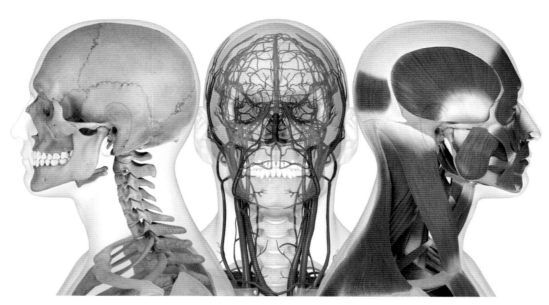

HUMAN
ANATOMY
THE **DEFINITIVE** VISUAL GUIDE

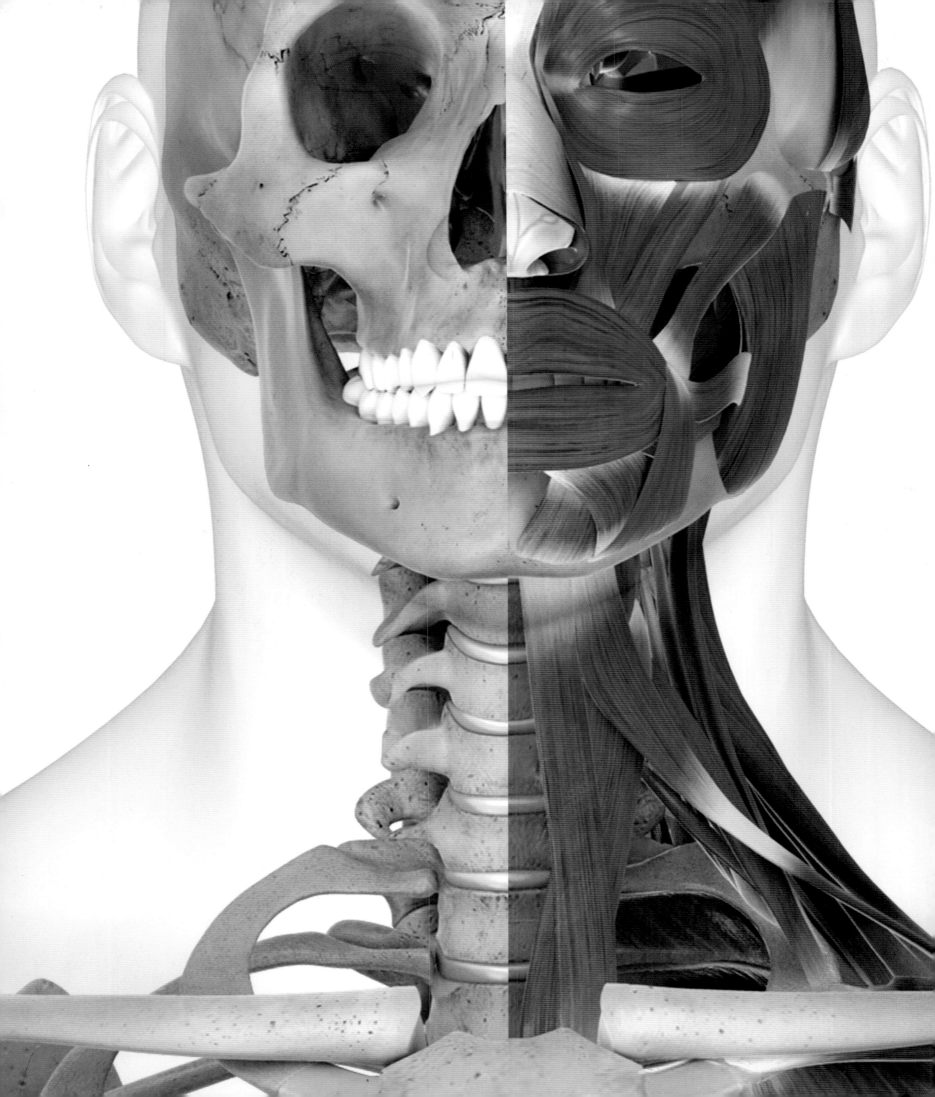

ALICE ROBERTS

HUMAN ANATOMY

THE **DEFINITIVE** VISUAL GUIDE

LONDON, NEW YORK, MELBOURNE,
MUNICH, AND DELHI

DK LONDON

Project Art Editor	**Project Editor**
Duncan Turner	Ruth O'Rourke-Jones
Jacket Designer	**Editor**
Duncan Turner	Lili Bryant
Pre-Production	**Jacket Editor**
Producer	Manisha Majithia
Vikki Nousiainen	**Managing Editor**
Managing Art Editor	Angeles Gavira
Michelle Baxter	**Production Editor**
Art Director	Rachel Ng
Philip Ormerod	**Publisher**
Associate Publishing	Sarah Larter
Director	**Publishing Director**
Liz Wheeler	Jonathan Metcalf

DK DELHI

Senior Editor	**Editor**
Anita Kakkar	Pallavi Singh
Art Editors	**Design**
Suhita Dharamjit,	**Assistant**
Amit Malhotra	Anjali Sachar
Deputy Managing	**Managing**
Art Editor	**Editor**
Sudakshina Basu	Rohan Sinha
DTP Designer	**Pre-Production Manager**
Vishal Bhatia	Balwant Singh
Production Manager	**Picture Researcher**
Pankaj Sharma	Sumedha Chopra

Illustrators

Medi-Mation (Creative Director: Rajeev Doshi)

Antbits Ltd (Richard Tibbitts)
Dotnamestudios (Andrew Kerr)
Deborah Maizels

Editor-in-Chief Professor Alice Roberts

Authors	**Consultants**
THE INTEGRATED BODY	**THE INTEGRATED BODY**
Linda Geddes	Professor Mark Hanson, University of Southampton
BODY SYSTEMS, IMAGING THE BODY	**BODY SYSTEMS, IMAGING THE BODY**
Professor Alice Roberts	Professor Harold Ellis, King's College, London Professor Susan Standring, King's College London

Content previously published in The Complete Human Body, 2010

This edition first published in Great Britain in 2014
by Dorling Kindersley Limited
80 Strand, London WC2R 0RL

Copyright © 2010, 2014
Dorling Kindersley Limited, London

A Penguin Random House Company

Foreword copyright © Alice Roberts

2 4 6 8 10 9 7 5 3 1

001 – 192972 – May/2014

A CIP catalogue record for this book is available from the British Library.

ISBN: 978 1 4093 4736 1

Printed and bound in China by South China.

See our complete catalogue at
www.dk.com

CONTENTS

006 Foreword

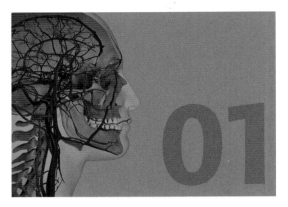

008
THE INTEGRATED BODY

010 Human genetic formula
012 Cell
014 Body composition
016 Body systems
018 Terminology and planes

020
BODY SYSTEMS

022 SKIN, HAIR, AND NAIL

024 SKELETAL SYSTEM OVERVIEW
026 Head and neck
036 Thorax
040 Spine
042 Abdomen and pelvis
046 Pelvis

048 Shoulder and upper arm
054 Lower arm and hand
056 Hand and wrist joints
058 Hip and thigh
062 Hip and knee
064 Lower leg and foot
066 Foot and ankle

068 MUSCULAR SYSTEM OVERVIEW
070 Head and neck
076 Thorax
082 Abdomen and pelvis
086 Shoulder and upper arm
094 Lower arm and hand
098 Hip and thigh
106 Lower leg and foot

110 NERVOUS SYSTEM OVERVIEW
112 Brain
118 Head and neck
120 Brain (transverse and coronal sections)
122 Head and neck (cranial nerves)
124 Eye
126 Ear
128 Neck
130 Thorax
132 Abdomen and pelvis
134 Shoulder and upper arm
138 Lower arm and hand
140 Hip and thigh
144 Lower leg and foot

146 RESPIRATORY SYSTEM OVERVIEW
148 Head and neck
150 Thorax
152 Lungs

154 CARDIOVASCULAR SYSTEM OVERVIEW
156 Head and neck
160 Thorax
162 Heart
166 Abdomen and pelvis
168 Shoulder and upper arm
172 Lower arm and hand
174 Hip and thigh
178 Lower leg and foot

180 LYMPHATIC AND IMMUNE SYSTEM OVERVIEW
182 Head and neck
184 Thorax
186 Abdomen and pelvis
188 Shoulder and upper arm
190 Hip and thigh

192 DIGESTIVE SYSTEM OVERVIEW
194 Head and neck
196 Thorax
198 Abdomen and pelvis
200 Stomach and intestines
202 Liver, pancreas, and gallbladder

204 URINARY SYSTEM OVERVIEW
206 Abdomen and pelvis

208 REPRODUCTIVE SYSTEM OVERVIEW
210 Thorax
212 Abdomen and pelvis

216 ENDOCRINE SYSTEM OVERVIEW
218 Head and neck

230 Lower arm and hand
232 Lower limb and foot

234 GLOSSARY
241 INDEX
255 ACKNOWLEDGMENTS

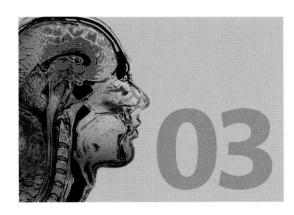

220
IMAGING THE BODY

222 Imaging techniques
224 Head and neck
226 Thorax
228 Abdomen and pelvis

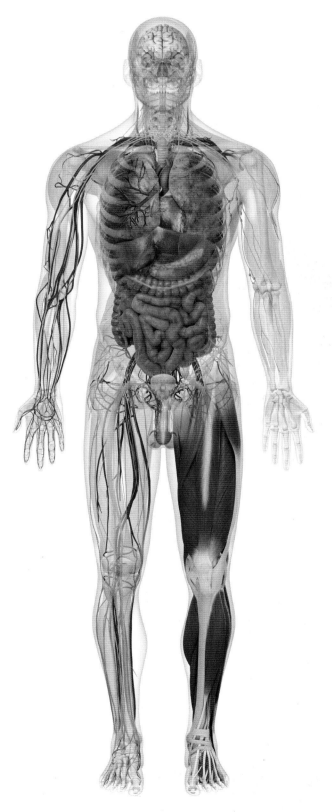

FOREWORD

Anatomy is a very visual subject, and illustrated anatomy books have been around for centuries. In the same way that a map must represent the physical features of a landscape, anatomical illustrations must convey the detailed layout of the human body. The mapmaker is concerned with the topography of a landscape, while the anatomist focuses on the topography of the body. The maps – whether of landscapes or the body – are collected into books known as atlases. The first anatomical atlases appeared in the Renaissance period, but students of anatomy today still rely heavily on visual media. Plenty of students still use atlases, alongside electronic resources.

Anatomical depictions have changed through time, reflecting the development of anatomical knowledge, changing styles and taste, and the constraints of different media. One of the earliest and most well-known atlases is Andreas Vesalius' *De humani corporis fabrica* (On the structure of the human body), published in 1543. The anatomical illustrations in this book took the form of a series of posed, dissected figures standing against a landscape. It was a book intended not just for medical students, but for a general readership. The heavy use of images to convey information made sense for this visual subject, and also helped to make anatomy accessible.

The late seventeenth century saw a striking change in anatomical depictions. Flayed figures, gracefully arranged against landscapes, gave way to brutally realistic illustrations of cadaveric specimens in the dissection room. The connection between anatomy and death was impossible to ignore in these pictures. The style of anatomy illustration has also been influenced by the methods available to capture and print images. As lithography replaced woodcut printing, it was possible to render anatomy in finer detail. Anatomical illustrators leapt on the potential offered by colour printing, using different colours to pick out arteries, veins, and nerves. More recently, the advent of photography meant that anatomy could be captured more objectively. It would be reasonable to suppose that photography would offer the best solution to the challenges facing the medical illustrator, but the task requires more than objectivity and fidelity. Images need to be uncluttered, and sometimes a simple line drawing can convey information better than a photograph of an actual dissection. The challenge facing the medical illustrator has always centred on what to keep in, and what to leave out.

The development of medical imaging, including the use of X rays, ultrasound, and MRI (magnetic resonance imaging), has had a huge impact on medicine, and has also had a profound effect on the way we visualize and conceptualize the body. Some anatomy atlases are still based on photographic or drawn representations of dissected, cadaveric specimens, and these have their place. But a new style has emerged, heavily influenced by medical imaging, featuring living anatomy. The supernatural, re-animated skeletons and muscle-men of the Renaissance anatomy atlases, and the later, somewhat brutal illustrations of dissected specimens, have been replaced with representations of the inner structure of a living woman or man.

Historically, and by necessity, anatomy has been a morbid subject. The general reader may understandably have been put off by opening an atlas to be confronted with images of dead flesh, slightly shrunken eyeballs resting in dissected sockets, and dead guts spilling out of opened abdomens. But the depiction of living anatomy, informed by medical imaging techniques, reveals anatomy in all its glory, without the gore.

The illustrations in this atlas are all about living anatomy. Most of the images in this book are founded on a 3D reconstruction of the anatomy of a whole body, drawn up in digital media and based on scans. We have grappled with the challenge of what to keep in, and what to leave out. It's overwhelming to see all the elements at the same time, so the anatomy of this idealized living human is stripped down, revealing the bones, muscles, nerves, blood vessels, and organs of the body in turn. The result is, I hope, an anatomy atlas which will be useful to any student of anatomy as well as appealing to anyone with an interest in the structure of the human body.

PROFESSOR ALICE ROBERTS

The body piece by piece
A series of MRI scans show horizontal slices through the body, starting with the head and working downwards, through the thorax and upper limbs, to the lower limbs and finally the feet.

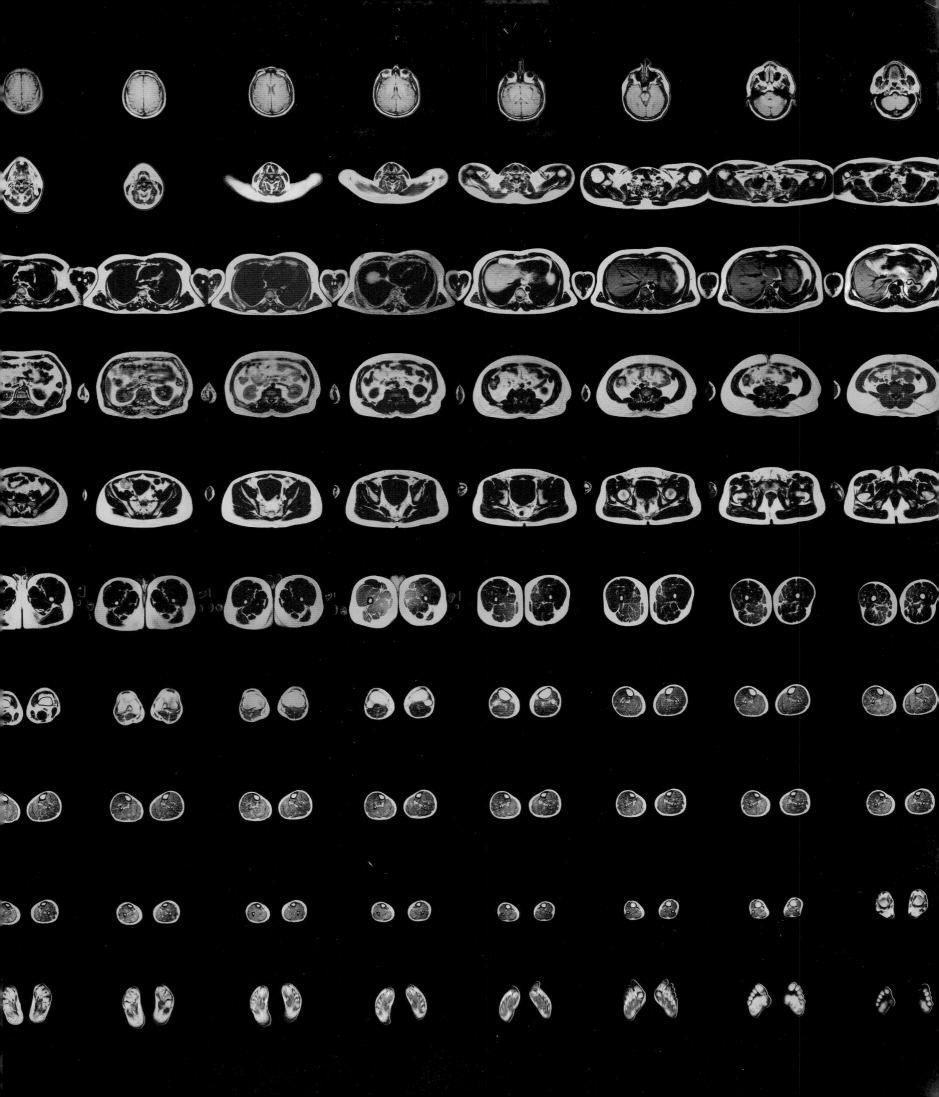

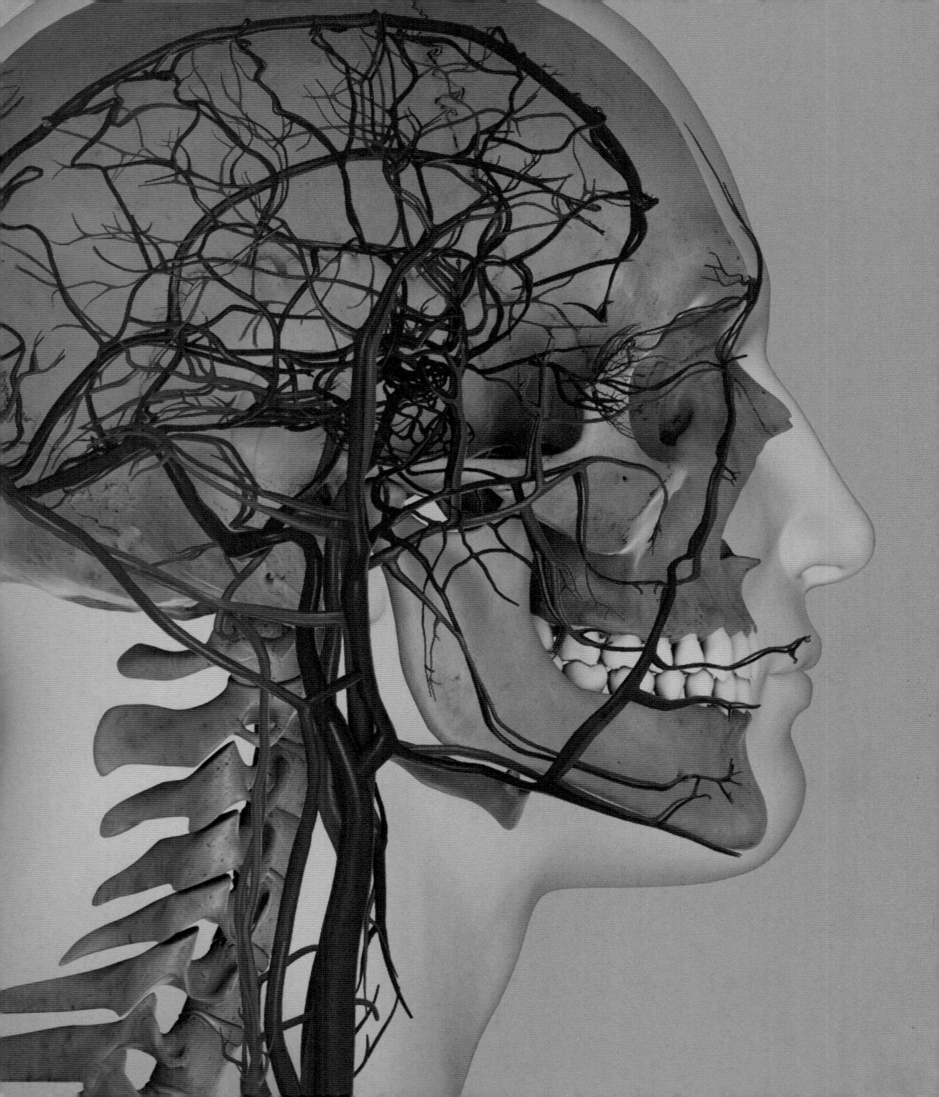

01 The Integrated Body

The human body comprises trillions of cells, each one a complex unit with intricate workings in itself. Cells are the building blocks of tissues, organs, and eventually, the integrated body systems that all interact – allowing us to function and survive.

010 Human genetic formula

012 Cell

014 Body composition

016 Body systems

018 Terminology and planes

HUMAN GENETIC FORMULA

DNA (deoxyribonucleic acid) is the blueprint for all life, from the humblest yeast to the human being. It provides a set of instructions on how to assemble the many thousands of different proteins that make us who we are. It also tightly regulates this assembly, ensuring that it does not run out of control.

THE MOLECULE OF LIFE

Although we all look different, the basic structure of our DNA is identical. It consists of chemical building blocks called bases, or nucleotides. What varies between individuals is the precise order in which these bases are connected into pairs. When base pairs are strung together they can form functional units called genes, which "spell out" the instructions for making a protein. Each gene encodes a single protein, although some complex proteins are encoded by more than one gene. Proteins have a wide range of vital functions in the body. They form structures such as skin or hair, carry signals around the body, and fight off infectious agents such as bacteria. Proteins also make up cells, the basic units of the body, and carry out the thousands of basic biochemical processes needed to sustain life. However, only about 1.5 per cent of our DNA encodes genes. The rest consists of regulatory sequences, structural DNA, or has no obvious purpose – so-called "junk DNA".

DNA double helix

In the vast majority of organisms, including humans, long strands of DNA twist around each other to form a right-handed spiral structure called a double helix. The helix consists of a sugar (deoxyribose) and phosphate backbone and complementary base pairs that stick together in the middle. Each twist of the helix contains around ten base pairs.

Cytosine

Guanine

Thymine

Adenine

DNA backbone
Formed of alternating units of phosphate and a sugar called deoxyribose

PACKAGING DNA

The human genome is composed of approximately 3 billion bases of DNA – about 2m (6½ft) of DNA in every cell if it was stretched from end to end. So our DNA must be packaged in order to fit inside each cell. DNA is concentrated into dense structures called chromosomes. Each cell has 23 pairs of chromosomes (46 in total) – one set from each parent. To package DNA, the double helix must first be coiled around histone proteins, forming a structure that looks like a string of beads. These histone "beads" wind up and lock together into densely coiled "chromatin", which, when a cell prepares to divide, further winds back on itself into tightly coiled chromosomes.

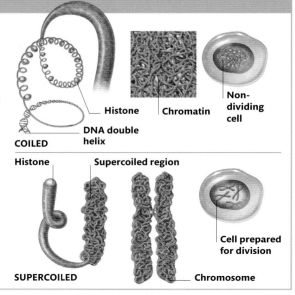

COILED

Histone

Chromatin

DNA double helix

Non-dividing cell

Histone

Supercoiled region

SUPERCOILED

Cell prepared for division

Chromosome

MAKING PROTEINS

Proteins consist of building blocks called amino acids, strung together in chains and folded. Every three base pairs of DNA codes for one amino acid, and the body makes 20 different amino acids – others are obtained from the diet. Protein synthesis occurs in two steps: transcription and translation. In transcription, the DNA double helix unwinds, exposing single-stranded DNA. Complementary sequences of a related molecule called RNA (ribonucleic acid) then create a copy of the DNA sequence that lock onto the exposed DNA bases to be translated into protein. This "messenger RNA" travels to ribosomes, where it is translated into strings of amino acids. These are then folded into the 3-D structure of a particular protein.

BASE PAIRS

DNA consists of building blocks called bases. There are four types: adenine (A), thymine (T), cytosine (C), and guanine (G). Each base is attached to a phosphate group and a deoxyribose sugar ring to form a nucleotide. In humans, bases pair up to form a double-stranded helix in which adenine pairs with thymine, and cytosine with guanine. The two strands are "complementary" to each other. Even if they are unwound and unzipped, they can realign and rejoin.

GENES

A gene is a unit of DNA needed to make a protein. Genes range in size from just a few hundred to millions of base pairs. They control our development, but are also switched on and off in response to environmental factors. For example, when an immune cell encounters a bacterium, genes are switched on that produce antibodies to destroy it. Gene expression is regulated by proteins that bind to regulatory sequences within each gene. Genes contain regions that are translated into protein (exons) and non-coding regions (introns).

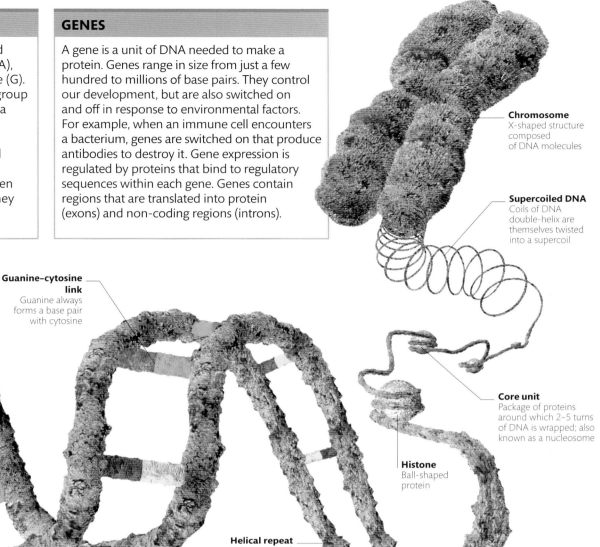

Chromosome
X-shaped structure composed of DNA molecules

Supercoiled DNA
Coils of DNA double-helix are themselves twisted into a supercoil

Guanine–cytosine link
Guanine always forms a base pair with cytosine

Core unit
Package of proteins around which 2–5 turns of DNA is wrapped; also known as a nucleosome

Histone
Ball-shaped protein

Adenine–thymine link
Adenine and thymine always form base pairs together

Helical repeat
Helix turns 360° for every 10.4 base pairs

THE HUMAN GENOME

Different organisms contain different genes, but a surprisingly large proportion of genes are shared between organisms. For example, roughly half of the genes found in humans are also found in bananas. However, it would not be possible to substitute the banana version of a gene for a human one because variations in the order of the base pairs within each gene also distinguish us. Humans possess more or less the same genes, but many of the differences between individuals can be explained by subtle variations within each gene. In humans, DNA differs by only about 0.2 per cent, while human DNA differs from chimpanzee DNA by around 5 per cent. Human genes are divided unevenly between 23 pairs of chromosomes, and each chromosome consists of gene-rich and gene-poor sections. When chromosomes are stained, differences in these regions show up as light and dark bands, giving chromosomes a striped appearance. We still don't know the exact number of protein-coding genes in the human genome, but researchers currently estimate between 20,000 and 25,000.

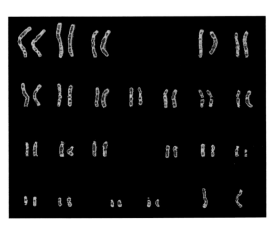

Karyotype

This is an organized profile of the chromosomes in someone's cells, arranged by size. Studying someone's karyotype enables doctors to determine whether any chromosomes are missing or abnormal.

GENETIC ENGINEERING

This form of gene manipulation enables us to substitute a defective gene with a functional one, or introduce new genes. Glow-in-the-dark mice were created by introducing a jellyfish gene that encodes a fluorescent protein into the mouse genome. Finding safe ways of delivering replacement genes to the correct cells in humans could lead to cures for many types of inherited diseases – so-called gene therapy.

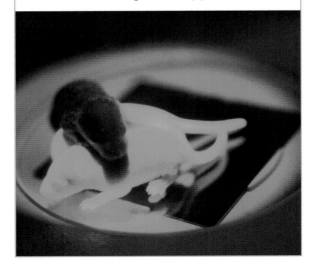

CELL

It is hard to comprehend what 75 trillion cells looks like, but observing yourself in a mirror would be a good start. That is how many cells exist in the average human body – and we replace millions of these cells every single day.

CELL ANATOMY

The cell is the basic functional unit of the human body. Cells are extremely small, typically only about 0.01mm across – even our largest cells are no bigger than the width of a human hair. They are also immensely versatile: some can form sheets like those in your skin or lining your mouth, while others can store or generate energy, such as fat and muscle cells. Despite their amazing diversity, there are certain features that all cells have in common, including an outer membrane, a control centre called a nucleus, and tiny powerhouses called mitochondria.

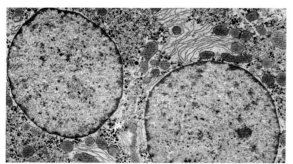

Liver cell
These cells make protein, cholesterol, and bile, and detoxify and modify substances from the blood. This requires lots of energy, so liver cells are packed with mitochondria (orange).

CELL METABOLISM

When a cell breaks down nutrients to generate energy for building new proteins or nucleic acids, it is known as cell metabolism. Cells use a variety of fuels to generate energy, but the most common one is glucose, which is transformed into adenosine triphosphate (ATP). This takes place in structures called mitochondria through a process called cellular respiration: enzymes within the mitochondria react with oxygen and glucose to produce ATP, carbon dioxide, and water. Energy is released when ATP is converted into adenoside diphosphate (ADP) via the loss of a phosphate group.

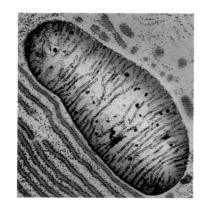

Mitochondrion
While the number of mitochondria varies between different cells, all have the same basic structure: an outer membrane and a highly folded inner membrane, where the production of energy actually takes place.

Nuclear membrane
A two-layered membrane with pores for substances to enter and leave the nucleus

Nucleus
The cell's control centre, containing chromatin and most of the cell's DNA

Nucleolus
The region at the centre of the nucleus; plays a vital role in ribosome production

Nucleoplasm
Fluid within the nucleus, in which nucleolus and chromosomes float

Microtubules
Part of cell's cytoskeleton, these aid movement of substances through the watery cytoplasm

Centriole
Composed of two cylinders of tubules; essential to cell reproduction

Microvilli
These projections increase the cell's surface area, aiding absorption of nutrients

Golgi complex
A structure that processes and repackages proteins produced in the rough endoplasmic reticulum for release at the cell membrane

Released secretions
Secretions are released from the cell by exytosis, in which a vesicle merges with the cell membrane and releases its contents

Secretory vesicle
Sac containing various substances, such as enzymes, that are produced by the cell and secreted at the cell membrane

Lysosome
Produces powerful enzymes that aid in digestion and excretion of substances and worn-out organelles

CELL TRANSPORT

Materials are constantly being transported in and out of the cell via the cell membrane. Such materials include fuel for generating energy, or building blocks for protein assembly. Some cells secrete signalling molecules to communicate with the rest of the body. The cell membrane is studded with proteins that help transport, allow cells to communicate, and identify a cell to other cells. The membrane is permeable to some molecules, but others need active transport through special channels in the membrane. Cells have three methods of transport: diffusion, facilitated diffusion, and active transport.

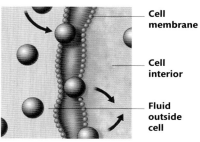

Diffusion
Molecules passively cross the membrane from areas of high to low concentration. Water and oxygen both cross by diffusion.

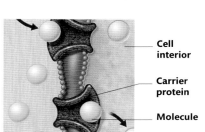

Facilitated diffusion
A carrier protein, or protein pore, binds with a molecule outside the cell, then changes shape and ejects the molecule into the cell.

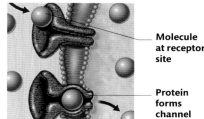

Active transport
Molecules bind to a receptor site on the cell membrane, triggering a protein, which changes into a channel that molecules travel through.

MAKING NEW BODY CELLS

While the cells lining the mouth are replaced every couple of days, some of the nerve cells in the brain have been there since before birth. Stem cells are specialized cells that constantly divide and give rise to new cells, such as blood cells. Cell division requires that a cell's DNA is accurately copied and then shared equally between two "daughter" cells, by a process called mitosis. The chromosomes are first replicated before being pulled to opposite ends of the cell. The cell then divides to produce two daughter cells, with the cytoplasm and organelles being shared between the two cells.

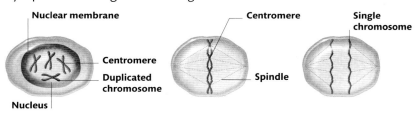

1 Preparation
The cell produces proteins and new organelles, and duplicates its DNA. The DNA condenses into X-shaped chromosomes.

2 Alignment
The chromosomes line up along a network of filaments – spindle – linked to a larger network, called the cytoskeleton.

3 Separation
The chromosomes are pulled apart and move to opposite ends of the cell. Each end has an identical set of chromosomes.

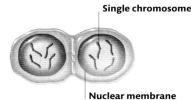

4 Splitting
The cell now splits into two, with the cytoplasm, cell membrane, and remaining organelles being shared roughly equally between the two daughter cells.

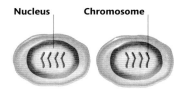

5 Offspring
Each daughter cell contains a complete copy of the DNA from the parent cell; this enables it to continue growing, and eventually divide itself.

Generic cell
At a cell's heart is the nucleus, where the genetic material is stored and the first stages of protein synthesis occur. Cells also contain other structures for assembling proteins, including ribosomes, the endoplasmic reticulum, and Golgi apparatus. The mitochondria provide the cell with energy.

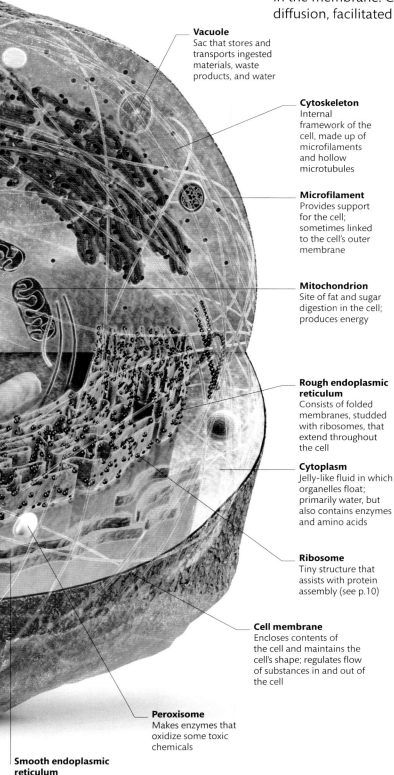

Vacuole
Sac that stores and transports ingested materials, waste products, and water

Cytoskeleton
Internal framework of the cell, made up of microfilaments and hollow microtubules

Microfilament
Provides support for the cell; sometimes linked to the cell's outer membrane

Mitochondrion
Site of fat and sugar digestion in the cell; produces energy

Rough endoplasmic reticulum
Consists of folded membranes, studded with ribosomes, that extend throughout the cell

Cytoplasm
Jelly-like fluid in which organelles float; primarily water, but also contains enzymes and amino acids

Ribosome
Tiny structure that assists with protein assembly (see p.10)

Cell membrane
Encloses contents of the cell and maintains the cell's shape; regulates flow of substances in and out of the cell

Peroxisome
Makes enzymes that oxidize some toxic chemicals

Smooth endoplasmic reticulum
Network of tubes and flat, curved sacs that helps to transport materials through the cell; site of calcium storage; main location of fat metabolism

BODY COMPOSITION

Cells are building blocks from which the human body is made. Some cells work alone – such as red blood cells, which carry oxygen – but many are organized into tissues. These tissues form organs, which in turn form specific body systems, where cells with various functions join forces to accomplish one or more tasks.

CELL TYPES

There are more than 200 types of cell in the body, each type specially adapted to its own particular function. Every cell contains the same genetic information, but not all of the genes are "switched on" in every cell. It is this pattern of gene expression that dictates the cell's appearance, its behavior, and its role in the body. A cell's fate is largely determined before birth, influenced by its position in the body and the cocktail of chemical messengers that it is exposed to in that environment. Early during development, stem cells begin to differentiate into three layers of specialized cells called the ectoderm, endoderm, and mesoderm. Cells of the ectoderm will form the skin and nails, the epithelial lining of the nose, mouth, and anus, the eyes, and the brain and spinal cord. Cells of the endoderm will become the inner linings of the digestive tract, the respiratory linings, and glandular organs. Mesoderm cells will develop into the muscles, and circulatory and excretory systems.

STEM CELLS

A few days after fertilization, an embryo consists of a ball of "embryonic stem cells" (ESCs). These cells have the potential to become any type of cell in the body. Scientists are trying to harness this property to grow replacement body parts. As the embryo grows, the stem cells become increasingly restricted in their potential and most are fully differentiated by the time we are born. Only a small number of stem cells remain in parts of the adult body, including in the bone marrow. Scientists believe that these cells could also be used to help cure disease.

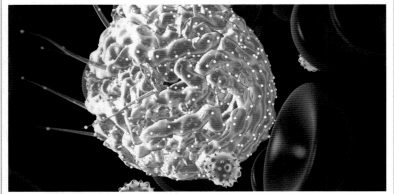

Adult stem cells
Adult stem cells, such as the large white cell in this image, are present in bone marrow, where they multiply and produce millions of blood cells, including red blood cells, also seen here.

Red blood cells
Unlike other cells, red blood cells lack a nucleus and organelles. Instead, they have oxygen-carrying protein (haemoglobin), which gives blood its red colour.

Epithelial cells
The skin cells and the cells lining the lungs and reproductive tracts are among the barrier cells, called epithelial cells, which line the cavities and surfaces of the body.

Adipose (fat) cells
These cells are highly adapted for storing fat – the bulk of their interior is taken up by a large droplet of semi-liquid fat. When we gain weight, they fill up with more fat.

Nerve cells
These electrically excitable cells transmit electrical signals down an extended stem called an axon. Found throughout the body, they enable us to feel sensations.

Photoreceptor cells
Located in the eye, these are two types – cone and rod (left). They have a light-sensitive pigment and generate electrical signals when struck by light, helping us to see.

Smooth muscle cells
One of three types of muscle cell, smooth muscle cells are spindle-shaped cells found in the arteries and the digestive tract that produce contractions.

Ovum (egg) cells
The largest cell in the female human body, eggs are female reproductive cells. Like sperm (below), they have just 23 chromosomes.

Sperm cells
Sperm are male reproductive cells, with a tail that enables them to swim up the female reproductive tract and fertilize an egg.

LEVELS OF ORGANISATION

The overall organization of the human body can be visualized as a hierarchy of levels. At its lowest are the body's basic chemical constituents, which form organic molecules, such as the DNA, the key to life. As the hierarchy ascends, the number of components in each of its levels – cells, tissues, organs, and systems – decreases, culminating in a single being at its apex. Cells are the smallest living units, with each adapted to carry out a specific role, but not in isolation. Groups of similar cells form tissues, which in turn form organs with a specific role. Organs with a common purpose are linked within a system, such as the cardiovascular system, shown right. These interdependent systems combine to produce a human body (see pp.16–17).

TISSUE TYPES

Cells often group together with their own kind to form tissues that carry out a specific function. However, not all cells within a tissue are necessarily identical. The four main types of tissue in the human body are muscle, connective tissue, nervous tissue, and epithelial tissue. Within these groups, different forms of these tissues can have very different appearances and functions. For example, blood, bone, and cartilage are all types of connective tissue, but so are fat layers, tendons, ligaments, and the fibrous tissue that holds organs and epithelial layers in place. Organs such as the heart and lungs are composed of several different kinds of tissue.

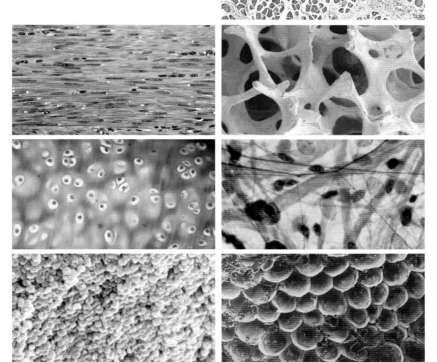

Skeletal muscle
This tissue performs voluntary limb movements. Its cells are arranged into bundles of fibres that connect to bones via tendons. They are packed with filaments that slide over one another to produce contractions.

Smooth muscle
Able to contract in long, wave-like motions involuntarily, smooth muscle is found in sheets on the walls of specific organs. It is vital for maintaining blood pressure and for pushing food through the system.

Cartilage
The high water-content makes this tissue rubbery yet stiff. It is composed of cells, called chondrocytes, set in a matrix of gel-like material, which the cells secrete. Cartilage is found in the bone joints and in the ear and nose.

Dense connective tissue
This contains fibroblast cells, which secrete the fibrous protein called type 1 collagen. The fibres are organized into a regular parallel pattern, making the tissue very strong. This tissue type occurs in the base layer of skin.

Epithelial tissue
This tissue forms a covering or lining for internal and external body surfaces. Some epithelial tissues can secrete substances such as digestive enzymes; others can absorb substances like food or water.

Spongy bone
Spongy bone is found in the centre of bones (see p.24) and is softer and weaker than compact bone. The lattice-like spaces in spongy bone are filled with bone marrow or connective tissue.

Loose connective tissue
This tissue type also contains cells called fibroblasts, but the fibres they secrete are loosely organized, making the tissue pliable. Loose connective tissue holds organs in place and provides support.

Adipose tissue
A type of connective tissue, adipose tissue is composed of fat cells called adipocytes, as well as some immune cells, fibroblast cells, and blood vessels. Its main task is to store energy, and to protect and insulate the body.

Nervous tissue
This forms the brain, spinal cord, and the nerves that control movement, transmit sensation, and regulate many body functions. It is mainly made up of networks of nerve cells (see opposite).

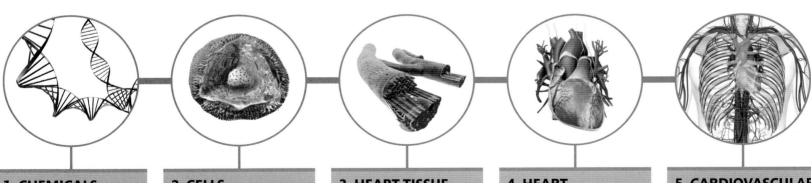

1. CHEMICALS
Key among the chemicals inside all cells is DNA (see pp.10–11). Its long molecules provide the instructions for making proteins. These, in turn, perform many roles, such as building cells.

2. CELLS
While cells may differ in size and shape, all have the same basic features: an outer membrane; organelles floating within jelly-like cytoplasm; and a nucleus containing DNA (see pp.12–13).

3. HEART TISSUE
One of the three types of muscle tissue, cardiac muscle is found only in the walls of the heart. Its cells contract together, as a network, to make the heart squeeze and pump blood.

4. HEART
Like other organs, the heart is made of several types of tissue, including cardiac muscle tissue. Among the others are connective and epithelial tissues, found in the chambers and valves.

5. CARDIOVASCULAR SYSTEM
The heart, blood, and blood vessels form the cardiovascular system. Its main tasks are to pump blood, deliver nutrients, and remove waste from the tissue cells.

BODY SYSTEMS

The human body can do many different things. It can digest food, think, move, even reproduce and create new life. Each of these tasks is performed by a different body system – a group of organs and tissues working together to complete that task. However, good health and body efficiency rely on the different body systems working together in harmony.

SYSTEM INTERACTION

Think about what your body is doing right now. You are breathing, your heart is beating, and your blood pressure is under control. You are also conscious and alert. If you were to start running, specialized cells called chemoreceptors would detect a change in your body's metabolic requirements and signal to the brain to release adrenaline. This would in turn signal the heart to beat faster, boosting blood circulation and providing more oxygen to the muscles. After a while, cells in the hypothalamus might detect an increase in temperature and send a signal to the skin to produce sweat, which would evaporate and cool you down.

The individual body systems are linked together by a vast network of positive and negative feedback loops. These use signalling molecules such as hormones and electrical impulses from nerves to maintain equilibrium. Here, the basic components and functions of each system are described, and examples of system interactions are examined.

BREATHING IN AND OUT

The mechanics of breathing rely upon an interaction between the respiratory and muscular systems. Together with three accessory muscles, the intercostal muscles and the diaphragm contract to increase the volume of the chest cavity. This draws air down into the lungs. A different set of muscles is used during forced exhalation. These rapidly compress the chest cavity, forcing air out of the lungs.

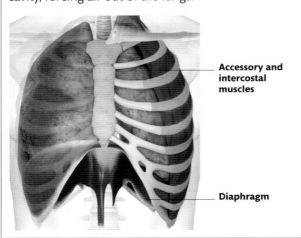

Accessory and intercostal muscles

Diaphragm

LYMPHATIC AND IMMUNE SYSTEM

The lymphatic system includes a network of vessels and nodes, which drain tissue fluid and return it to the veins. Its main functions are to maintain fluid balance within the cardiovascular system and to distribute immune cells around the body. Movement of lymphatic fluid relies on the muscles within the muscular system.

ENDOCRINE SYSTEM

The endocrine system communicates between the other systems, enabling them to be monitored and controlled. It uses chemical messengers, called hormones, which are secreted into the blood by specialized glands.

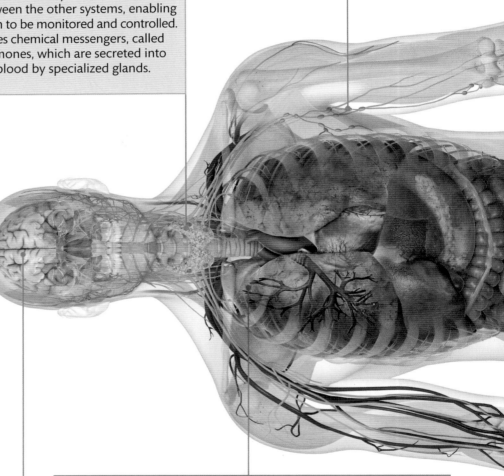

RESPIRATORY SYSTEM

Every cell in the body needs oxygen and must dispel carbon dioxide in order to function. The respiratory system ensures this by breathing air into the lungs, where the exchange of these molecules occurs between the air and blood. The cardiovascular system transports oxygen and carbon dioxide between the cells and the lungs.

NERVOUS SYSTEM

The brain, spinal cord, and nerves collect, process, and disseminate information from the body's internal and external environments. The nervous system communicates through networks of nerve cells, which connect with other systems. The brain controls and monitors all the other systems to ensure they are performing normally.

DIGESTIVE SYSTEM

As well as oxygen, every cell needs energy in order to function. The digestive system processes and breaks down the food we eat so that a variety of nutrients can be absorbed from the intestines into the circulatory system. These are then delivered to the cells of every body system in order to provide them with energy.

MUSCULAR SYSTEM

The muscular system is made up of three types of muscle: skeletal, smooth, and cardiac. It is responsible for generating movement – both of the limbs and within the other body systems. For example, smooth muscle aids the digestive system by helping to propel food down the oesophagus and through the stomach, intestines, and rectum. The respiratory system needs the thoracic muscles to contract to fill the lungs with air (see opposite).

SKELETAL SYSTEM

This system uses bones, cartilage, ligaments, and tendons to provide the body with structural support and protection. It encases much of the nervous system within a skull and vertebrae, and the vital organs of the respiratory and circulatory systems within the ribcage. The skeletal system also supports our immune and the circulatory systems by manufacturing red and white blood cells.

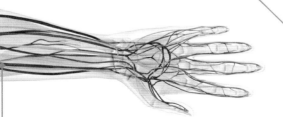

REPRODUCTIVE SYSTEM

Although the reproductive system is not essential for maintaining life, it is needed to propagate it. Both the testes of the male and the ovaries of the female produce gametes in the form of sperm and eggs, which fuse to create an embryo. The testes and ovaries also produce hormones including oestrogen and testosterone, therefore form part of the endocrine system.

CARDIOVASCULAR SYSTEM

The cardiovascular system uses blood to carry oxygen from the respiratory system and nutrients from the digestive system to cells of all the body's systems. It also removes waste products from these cells. At the centre of the cardiovascular system lies the muscular heart, which pumps blood through the blood vessels.

URINARY SYSTEM

The urinary system filters and removes many of the waste products generated by cells of the body. It does this by filtering blood through the kidneys and producing urine, which is collected in the bladder and then excreted through the urethra. The kidneys also help regulate blood pressure within the cardiovascular system by ensuring that the correct amount of water is reabsorbed by the blood.

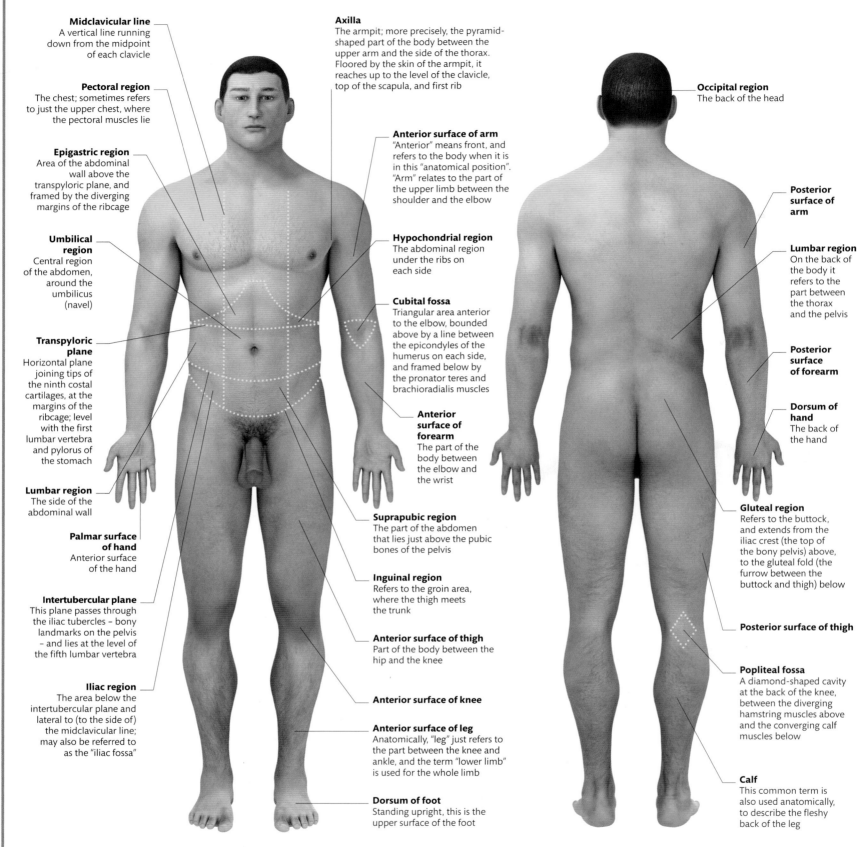

Midclavicular line
A vertical line running down from the midpoint of each clavicle

Pectoral region
The chest; sometimes refers to just the upper chest, where the pectoral muscles lie

Epigastric region
Area of the abdominal wall above the transpyloric plane, and framed by the diverging margins of the ribcage

Umbilical region
Central region of the abdomen, around the umbilicus (navel)

Transpyloric plane
Horizontal plane joining tips of the ninth costal cartilages, at the margins of the ribcage; level with the first lumbar vertebra and pylorus of the stomach

Lumbar region
The side of the abdominal wall

Palmar surface of hand
Anterior surface of the hand

Intertubercular plane
This plane passes through the iliac tubercles – bony landmarks on the pelvis – and lies at the level of the fifth lumbar vertebra

Iliac region
The area below the intertubercular plane and lateral to (to the side of) the midclavicular line; may also be referred to as the "iliac fossa"

Axilla
The armpit; more precisely, the pyramid-shaped part of the body between the upper arm and the side of the thorax. Floored by the skin of the armpit, it reaches up to the level of the clavicle, top of the scapula, and first rib

Anterior surface of arm
"Anterior" means front, and refers to the body when it is in this "anatomical position". "Arm" relates to the part of the upper limb between the shoulder and the elbow

Hypochondrial region
The abdominal region under the ribs on each side

Cubital fossa
Triangular area anterior to the elbow, bounded above by a line between the epicondyles of the humerus on each side, and framed below by the pronator teres and brachioradialis muscles

Anterior surface of forearm
The part of the body between the elbow and the wrist

Suprapubic region
The part of the abdomen that lies just above the pubic bones of the pelvis

Inguinal region
Refers to the groin area, where the thigh meets the trunk

Anterior surface of thigh
Part of the body between the hip and the knee

Anterior surface of knee

Anterior surface of leg
Anatomically, "leg" just refers to the part between the knee and ankle, and the term "lower limb" is used for the whole limb

Dorsum of foot
Standing upright, this is the upper surface of the foot

Occipital region
The back of the head

Posterior surface of arm

Lumbar region
On the back of the body it refers to the part between the thorax and the pelvis

Posterior surface of forearm

Dorsum of hand
The back of the hand

Gluteal region
Refers to the buttock, and extends from the iliac crest (the top of the bony pelvis) above, to the gluteal fold (the furrow between the buttock and thigh) below

Posterior surface of thigh

Popliteal fossa
A diamond-shaped cavity at the back of the knee, between the diverging hamstring muscles above and the converging calf muscles below

Calf
This common term is also used anatomically, to describe the fleshy back of the leg

Anterior surface regions

The anterior surface of the body is divided into general anatomical areas by imaginary lines drawn on the body. The location of many of these lines is defined by reference to underlying features such as muscles or bony prominences; for example, the cubital fossa is defined by reference to epicondyles of the humerus, and the pronator teres and brachioradialis muscles. Many of the regions may be divided into smaller areas. For instance, the upper part of the anterior thigh contains the femoral triangle.

Posterior surface regions

As with the anterior surface, the posterior surface can also be divided into anatomical regions. The anterior surface of the abdomen is divided by planes and mapped into nine regions – allowing doctors to describe precisely where areas of tenderness or lumps are felt on abdominal examination. The back is not divided into as many regions. This illustration shows some of the terms used for the broader regions of back of the body.

TERMINOLOGY AND PLANES

Anatomical language allows us to describe the structure of the body accurately and unambiguously. The illustrations here show the main regions of the anterior (front) and posterior (back) surfaces of the body. Sometimes it is easier to understand anatomy by dividing the body into two dimensional slices. The orientation of these planes through the body also have specific anatomical names. There are also terms to describe the relative position of structures within the body.

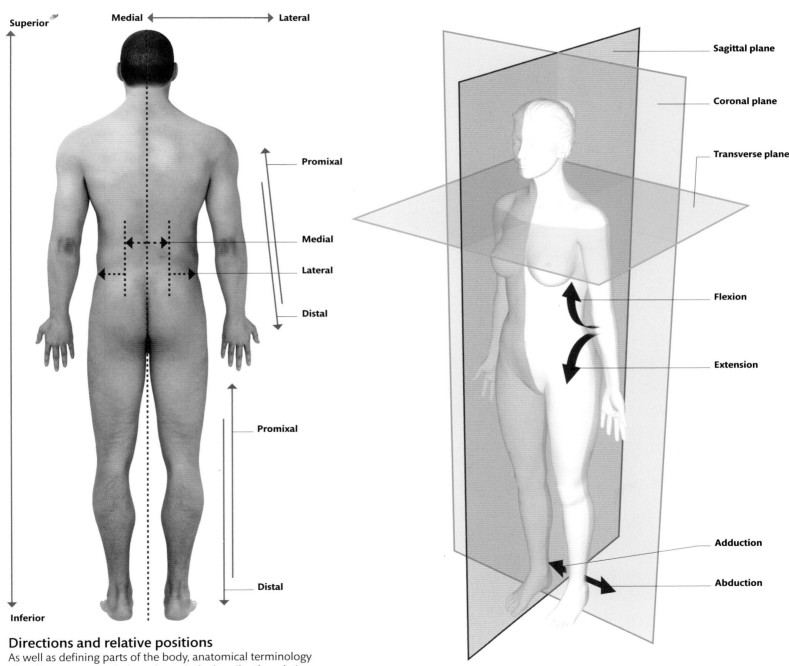

Directions and relative positions

As well as defining parts of the body, anatomical terminology also allows us to precisely and concisely describe the relative positions of various structures. These terms always refer back to relative positions of structures when the body is in the "anatomical position" (shown above). Medial and lateral describe positions of structures towards the midline, or towards the side of the body, respectively. Superior and inferior refer to vertical position – towards the top or bottom of the body. Proximal and distal are useful terms, describing a relative position towards the centre or periphery of the body.

Anatomical terms for planes and movement

The diagram above shows the three planes – sagittal, coronal, and transverse – cutting through a body. It also illustrates some medical terms that are used to describe certain movements of body parts: flexion decreases the angle of a joint, such as the elbow, while extension increases it; adduction draws a limb closer to the sagittal plane, while abduction moves it further away from that plane.

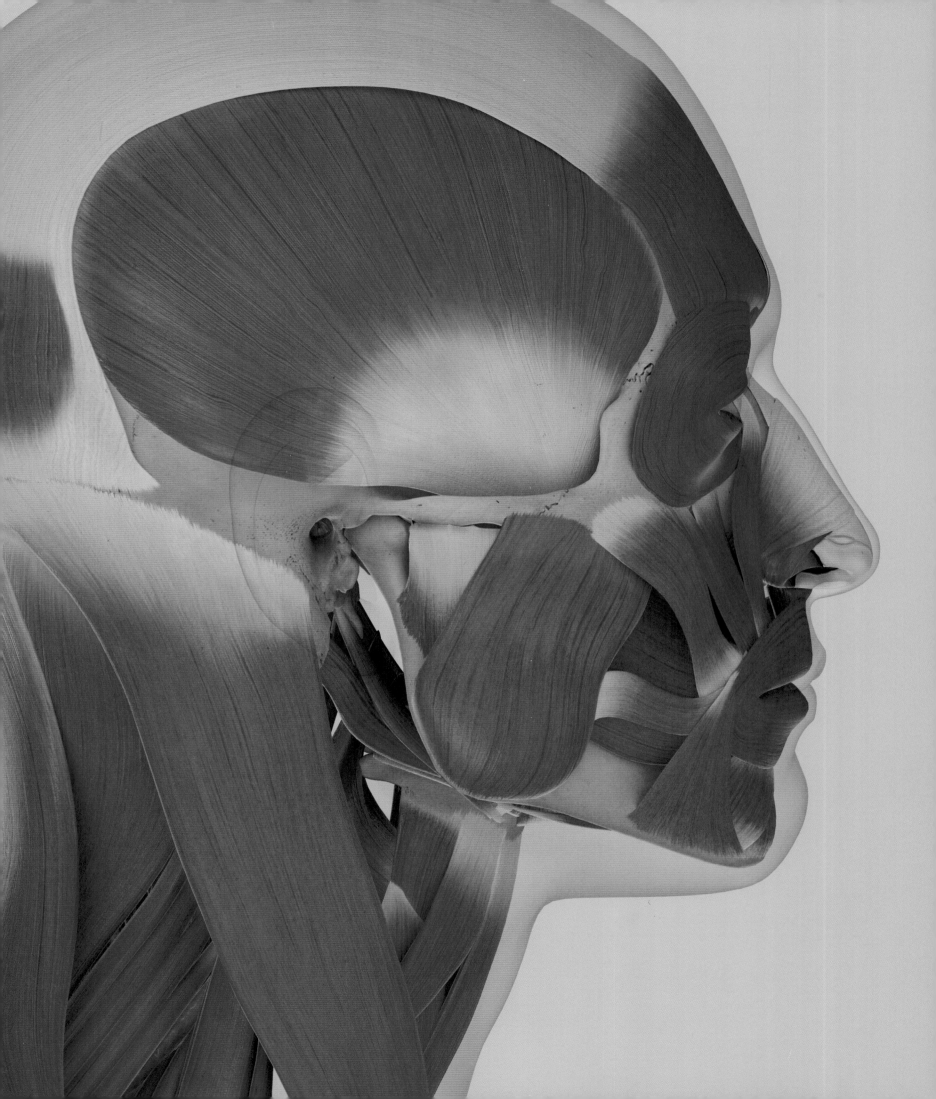

02 Body Systems

The human body is made up of eleven functional systems. No one system works in isolation, for example the endocrine and nervous systems work closely to keep the body regulated, while the respiratory and cardiovascular systems combine to deliver vital oxygen to cells. To build the clearest picture of how the body is put together it is, however, helpful to strip back our anatomy and consider it system by system. This chapter gives an overview of the basic structure of each system before looking at each region in detail.

022 Skin, hair, and nail

024 Skeletal system

068 Muscular system

110 Nervous system

146 Respiratory system

154 Cardiovascular system

180 Lymphatic and immune system

192 Digestive system

204 Urinary system

208 Reproductive system

216 Endocrine system

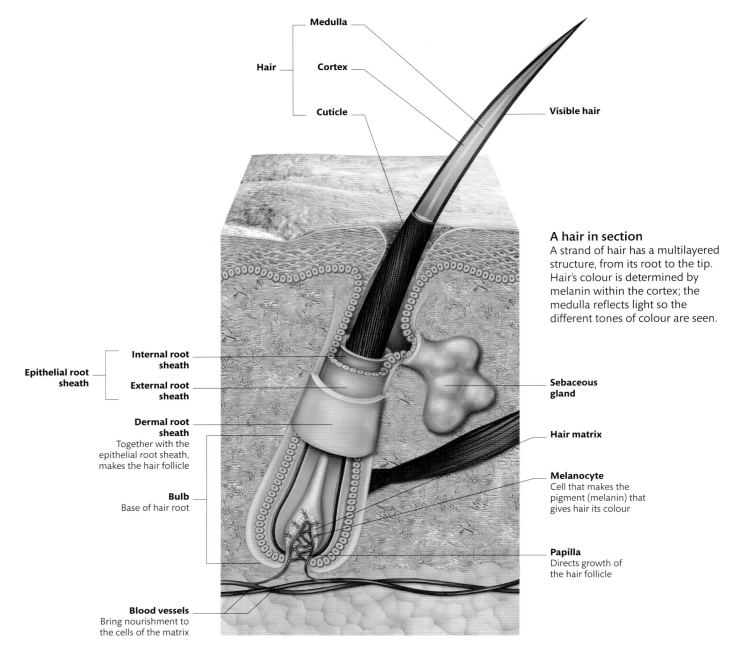

Medulla

Hair

Cortex

Cuticle

Visible hair

A hair in section
A strand of hair has a multilayered structure, from its root to the tip. Hair's colour is determined by melanin within the cortex; the medulla reflects light so the different tones of colour are seen.

Internal root sheath

Epithelial root sheath

External root sheath

Dermal root sheath
Together with the epithelial root sheath, makes the hair follicle

Bulb
Base of hair root

Blood vessels
Bring nourishment to the cells of the matrix

Sebaceous gland

Hair matrix

Melanocyte
Cell that makes the pigment (melanin) that gives hair its colour

Papilla
Directs growth of the hair follicle

SECTION THROUGH A HAIR

SKIN, HAIR, AND NAIL

The skin is our largest organ, weighing about 4kg (9lb) and covering an area of about 2 square metres (21 square feet). It forms a tough, waterproof layer, which protects us from the elements. However, it offers much more than protection: the skin lets us appreciate the texture and temperature of our environment; it regulates body temperature; it allows excretion in sweat, communication through blushing, gripping due to ridges on our fingertips, and vitamin D production in sunlight. Thick head hairs and fine body hairs help to keep us warm and dry. All visible hair is in fact dead; hairs are only alive at their root. Constantly growing and self-repairing, nails protect fingers and toes but also enhance their sensitivity.

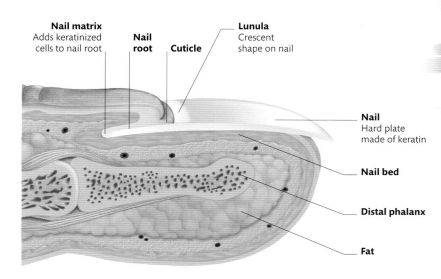

Nail matrix
Adds keratinized cells to nail root

Nail root

Cuticle

Lunula
Crescent shape on nail

Nail
Hard plate made of keratin

Nail bed

Distal phalanx

Fat

SECTION THROUGH A NAIL

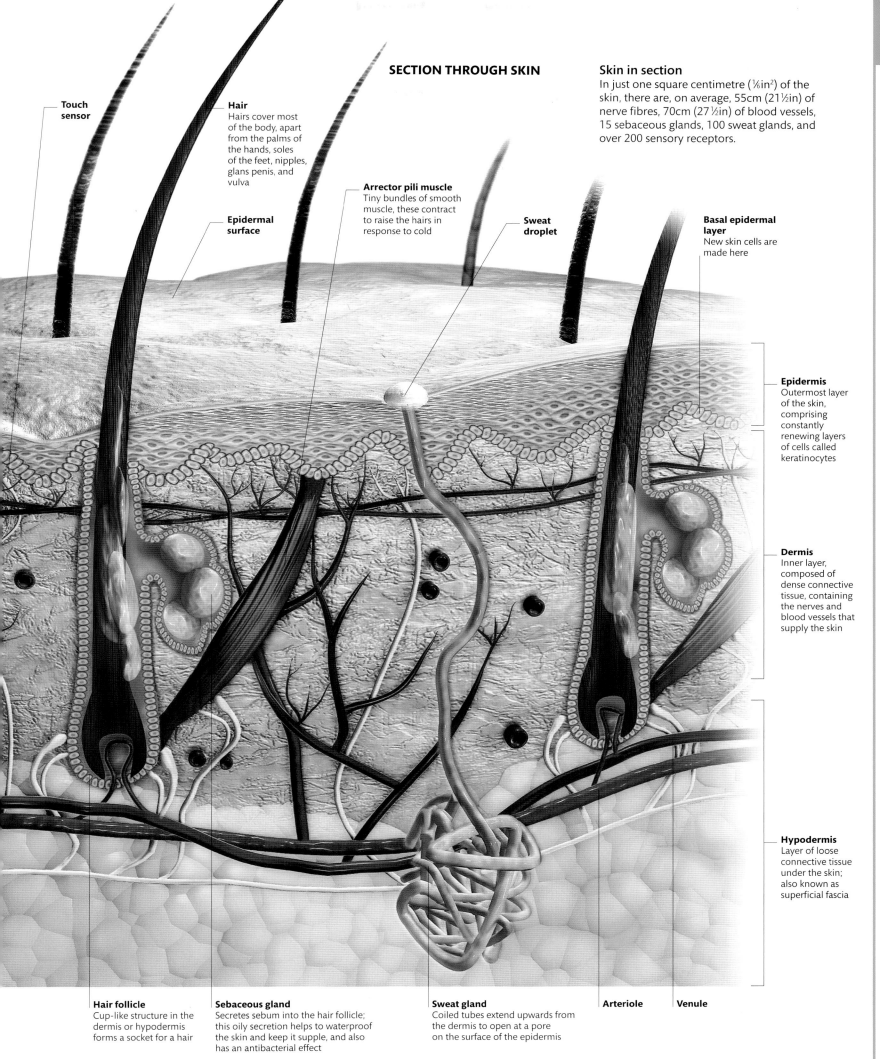

SECTION THROUGH SKIN

Skin in section
In just one square centimetre (⅛in²) of the skin, there are, on average, 55cm (21½in) of nerve fibres, 70cm (27½in) of blood vessels, 15 sebaceous glands, 100 sweat glands, and over 200 sensory receptors.

Touch sensor

Hair
Hairs cover most of the body, apart from the palms of the hands, soles of the feet, nipples, glans penis, and vulva

Epidermal surface

Arrector pili muscle
Tiny bundles of smooth muscle, these contract to raise the hairs in response to cold

Sweat droplet

Basal epidermal layer
New skin cells are made here

Epidermis
Outermost layer of the skin, comprising constantly renewing layers of cells called keratinocytes

Dermis
Inner layer, composed of dense connective tissue, containing the nerves and blood vessels that supply the skin

Hypodermis
Layer of loose connective tissue under the skin; also known as superficial fascia

Hair follicle
Cup-like structure in the dermis or hypodermis forms a socket for a hair

Sebaceous gland
Secretes sebum into the hair follicle; this oily secretion helps to waterproof the skin and keep it supple, and also has an antibacterial effect

Sweat gland
Coiled tubes extend upwards from the dermis to open at a pore on the surface of the epidermis

Arteriole

Venule

SKELETAL SYSTEM
OVERVIEW

The human skeleton gives the body its shape, supports the weight of all our other tissues, provides attachment for muscles, and forms a system of linked levers that the muscles can move. It also protects delicate organs and tissues, such as the brain within the skull, the spinal cord within the arches of the vertebrae, and the heart and lungs within the ribcage. The skeletal system differs between the sexes. This is most obvious in the pelvis, which is usually wider in a woman than in a man. The skull also varies, with men having a larger brow and more prominent areas for muscle attachment on the back of the head. The entire skeleton tends to be larger and more robust in a man.

BONE STRUCTURE

Most of the human skeleton develops first as cartilage, which is later replaced by bone throughout fetal development and childhood. Both bone and cartilage are connective tissues. Bone tissue consists of cells that are embedded in a mineralized matrix, making it extremely hard and strong. Bone is full of blood vessels and repairs easily.

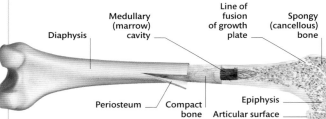

Diaphysis | Medullary (marrow) cavity | Line of fusion of growth plate | Spongy (cancellous) bone

Periosteum | Compact bone | Epiphysis | Articular surface

Long bone
Long bones are found in the limbs, and include the femur (shown above), humerus, radius, ulna, tibia, fibula, metatarsals, metacarpals, and phalanges. A long bone has flared ends (epiphyses), which narrow to form a neck (metaphysis), tapering down into a cylindrical shaft (diaphysis). Cartilage growth plates near the ends of bones allow rapid growth in childhood, but disappear by adulthood.

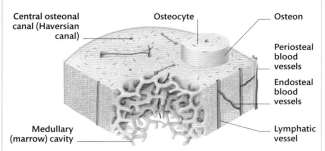

Central osteonal canal (Haversian canal) | Osteocyte | Osteon

Periosteal blood vessels

Endosteal blood vessels

Medullary (marrow) cavity | Lymphatic vessel

Compact bone
Also called cortical bone, compact bone is made up of osteons: concentric cylinders of bone tissue, each around 0.1–0.4mm in diameter, with a central vascular canal. Bone is full of blood vessels: those in the osteons connect to blood vessels within the medullary cavity of the bone as well as to vessels in the periosteum on the outside.

Cranium
Contains and protects the brain and the organs of special sense – the eyes, ears, and nose – and provides the supporting framework of the face

Vertebral column
Comprises stacked vertebrae and forms a strong, flexible backbone for the skeleton

Mandible
A single bone, the jaw contains the lower teeth and provides attachment for the chewing muscles

Manubrium

Clavicle

Sternum

Gladiolus

Scapula

Xiphoid process

Humerus

Costal cartilages
Attach the upper ribs to the sternum, and lower ribs to each other, and give the ribcage flexibility

Ribs

Ulna

Pelvis
Oddly shaped bone also called the innominate bone ("bone without a name")

Radius

Sacrum
Formed from five fused vertebrae; it provides a strong connection between the pelvis and the spine

Femur
The largest bone in the body at around 45cm (18in) long

Patella
The kneecap. This bone lies embedded in the tendon of the quadriceps muscle

Tibia
The shinbone; its sharp anterior edge can be felt along the front of the shin

Tarsals
A group of seven bones, including the talus; contributes to the ankle joint, and the heel-bone or calcaneus

Fibula
Contributes to the ankle joint and provides a surface for muscle attachment

Metatarsals
Five bones in the foot; the equivalent of the metacarpals in the hand

Phalanges
Fourteen phalanges form the toes of each foot

ANTERIOR (FRONT)

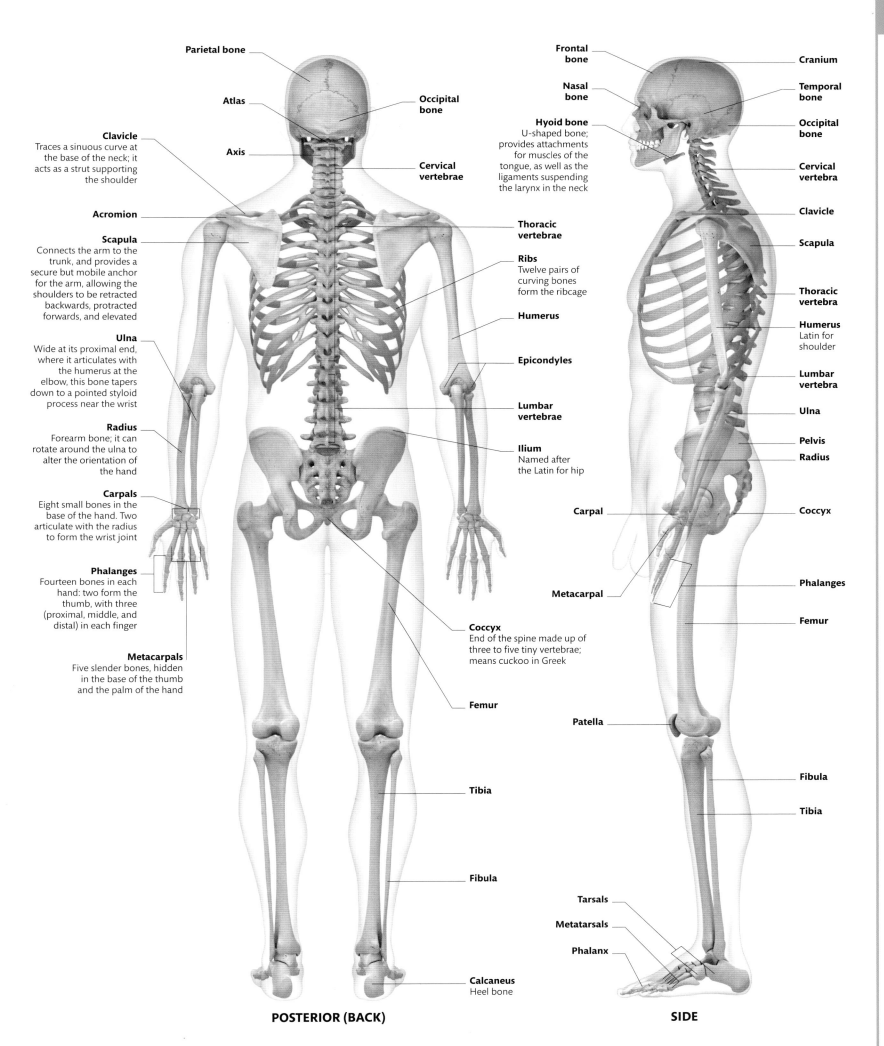

Parietal bone

Atlas

Occipital bone

Axis

Cervical vertebrae

Clavicle
Traces a sinuous curve at the base of the neck; it acts as a strut supporting the shoulder

Acromion

Scapula
Connects the arm to the trunk, and provides a secure but mobile anchor for the arm, allowing the shoulders to be retracted backwards, protracted forwards, and elevated

Ulna
Wide at its proximal end, where it articulates with the humerus at the elbow, this bone tapers down to a pointed styloid process near the wrist

Radius
Forearm bone; it can rotate around the ulna to alter the orientation of the hand

Carpals
Eight small bones in the base of the hand. Two articulate with the radius to form the wrist joint

Phalanges
Fourteen bones in each hand: two form the thumb, with three (proximal, middle, and distal) in each finger

Metacarpals
Five slender bones, hidden in the base of the thumb and the palm of the hand

Thoracic vertebrae

Ribs
Twelve pairs of curving bones form the ribcage

Humerus

Epicondyles

Lumbar vertebrae

Ilium
Named after the Latin for hip

Coccyx
End of the spine made up of three to five tiny vertebrae; means cuckoo in Greek

Femur

Tibia

Fibula

Calcaneus
Heel bone

POSTERIOR (BACK)

Frontal bone

Nasal bone

Hyoid bone
U-shaped bone; provides attachments for muscles of the tongue, as well as the ligaments suspending the larynx in the neck

Cranium

Temporal bone

Occipital bone

Cervical vertebra

Clavicle

Scapula

Thoracic vertebra

Humerus
Latin for shoulder

Lumbar vertebra

Ulna

Pelvis

Radius

Carpal

Coccyx

Metacarpal

Phalanges

Femur

Patella

Fibula

Tibia

Tarsals

Metatarsals

Phalanx

SIDE

HEAD AND NECK

The skull comprises the cranium and mandible. It houses and protects the brain and the eyes, ears, nose, and mouth. It encloses the first parts of the airway and of the alimentary tract, and provides attachment for the muscles of the head and neck. The cranium itself comprises more than 20 bones that meet each other at fibrous joints called sutures. In addition to the main bones labelled on these pages, there are sometimes extra bones along the sutures. In a young adult skull, the sutures are visible as tortuous lines between the cranial bones; they gradually fuse with age. The mandible of a newborn baby is in two halves, with a fibrous joint in the middle. The joint fuses during early infancy, so that the mandible becomes a single bone.

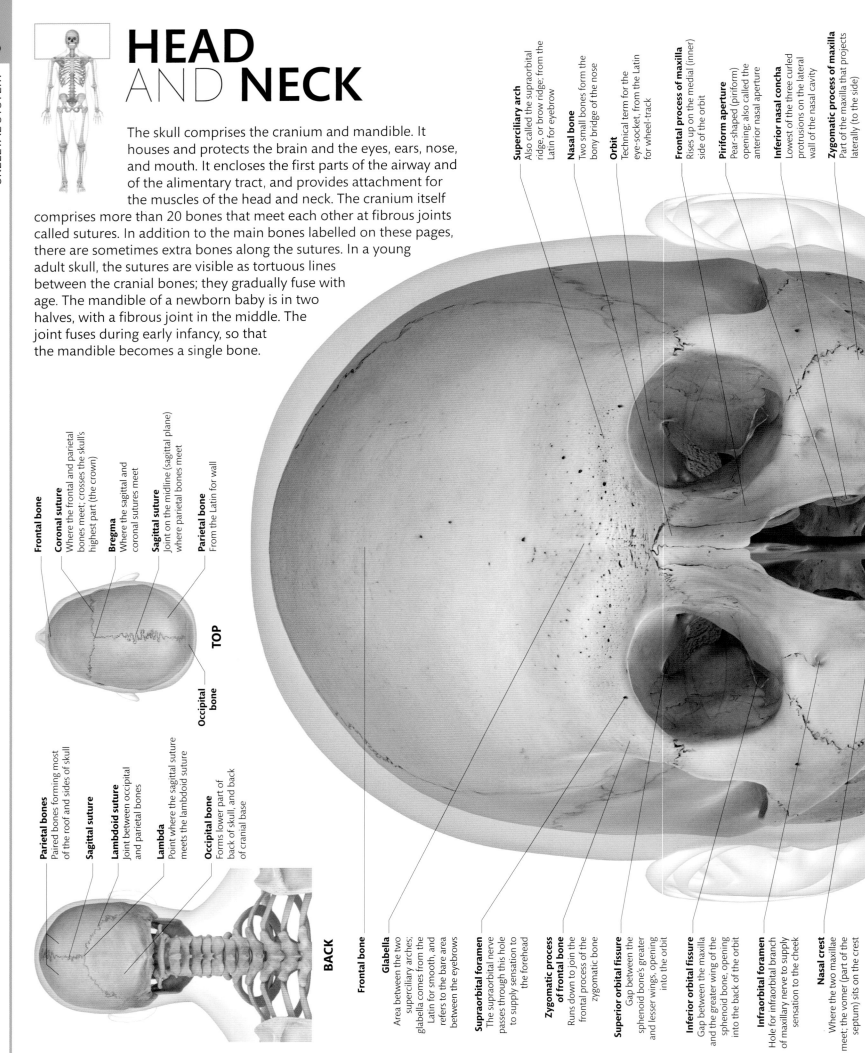

TOP

Frontal bone

Coronal suture
Where the frontal and parietal bones meet; crosses the skull's highest part (the crown)

Bregma
Where the sagittal and coronal sutures meet

Sagittal suture
Joint on the midline (sagittal plane) where parietal bones meet

Parietal bone
From the Latin for wall

BACK

Parietal bones
Paired bones forming most of the roof and sides of skull

Sagittal suture

Lambdoid suture
Joint between occipital and parietal bones

Lambda
Point where the sagittal suture meets the lambdoid suture

Occipital bone
Forms lower part of back of skull, and back of cranial base

Occipital bone

Frontal bone

Glabella
Area between the two superciliary arches; glabella comes from the Latin for smooth, and refers to the bare area between the eyebrows

Supraorbital foramen
The supraorbital nerve passes through this hole to supply sensation to the forehead

Zygomatic process of frontal bone
Runs down to join the frontal process of the zygomatic bone

Superior orbital fissure
Gap between the sphenoid bone's greater and lesser wings, opening into the orbit

Inferior orbital fissure
Gap between the maxilla and the greater wing of the sphenoid bone, opening into the back of the orbit

Infraorbital foramen
Hole for infraorbital branch of maxillary nerve to supply sensation to the cheek

Nasal crest
Where the two maxillae meet; the vomer (part of the septum) sits on the crest

Superciliary arch
Also called the supraorbital ridge, or brow ridge; from the Latin for eyebrow

Nasal bone
Two small bones form the bony bridge of the nose

Orbit
Technical term for the eye-socket, from the Latin for wheel-track

Frontal process of maxilla
Rises up on the medial (inner) side of the orbit

Piriform aperture
Pear-shaped (piriform) opening; also called the anterior nasal aperture

Inferior nasal concha
Lowest of the three curled protrusions on the lateral wall of the nasal cavity

Zygomatic process of maxilla
Part of the maxilla that projects laterally (to the side)

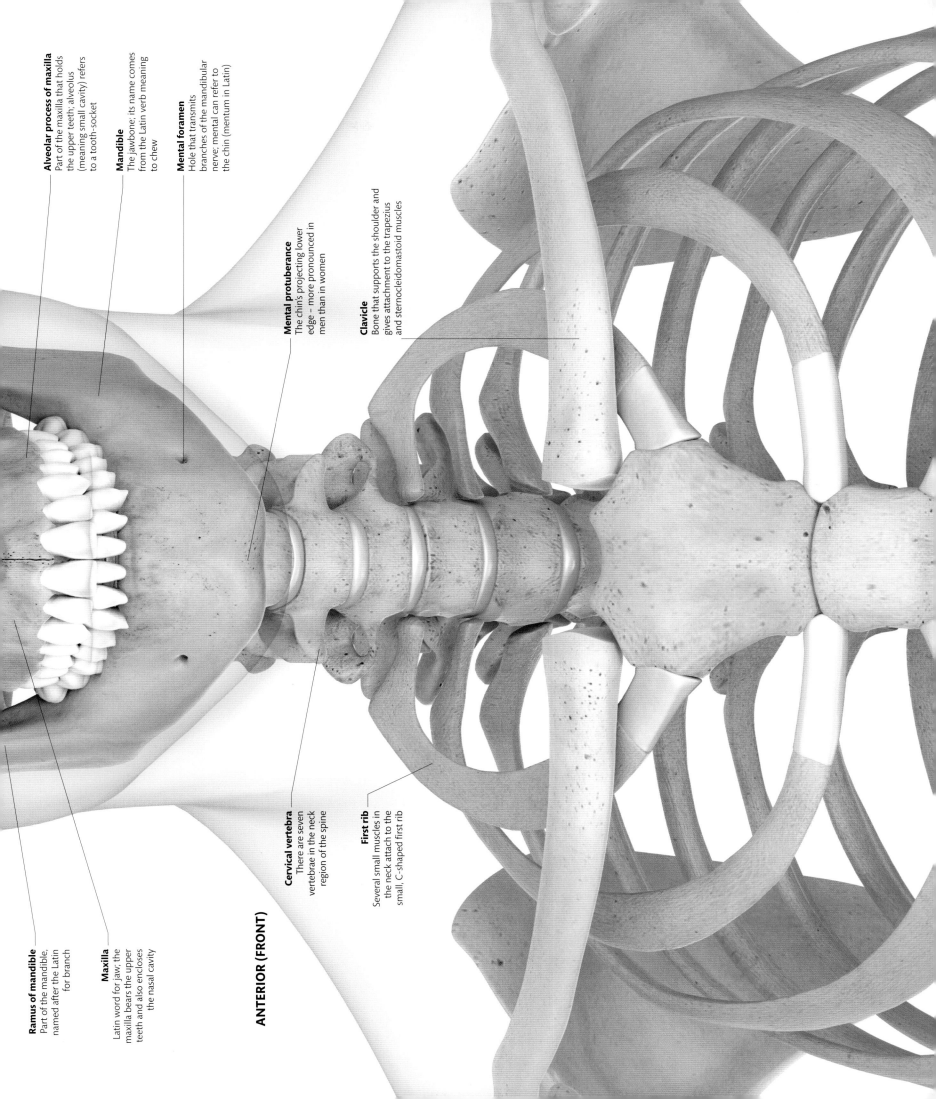

Ramus of mandible
Part of the mandible, named after the Latin for branch

Maxilla
Latin word for jaw; the maxilla bears the upper teeth and also encloses the nasal cavity

ANTERIOR (FRONT)

Alveolar process of maxilla
Part of the maxilla that holds the upper teeth; alveolus (meaning small cavity) refers to a tooth-socket

Mandible
The jawbone; its name comes from the Latin verb meaning to chew

Mental foramen
Hole that transmits branches of the mandibular nerve; mental can refer to the chin (mentum in Latin)

Mental protuberance
The chin's projecting lower edge – more pronounced in men than in women

Clavicle
Bone that supports the shoulder and gives attachment to the trapezius and sternocleidomastoid muscles

Cervical vertebra
There are seven vertebrae in the neck region of the spine

First rib
Several small muscles in the neck attach to the small, C-shaped first rib

HEAD AND NECK

The cervical spine includes seven vertebrae, the top two of which have specific names. The first vertebra, which supports the skull, is called the atlas, after the Greek god who carried the sky on his shoulders. Nodding movements of the head occur at the joint between the atlas and the skull. The second cervical vertebra is the axis, so-called because when you shake your head from side to side, the atlas rotates on the axis. In this side view, we can also see more of the bones that make up the cranium, as well as the temporomandibular (jaw) joint between the mandible and the skull. The hyoid bone is also visible. This small bone is a very important anchor for the muscles that form the tongue and the floor of the mouth, as well as muscles that attach to the larynx and pharynx.

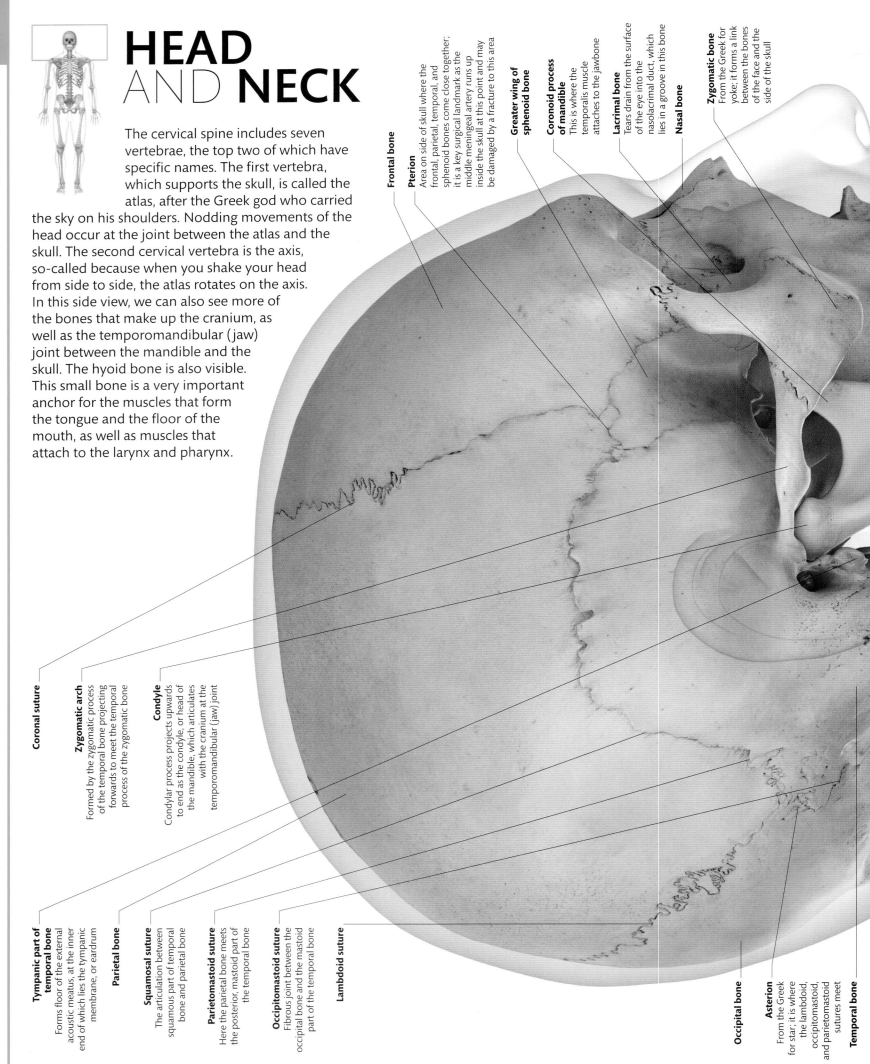

Frontal bone

Pterion
Area on side of skull where the frontal, parietal, temporal, and sphenoid bones come close together; it is a key surgical landmark as the middle meningeal artery runs up inside the skull at this point and may be damaged by a fracture to this area

Greater wing of sphenoid bone

Coronoid process of mandible
This is where the temporalis muscle attaches to the jawbone

Lacrimal bone
Tears drain from the surface of the eye into the nasolacrimal duct, which lies in a groove in this bone

Nasal bone

Zygomatic bone
From the Greek for yoke; it forms a link between the bones of the face and the side of the skull

Coronal suture

Zygomatic arch
Formed by the zygomatic process of the temporal bone projecting forwards to meet the temporal process of the zygomatic bone

Condyle
Condylar process projects upwards to end as the condyle, or head of the mandible, which articulates with the cranium at the temporomandibular (jaw) joint

Tympanic part of temporal bone
Forms floor of the external acoustic meatus, at the inner end of which lies the tympanic membrane, or eardrum

Parietal bone

Squamosal suture
The articulation between squamous part of temporal bone and parietal bone

Parietomastoid suture
Here the parietal bone meets the posterior, mastoid part of the temporal bone

Occipitomastoid suture
Fibrous joint between the occipital bone and the mastoid part of the temporal bone

Lambdoid suture

Occipital bone

Asterion
From the Greek for star; it is where the lambdoid, occipitomastoid, and parietomastoid sutures meet

Temporal bone

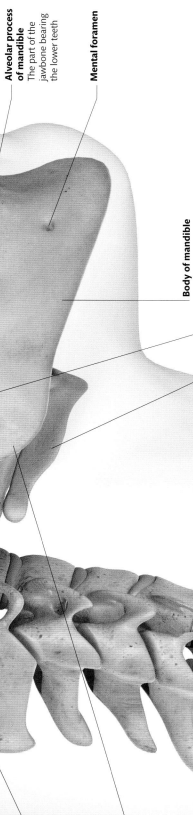

Maxilla

Alveolar process of mandible
The part of the jawbone bearing the lower teeth

Mental foramen

Body of mandible

Ramus of mandible

Hyoid bone
Takes its name from the Greek for U-shaped; it is a separate bone, lying just under the mandible, which provides an anchor for muscles forming the floor of the mouth and the tongue; the larynx hangs below it

Styloid process
Named after the Greek for pillar, this pointed projection sticks out under the skull and forms an anchor for several slender muscles and ligaments

Mastoid process
The name of this conical projection under the skull comes from the Greek for breast

Angle of mandible
Where the body of the mandible turns a corner to become the ramus

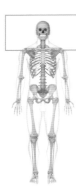

HEAD AND NECK

The most striking features of the skull viewed from these angles are the holes in it. In the middle, there is one large hole – the foramen magnum – through which the brainstem emerges to become the spinal cord. But there are also many smaller holes, most of them paired. Through these holes, the cranial nerves from the brain escape to supply the muscles, skin, and mucosa, and the glands of the head and neck. Blood vessels also pass through some holes, on their way to and from the brain. At the front, we can also see the upper teeth sitting in their sockets in the maxillae, and the bony, hard palate.

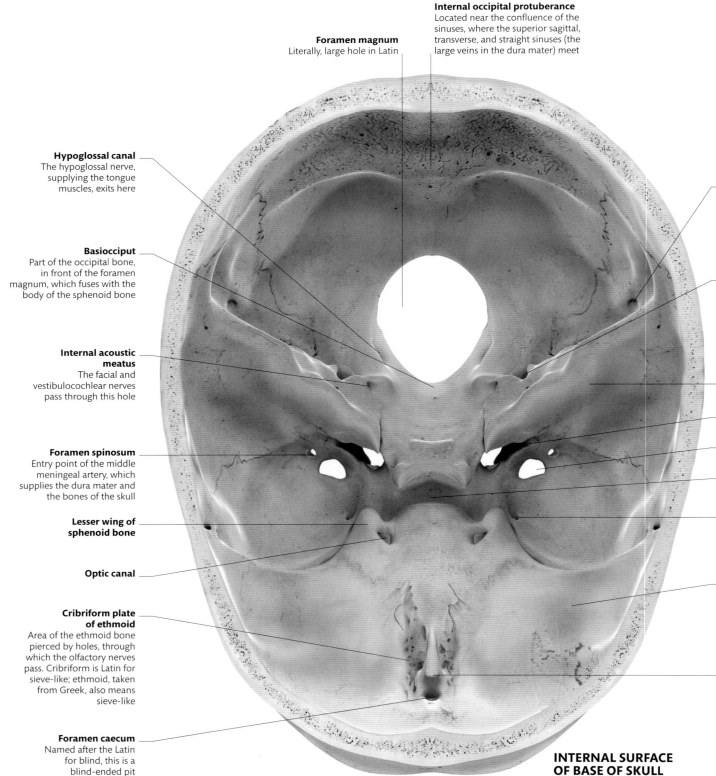

Internal occipital protuberance
Located near the confluence of the sinuses, where the superior sagittal, transverse, and straight sinuses (the large veins in the dura mater) meet

Foramen magnum
Literally, large hole in Latin

Hypoglossal canal
The hypoglossal nerve, supplying the tongue muscles, exits here

Basiocciput
Part of the occipital bone, in front of the foramen magnum, which fuses with the body of the sphenoid bone

Internal acoustic meatus
The facial and vestibulocochlear nerves pass through this hole

Foramen spinosum
Entry point of the middle meningeal artery, which supplies the dura mater and the bones of the skull

Lesser wing of sphenoid bone

Optic canal

Cribriform plate of ethmoid
Area of the ethmoid bone pierced by holes, through which the olfactory nerves pass. Cribriform is Latin for sieve-like; ethmoid, taken from Greek, also means sieve-like

Foramen caecum
Named after the Latin for blind, this is a blind-ended pit

Mastoid foramen
An emissary (valveless) vein passes out through this hole

Jugular foramen
The internal jugular vein and the glossopharyngeal, vagus, and accessory nerves emerge from this hole

Petrous part of temporal bone

Foramen lacerum

Foramen ovale

Pituitary fossa

Foramen rotundum
The maxillary division of the trigeminal nerve passes through this round hole

Orbital part of frontal bone
Part of the frontal bone that forms the roof of the orbit, and also the floor at the front of the cranial cavity

Crista galli
Vertical crest on the ethmoid bone that takes its name from the Latin for cock's comb; it provides attachment for the falx cerebri – the membrane between the two cerebral hemispheres

INTERNAL SURFACE OF BASE OF SKULL

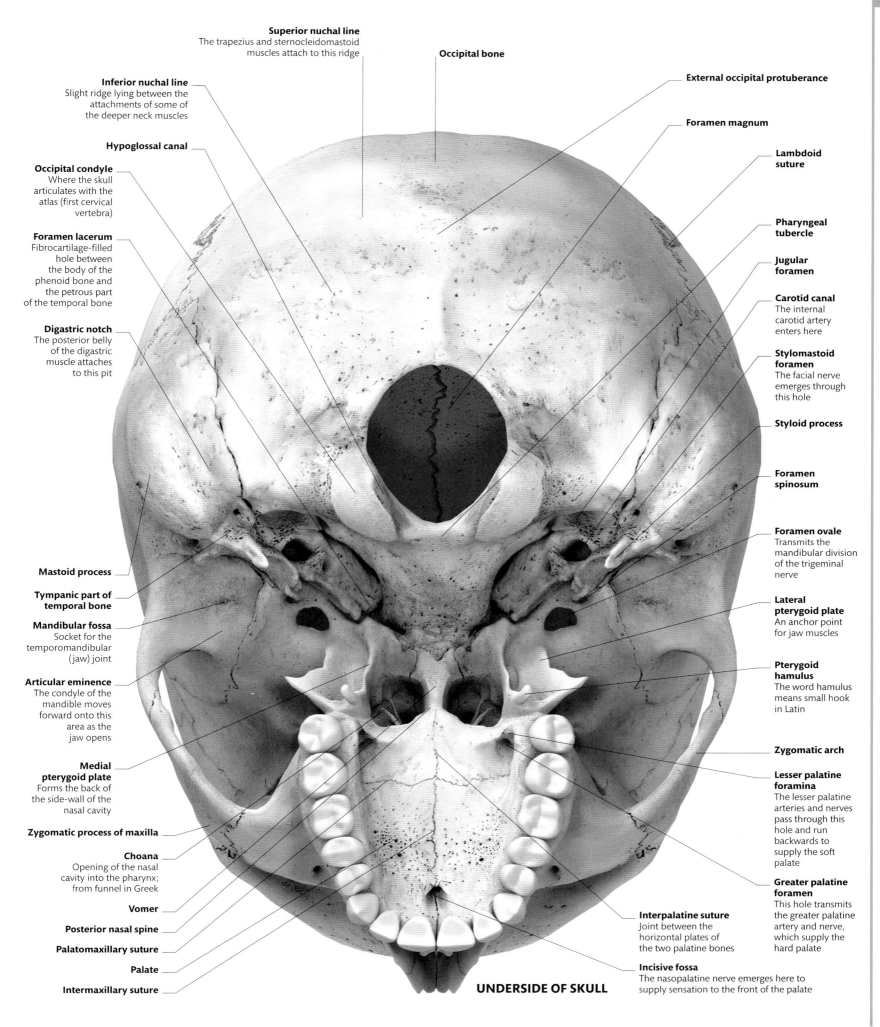

Superior nuchal line
The trapezius and sternocleidomastoid muscles attach to this ridge

Inferior nuchal line
Slight ridge lying between the attachments of some of the deeper neck muscles

Hypoglossal canal

Occipital condyle
Where the skull articulates with the atlas (first cervical vertebra)

Foramen lacerum
Fibrocartilage-filled hole between the body of the phenoid bone and the petrous part of the temporal bone

Digastric notch
The posterior belly of the digastric muscle attaches to this pit

Mastoid process

Tympanic part of temporal bone

Mandibular fossa
Socket for the temporomandibular (jaw) joint

Articular eminence
The condyle of the mandible moves forward onto this area as the jaw opens

Medial pterygoid plate
Forms the back of the side-wall of the nasal cavity

Zygomatic process of maxilla

Choana
Opening of the nasal cavity into the pharynx; from funnel in Greek

Vomer

Posterior nasal spine

Palatomaxillary suture

Palate

Intermaxillary suture

Occipital bone

External occipital protuberance

Foramen magnum

Lambdoid suture

Pharyngeal tubercle

Jugular foramen

Carotid canal
The internal carotid artery enters here

Stylomastoid foramen
The facial nerve emerges through this hole

Styloid process

Foramen spinosum

Foramen ovale
Transmits the mandibular division of the trigeminal nerve

Lateral pterygoid plate
An anchor point for jaw muscles

Pterygoid hamulus
The word hamulus means small hook in Latin

Zygomatic arch

Lesser palatine foramina
The lesser palatine arteries and nerves pass through this hole and run backwards to supply the soft palate

Greater palatine foramen
This hole transmits the greater palatine artery and nerve, which supply the hard palate

Interpalatine suture
Joint between the horizontal plates of the two palatine bones

Incisive fossa
The nasopalatine nerve emerges here to supply sensation to the front of the palate

UNDERSIDE OF SKULL

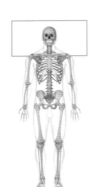

HEAD AND NECK

This section – right through the middle of the skull – lets us in on some secrets. We can clearly appreciate the size of the cranial cavity, which is almost completely filled by the brain, with just a small gap for membranes, fluid, and blood vessels. Some of those blood vessels leave deep grooves on the inner surface of the skull: we can trace the course of the large venous sinuses and the branches of the middle meningeal artery. We can also see that the skull bones are not solid, but contain trabecular bone (or diploe), which itself contains red marrow. Some skull bones also contain air spaces, like the sphenoidal sinus visible here. We can also appreciate the large size of the nasal cavity, hidden away inside the skull.

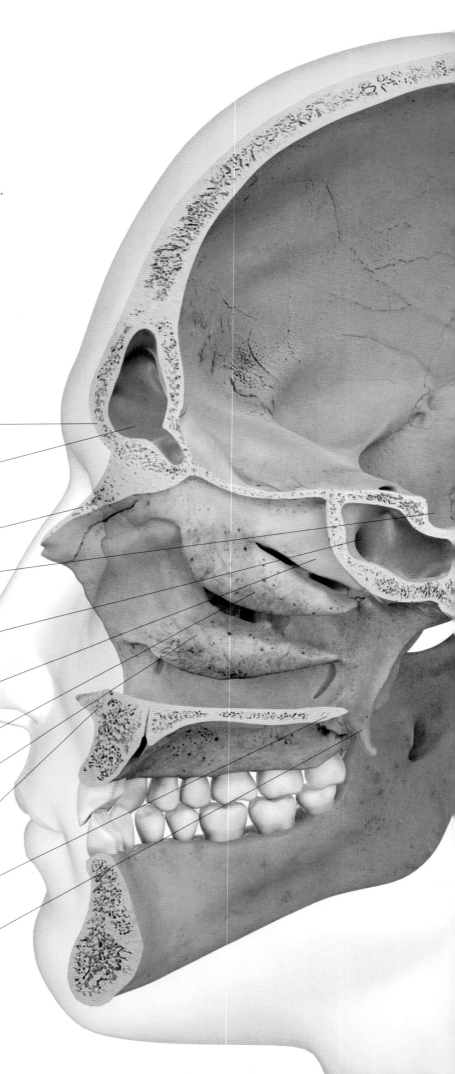

Frontal bone
Forms the anterior cranial fossa, where the frontal lobes of the brain lie, inside the skull

Frontal sinus
One of the paranasal air sinuses that drain into the nasal cavity, this is an air space within the frontal bone

Nasal bone

Pituitary fossa
Fossa is the Latin word for ditch; the pituitary gland occupies this small cavity on the upper surface of the sphenoid bone

Sphenoidal sinus
Another paranasal air sinus; it lies within the body of the sphenoid bone

Superior nasal concha
Part of the ethmoid bone, which forms the roof and upper side-walls of the nasal cavity

Middle nasal concha
Like the superior nasal concha, this is also part of the ethmoid bone

Inferior nasal concha
A separate bone, attached to the inner surface of the maxilla; the conchae increase the surface area of the nasal cavity

Anterior nasal crest

Palatine bone
Joins to the maxilla and forms the back of the hard palate

Pterygoid process
Sticking down from the greater wing of the sphenoid bone, this process flanks the back of the nasal cavity and provides attachment for muscles of the palate and jaw

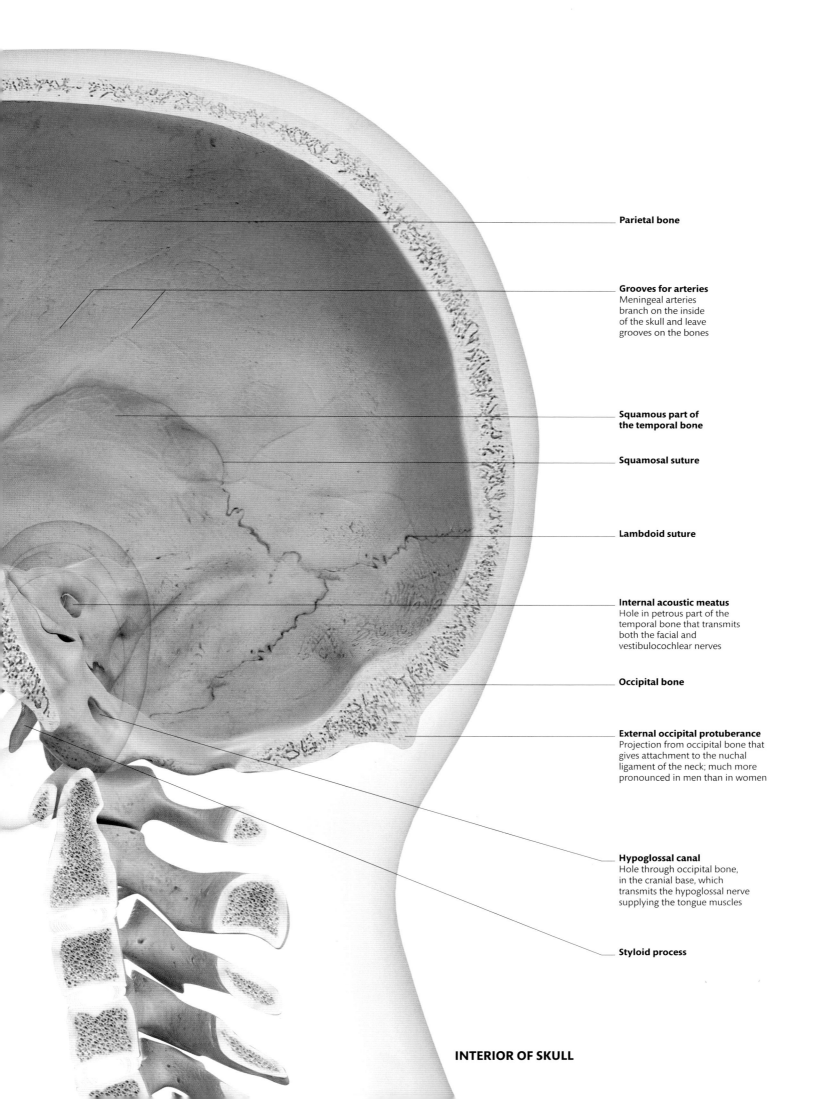

Parietal bone

Grooves for arteries
Meningeal arteries
branch on the inside
of the skull and leave
grooves on the bones

**Squamous part of
the temporal bone**

Squamosal suture

Lambdoid suture

Internal acoustic meatus
Hole in petrous part of the
temporal bone that transmits
both the facial and
vestibulocochlear nerves

Occipital bone

External occipital protuberance
Projection from occipital bone that
gives attachment to the nuchal
ligament of the neck; much more
pronounced in men than in women

Hypoglossal canal
Hole through occipital bone,
in the cranial base, which
transmits the hypoglossal nerve
supplying the tongue muscles

Styloid process

INTERIOR OF SKULL

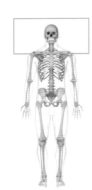

HEAD AND NECK

In this view of the skull, we can clearly see that it is not a single bone, and we can also see how the various cranial bones fit together to produce the shape we are more familiar with. The butterfly-shaped sphenoid bone is right in the middle of the action – it forms part of the skull base, the orbits, and the side walls of the skull, and it articulates with many of the other bones of the skull. The temporal bones also form part of the skull's base and side walls. The extremely dense petrous parts of the temporal bones contain and protect the delicate workings of the ear, including the tiny ossicles (malleus, incus, and stapes) that transmit vibrations from the eardrum to the inner ear.

FIBROUS JOINTS

In places, the connective tissue between developing bones solidifies to create fibrous joints. Linked by microscopic fibres of collagen, these fixed joints anchor the edges of adjacent bones, or bone and tooth, so that they are locked together. Such joints include the sutures of the skull, the teeth sockets (gomphoses), and the lower joint between the tibia and fibula.

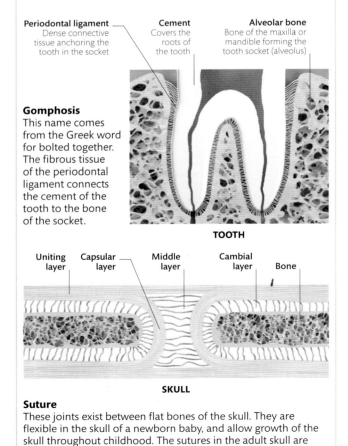

Periodontal ligament
Dense connective tissue anchoring the tooth in the socket

Cement
Covers the roots of the tooth

Alveolar bone
Bone of the maxilla or mandible forming the tooth socket (alveolus)

Gomphosis
This name comes from the Greek word for bolted together. The fibrous tissue of the periodontal ligament connects the cement of the tooth to the bone of the socket.

TOOTH

Uniting layer | Capsular layer | Middle layer | Cambial layer | Bone

SKULL

Suture
These joints exist between flat bones of the skull. They are flexible in the skull of a newborn baby, and allow growth of the skull throughout childhood. The sutures in the adult skull are interlocking, practically immovable joints, and eventually fuse completely in later adulthood.

Parietal bone

Frontal bone
Forms joints with the parietal and sphenoid bones on the top and sides of the skull, and with the maxilla, nasal, lacrimal, and ethmoid bones below

Occipital bone

Parietal bone
Forms the roof and side of the skull

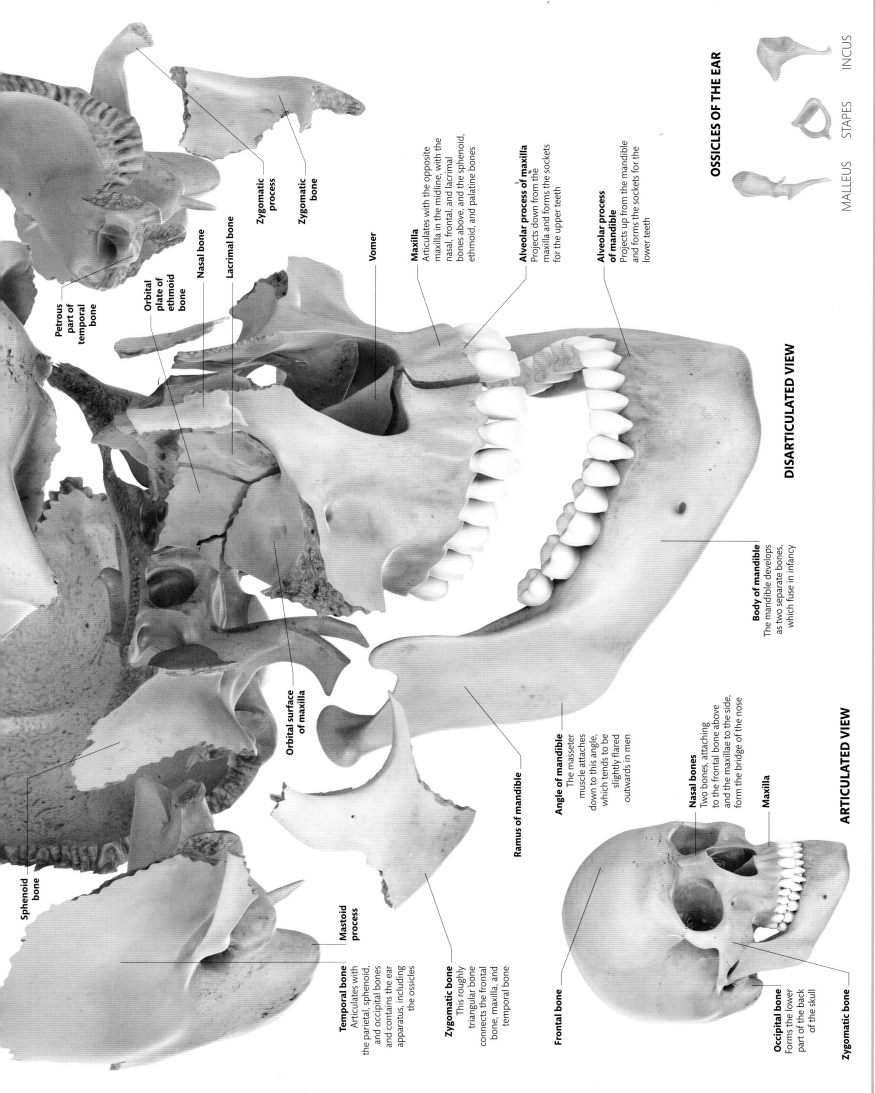

OSSICLES OF THE EAR

MALLEUS   STAPES   INCUS

DISARTICULATED VIEW

Petrous part of temporal bone

Orbital plate of ethmoid bone

Nasal bone

Lacrimal bone

Zygomatic process

Zygomatic bone

Vomer

Maxilla
Articulates with the opposite maxilla in the midline, with the nasal, frontal, and lacrimal bones above, and the sphenoid, ethmoid, and palatine bones

Alveolar process of maxilla
Projects down from the maxilla and forms the sockets for the upper teeth

Alveolar process of mandible
Projects up from the mandible and forms the sockets for the lower teeth

Body of mandible
The mandible develops as two separate bones, which fuse in infancy

Sphenoid bone

Mastoid process

Temporal bone
Articulates with the parietal, sphenoid, and occipital bones and contains the ear apparatus, including the ossicles

Zygomatic bone
This roughly triangular bone connects the frontal bone, maxilla, and temporal bone

Orbital surface of maxilla

Ramus of mandible

Angle of mandible
The masseter muscle attaches down to this angle, which tends to be slightly flared outwards in men

Nasal bones
Two bones, attaching to the frontal bone above and the maxillae to the side, form the bridge of the nose

Maxilla

Frontal bone

Occipital bone
Forms the lower part of the back of the skull

Zygomatic bone

ARTICULATED VIEW

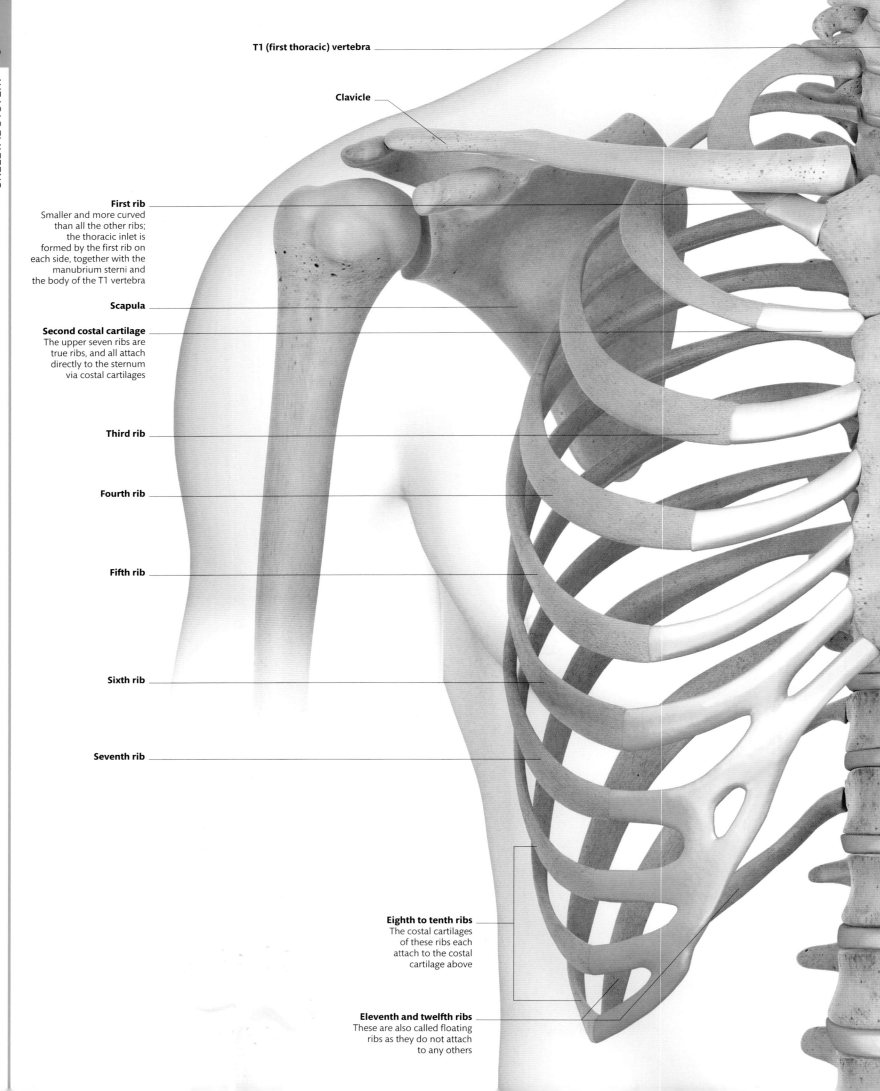

T1 (first thoracic) vertebra

Clavicle

First rib
Smaller and more curved than all the other ribs; the thoracic inlet is formed by the first rib on each side, together with the manubrium sterni and the body of the T1 vertebra

Scapula

Second costal cartilage
The upper seven ribs are true ribs, and all attach directly to the sternum via costal cartilages

Third rib

Fourth rib

Fifth rib

Sixth rib

Seventh rib

Eighth to tenth ribs
The costal cartilages of these ribs each attach to the costal cartilage above

Eleventh and twelfth ribs
These are also called floating ribs as they do not attach to any others

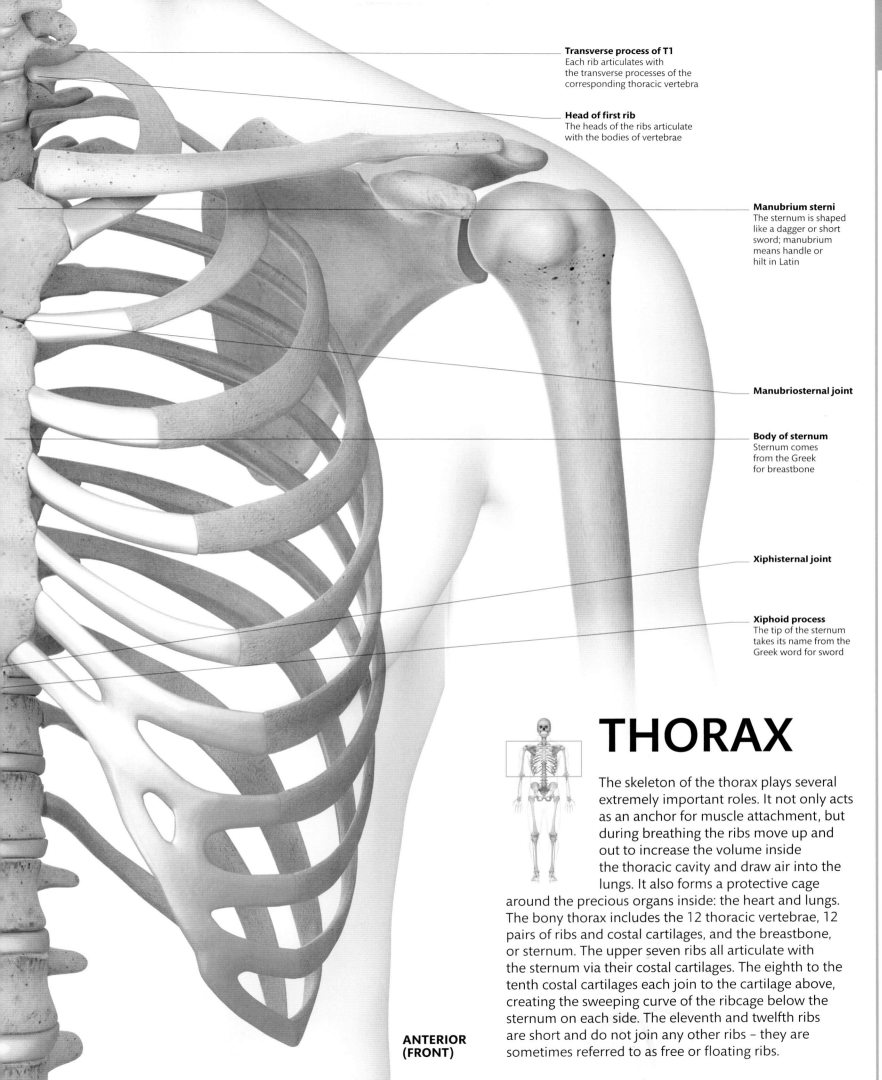

Transverse process of T1
Each rib articulates with the transverse processes of the corresponding thoracic vertebra

Head of first rib
The heads of the ribs articulate with the bodies of vertebrae

Manubrium sterni
The sternum is shaped like a dagger or short sword; manubrium means handle or hilt in Latin

Manubriosternal joint

Body of sternum
Sternum comes from the Greek for breastbone

Xiphisternal joint

Xiphoid process
The tip of the sternum takes its name from the Greek word for sword

**ANTERIOR
(FRONT)**

THORAX

The skeleton of the thorax plays several extremely important roles. It not only acts as an anchor for muscle attachment, but during breathing the ribs move up and out to increase the volume inside the thoracic cavity and draw air into the lungs. It also forms a protective cage around the precious organs inside: the heart and lungs. The bony thorax includes the 12 thoracic vertebrae, 12 pairs of ribs and costal cartilages, and the breastbone, or sternum. The upper seven ribs all articulate with the sternum via their costal cartilages. The eighth to the tenth costal cartilages each join to the cartilage above, creating the sweeping curve of the ribcage below the sternum on each side. The eleventh and twelfth ribs are short and do not join any other ribs – they are sometimes referred to as free or floating ribs.

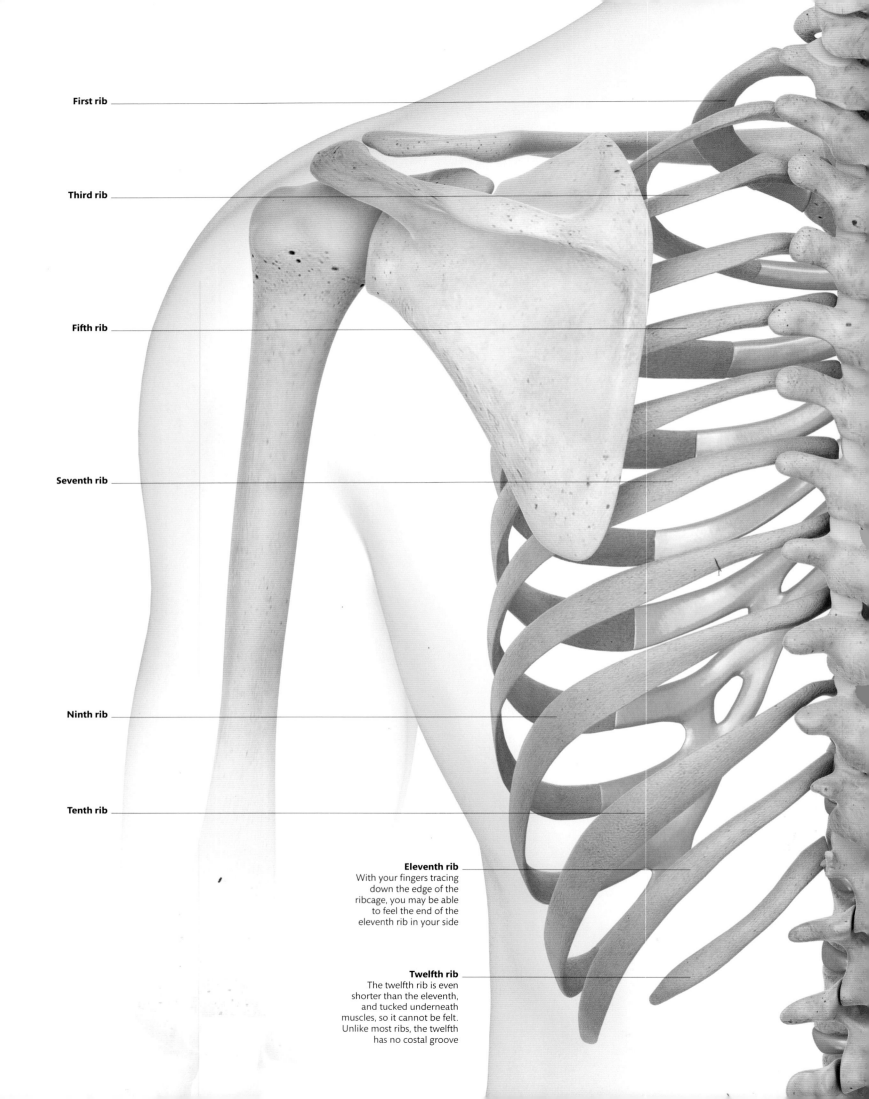

First rib

Third rib

Fifth rib

Seventh rib

Ninth rib

Tenth rib

Eleventh rib
With your fingers tracing
down the edge of the
ribcage, you may be able
to feel the end of the
eleventh rib in your side

Twelfth rib
The twelfth rib is even
shorter than the eleventh,
and tucked underneath
muscles, so it cannot be felt.
Unlike most ribs, the twelfth
has no costal groove

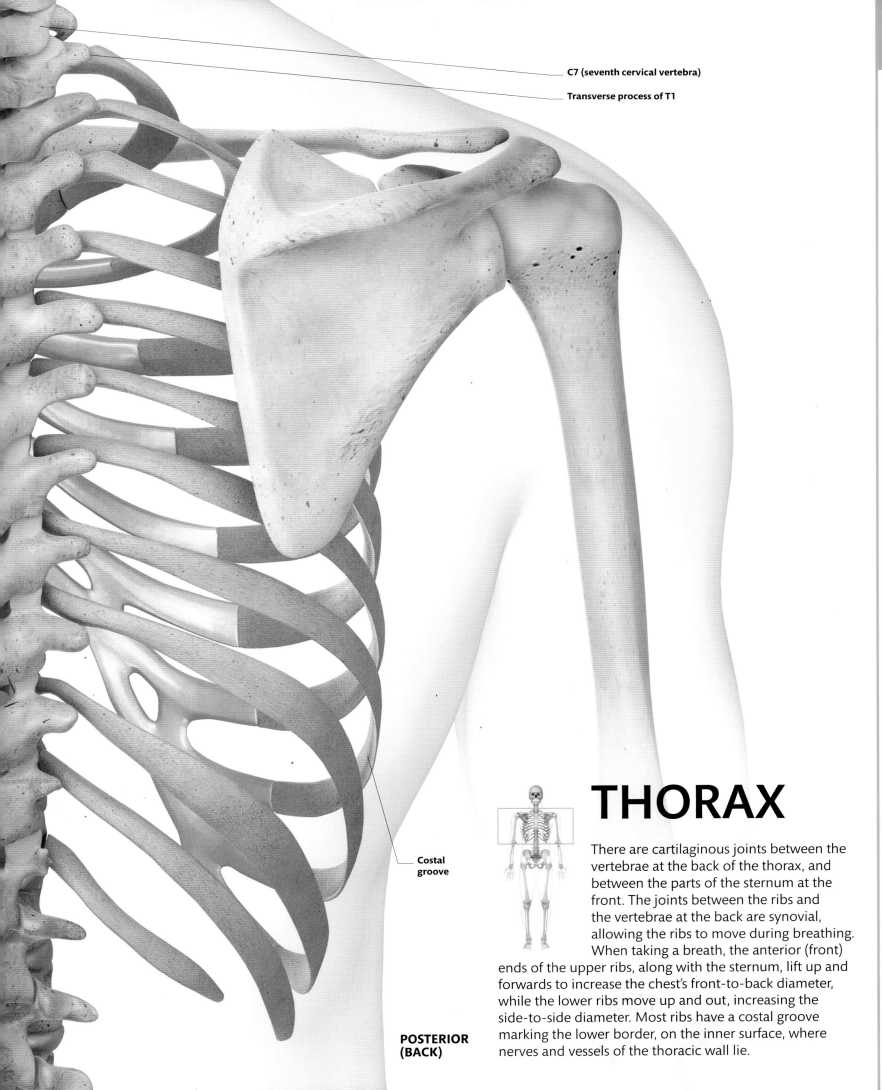

C7 (seventh cervical vertebra)

Transverse process of T1

Costal groove

THORAX

There are cartilaginous joints between the vertebrae at the back of the thorax, and between the parts of the sternum at the front. The joints between the ribs and the vertebrae at the back are synovial, allowing the ribs to move during breathing. When taking a breath, the anterior (front) ends of the upper ribs, along with the sternum, lift up and forwards to increase the chest's front-to-back diameter, while the lower ribs move up and out, increasing the side-to-side diameter. Most ribs have a costal groove marking the lower border, on the inner surface, where nerves and vessels of the thoracic wall lie.

POSTERIOR (BACK)

SPINE

The spine, or vertebral column, occupies a central position in the skeleton, and plays several extremely important roles: it supports the trunk, encloses and protects the spinal cord, provides sites for muscle attachment, and contains blood-forming bone marrow. The entire vertebral column is about 70cm (28in) long in men, and 60cm (24in) long in women. About a quarter of this length is made up by the cartilaginous intervertebral discs between the vertebrae. The number of vertebrae varies from 32 to 35, mostly due to variation in the number of small vertebrae that make up the coccyx. Although there is a general pattern for a vertebra – most possess a body, a neural arch, and spinous and transverse processes – there are recognizable features that mark out the vertebrae of each section of the spine.

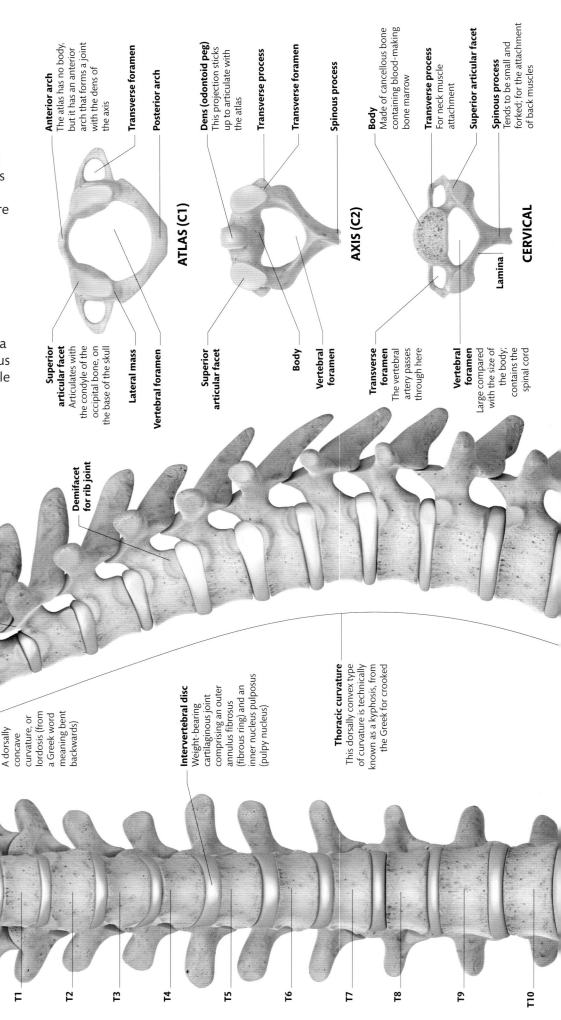

ATLAS (C1)

Anterior arch
The atlas has no body, but it has an anterior arch that forms a joint with the dens of the axis

Transverse foramen

Posterior arch

Superior articular facet
Articulates with the condyle of the occipital bone, on the base of the skull

Lateral mass

Vertebral foramen

AXIS (C2)

Dens (odontoid peg)
This projection sticks up to articulate with the atlas

Transverse process

Transverse foramen

Spinous process

Superior articular facet

Body

Vertebral foramen

CERVICAL

Body
Made of cancellous bone containing blood-making bone marrow

Transverse process
For neck muscle attachment

Superior articular facet

Spinous process
Tends to be small and forked; for the attachment of back muscles

Transverse foramen
The vertebral artery passes through here

Vertebral foramen
Large compared with the size of the body; contains the spinal cord

Lamina

Intervertebral foramen
These are the holes between adjacent vertebrae through which spinal nerves emerge

Superior articular process

Demifacet for rib joint

Cervical curvature
A dorsally concave curvature, or lordosis (from a Greek word meaning bent backwards)

Intervertebral disc
Weight-bearing cartilaginous joint comprising an outer annulus fibrosus (fibrous ring) and an inner nucleus pulposus (pulpy nucleus)

Thoracic curvature
This dorsally convex type of curvature is technically known as a kyphosis, from the Greek for crooked

C1 (atlas)
C2 (axis)
C3
C4
C5
C6
C7
T1
T2
T3
T4
T5
T6
T7
T8
T9
T10

Cervical spine
(Seven vertebrae make up the spine in the neck)

Thoracic spine
(Twelve vertebrae, providing attachment for twelve pairs of ribs)

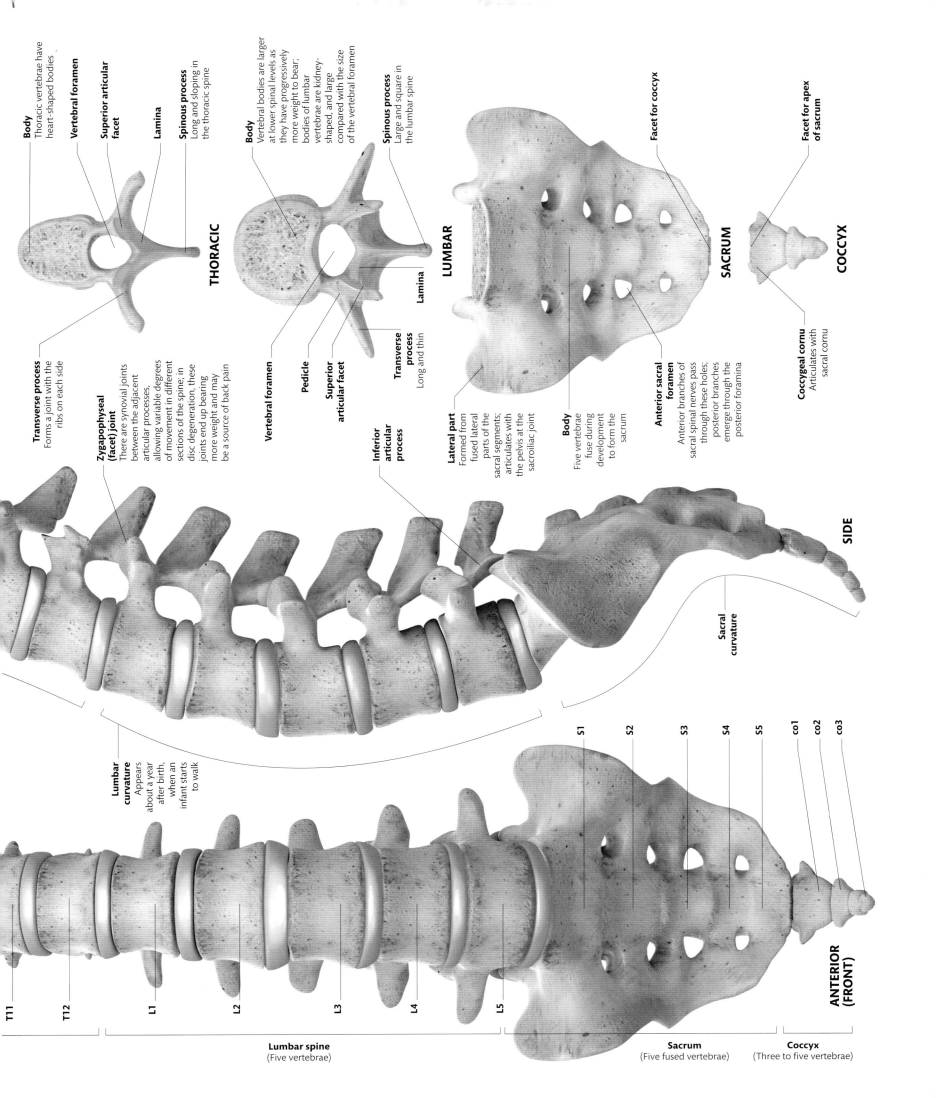

Body
Thoracic vertebrae have heart-shaped bodies

Vertebral foramen

Superior articular facet

Lamina

Spinous process
Long and sloping in the thoracic spine

THORACIC

Transverse process
Forms a joint with the ribs on each side

Zygapophyseal (facet) joint
There are synovial joints between the adjacent articular processes, allowing variable degrees of movement in different sections of the spine; in disc degeneration, these joints end up bearing more weight and may be a source of back pain

Body
Vertebral bodies are larger at lower spinal levels as they have progressively more weight to bear; bodies of lumbar vertebrae are kidney-shaped, and large compared with the size of the vertebral foramen

Spinous process
Large and square in the lumbar spine

Lamina

LUMBAR

Vertebral foramen

Pedicle

Superior articular facet

Transverse process
Long and thin

Inferior articular process

Lateral part
Formed from fused lateral parts of the sacral segments; articulates with the pelvis at the sacroiliac joint

Body
Five vertebrae fuse during development to form the sacrum

Anterior sacral foramen
Anterior branches of sacral spinal nerves pass through these holes; posterior branches emerge through the posterior foramina

SACRUM

Facet for coccyx

Facet for apex of sacrum

COCCYX

Coccygeal cornu
Articulates with sacral cornu

SIDE

Sacral curvature

Lumbar curvature
Appears about a year after birth, when an infant starts to walk

S1
S2
S3
S4
S5
co1
co2
co3

T11
T12
L1
L2
L3
L4
L5

ANTERIOR (FRONT)

Lumbar spine
(Five vertebrae)

Sacrum
(Five fused vertebrae)

Coccyx
(Three to five vertebrae)

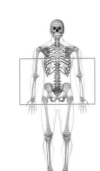

ABDOMEN AND PELVIS

The bony boundaries of the abdomen include the five lumbar vertebrae at the back, the lower margin of the ribs above, and the pubic bones and iliac crest of the pelvic bones below. The abdominal cavity itself extends up under the ribcage, as high as the gap between the fifth and sixth ribs, due to the domed shape of the diaphragm. This means that some abdominal organs, such as the liver, stomach, and spleen are, in fact, largely tucked up under the ribs. The pelvis is a basin shape, and is enclosed by the two pelvic (or innominate) bones, at the front and sides, and by the sacrum at the back. Each pelvic bone is made of three fused bones: the ilium at the rear, the ischium at the lower front, and the pubis above it.

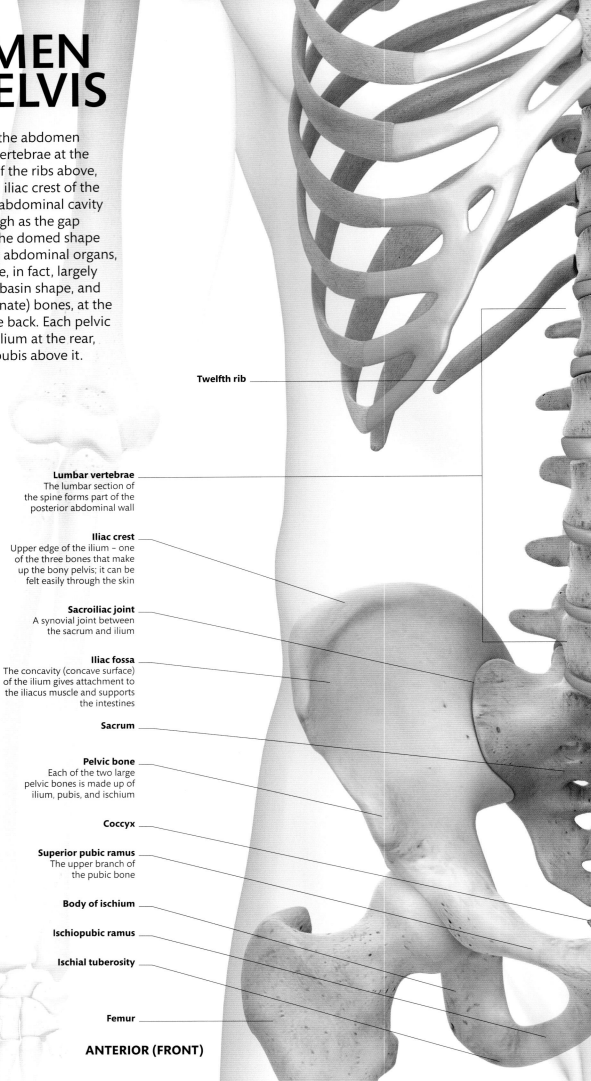

Twelfth rib

Lumbar vertebrae
The lumbar section of the spine forms part of the posterior abdominal wall

Iliac crest
Upper edge of the ilium – one of the three bones that make up the bony pelvis; it can be felt easily through the skin

Sacroiliac joint
A synovial joint between the sacrum and ilium

Iliac fossa
The concavity (concave surface) of the ilium gives attachment to the iliacus muscle and supports the intestines

Sacrum

Pelvic bone
Each of the two large pelvic bones is made up of ilium, pubis, and ischium

Coccyx

Superior pubic ramus
The upper branch of the pubic bone

Body of ischium

Ischiopubic ramus

Ischial tuberosity

Femur

ANTERIOR (FRONT)

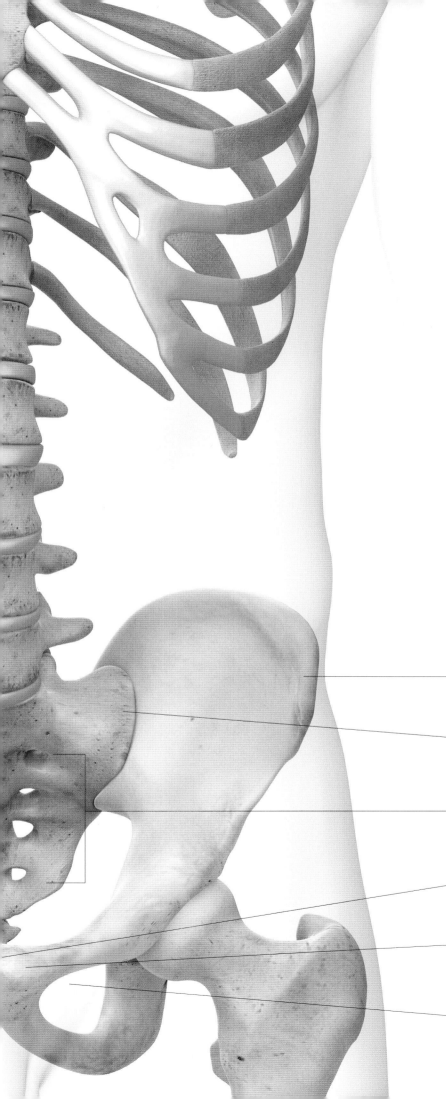

CARTILAGINOUS JOINTS

Semi-movable cartilaginous joints are formed of bones separated by a disc of resilient and compressible fibrocartilage, which allows limited movement. Cartilaginous joints include the junctions between ribs and costal cartilages, joints between the components of the sternum, and the pubic symphysis. The intervertebral discs are also specialized cartilaginous joints.

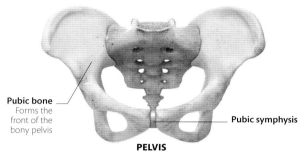

Pubic bone
Forms the front of the bony pelvis

Pubic symphysis

PELVIS

Pubic symphysis
At the front of the bony pelvis, the two pubic bones meet each other. The articular surface of each is covered with hyaline cartilage, with a pad of fibrocartilage joining them in the middle.

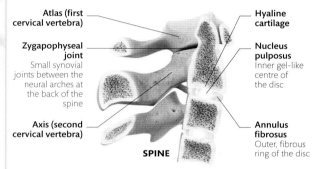

Atlas (first cervical vertebra)

Zygapophyseal joint
Small synovial joints between the neural arches at the back of the spine

Axis (second cervical vertebra)

Hyaline cartilage

Nucleus pulposus
Inner gel-like centre of the disc

Annulus fibrosus
Outer, fibrous ring of the disc

SPINE

Intervertebral disc
Each fibrocartilage pad or disc between vertebrae is organized into an outer annulus fibrosus and an inner nucleus pulposus.

Anterior superior iliac spine
This is the anterior (front) end of the iliac crest

Ala of sacrum
The bony masses to the sides of the sacrum are called the alae, which means wings in Latin

Anterior sacral foramina
Anterior (frontal) branches of the sacral spinal nerves pass out through these holes

Pubic symphysis
A cartilaginous joint between the two pubic bones

Pubic tubercle
This small bony projection provides an attachment point for the inguinal ligament

Obturator foramen
This hole is largely closed over by a membrane, with muscles attaching on either side; its name comes from the Latin for stopped up

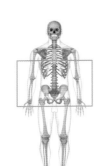

ABDOMEN
AND PELVIS

The orientation of the facet joints (the joints between the vertebrae) of the lumbar spine means that rotation of the vertebrae is limited, but flexion and extension can occur freely. There is, however, rotation at the lumbosacral joint, which allows the pelvis to swing during walking. The sacroiliac joints are unusual in that they are synovial joints (which are usually very movable), yet they are particularly limited in their movement. This is because strong sacroiliac ligaments around the joints bind the ilium (part of the pelvic bone) tightly to the sacrum on each side. Lower down, the sacrospinous and sacrotuberous ligaments, stretching from the sacrum and coccyx to the ilium, provide additional support and stability.

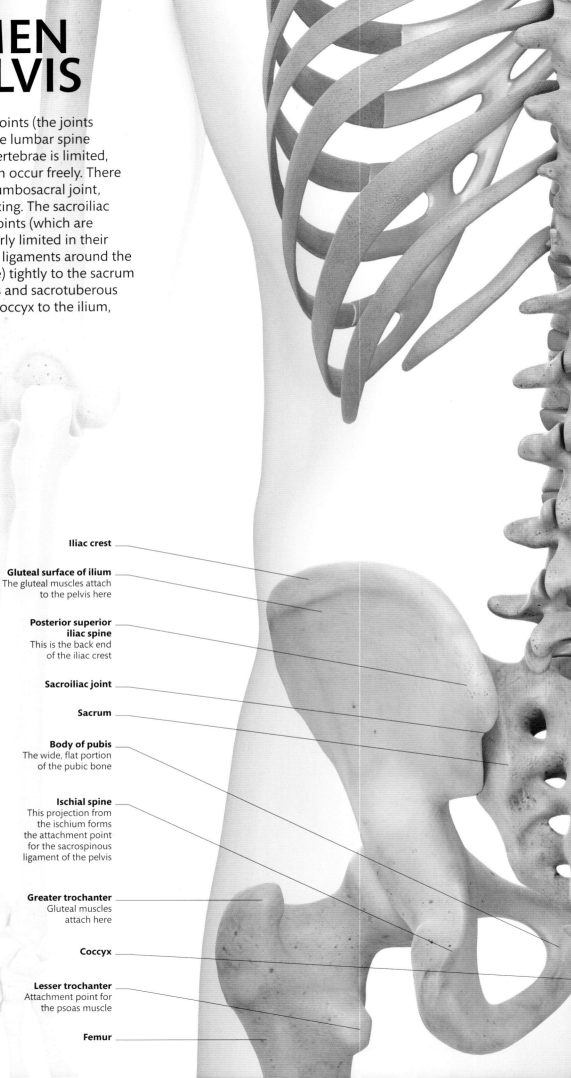

Iliac crest

Gluteal surface of ilium
The gluteal muscles attach
to the pelvis here

**Posterior superior
iliac spine**
This is the back end
of the iliac crest

Sacroiliac joint

Sacrum

Body of pubis
The wide, flat portion
of the pubic bone

Ischial spine
This projection from
the ischium forms
the attachment point
for the sacrospinous
ligament of the pelvis

Greater trochanter
Gluteal muscles
attach here

Coccyx

Lesser trochanter
Attachment point for
the psoas muscle

Femur

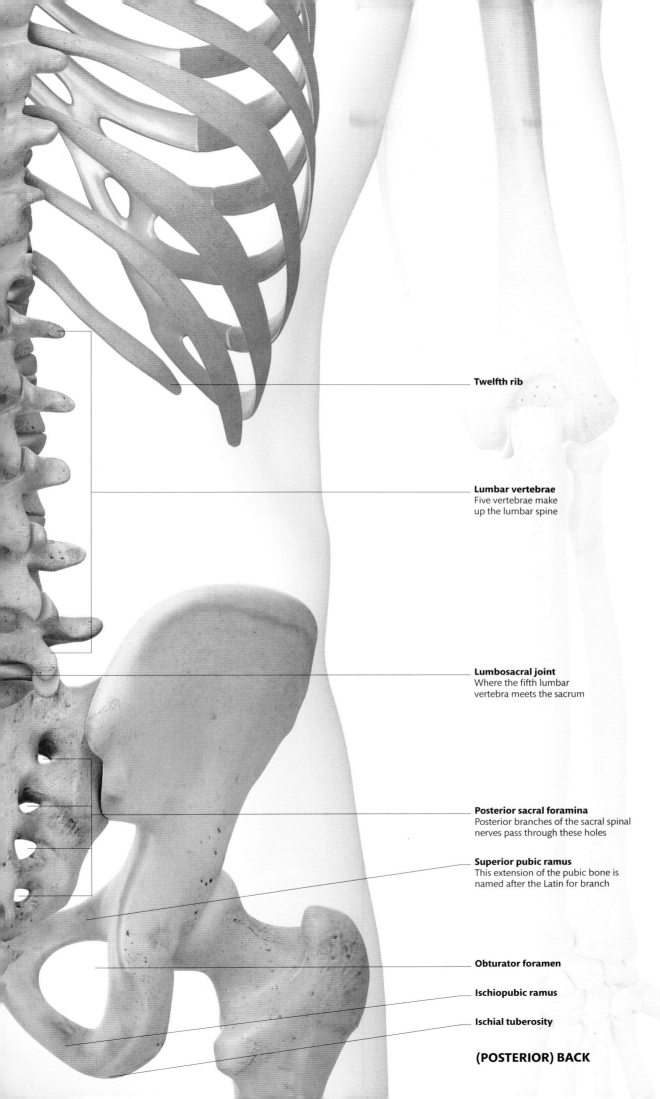

Twelfth rib

Lumbar vertebrae
Five vertebrae make
up the lumbar spine

Lumbosacral joint
Where the fifth lumbar
vertebra meets the sacrum

Posterior sacral foramina
Posterior branches of the sacral spinal
nerves pass through these holes

Superior pubic ramus
This extension of the pubic bone is
named after the Latin for branch

Obturator foramen

Ischiopubic ramus

Ischial tuberosity

(POSTERIOR) BACK

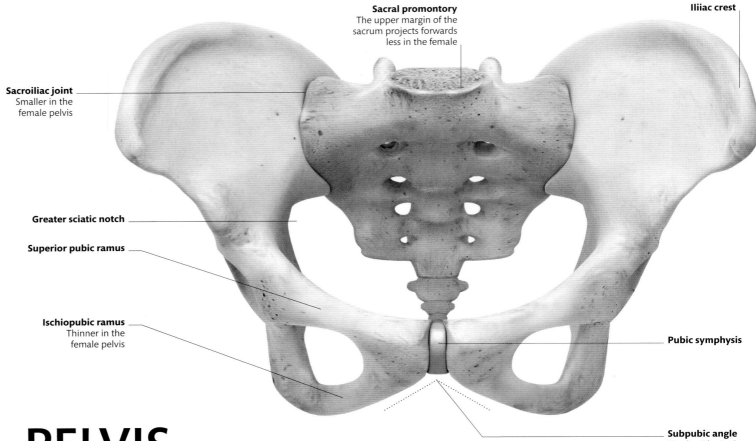

Sacral promontory
The upper margin of the sacrum projects forwards less in the female

Iliac crest

Sacroiliac joint
Smaller in the female pelvis

Greater sciatic notch

Superior pubic ramus

Ischiopubic ramus
Thinner in the female pelvis

Pubic symphysis

Subpubic angle
Much wider in the female pelvis

FEMALE PELVIS ANTERIOR (FRONT)

PELVIS

The bony pelvis is the part of the skeleton that is most different between the sexes, because the pelvis in the female has to accommodate the birth canal, unlike the male pelvis. Comparing the pelvic bones of a man and a woman, there are obvious differences between the two. The shape of the ring formed by the sacrum and the two pelvic bones – the pelvic brim – tends to be a wide oval in the woman and much narrower and heart-shaped in a man. The subpubic angle, underneath the joint between the two pubic bones, is much narrower in a man compared with a woman. As with the rest of the skeleton, the pelvic bone also tends to be more chunky or robust in a man, with more obvious ridges where muscles attach.

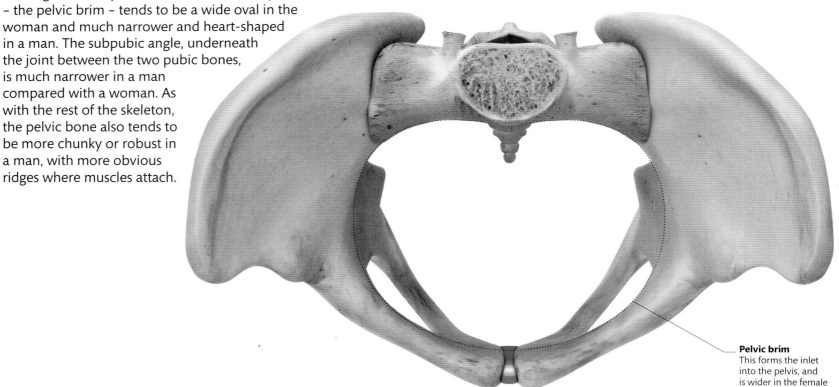

Pelvic brim
This forms the inlet into the pelvis, and is wider in the female

FEMALE PELVIS VIEWED FROM ABOVE

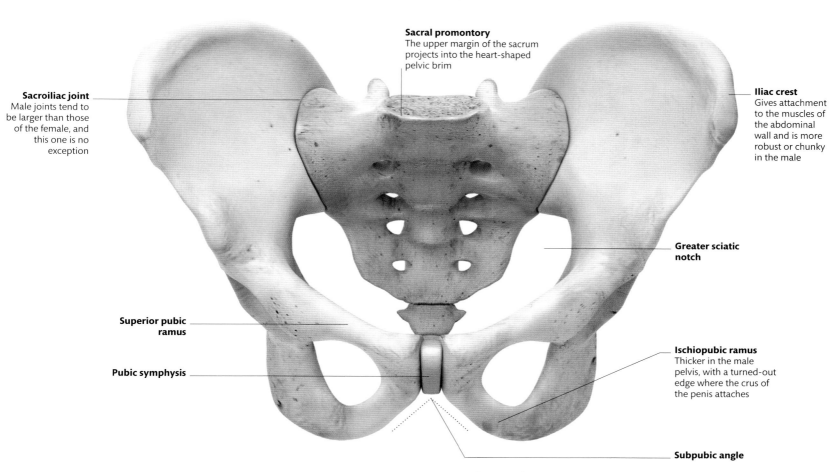

Sacral promontory
The upper margin of the sacrum projects into the heart-shaped pelvic brim

Sacroiliac joint
Male joints tend to be larger than those of the female, and this one is no exception

Iliac crest
Gives attachment to the muscles of the abdominal wall and is more robust or chunky in the male

Greater sciatic notch

Superior pubic ramus

Ischiopubic ramus
Thicker in the male pelvis, with a turned-out edge where the crus of the penis attaches

Pubic symphysis

Subpubic angle

MALE PELVIS ANTERIOR (FRONT)

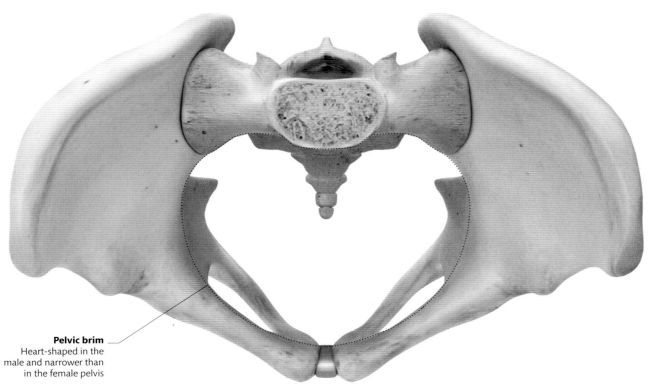

Pelvic brim
Heart-shaped in the male and narrower than in the female pelvis

MALE PELVIS VIEWED FROM ABOVE

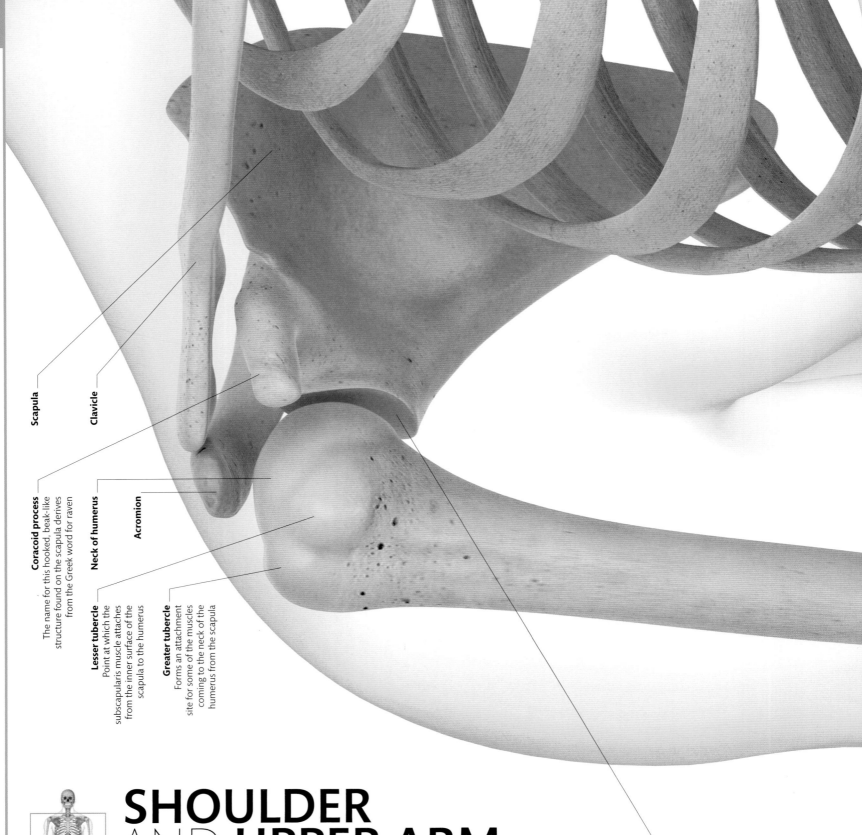

Scapula

Clavicle

Coracoid process
The name for this hooked, beak-like structure found on the scapula derives from the Greek word for raven

Neck of humerus

Acromion

Lesser tubercle
Point at which the subscapularis muscle attaches from the inner surface of the scapula to the humerus

Greater tubercle
Forms an attachment site for some of the muscles coming to the neck of the humerus from the scapula

Glenoid fossa
Shallow area which articulates with the head of the humerus, forming part of the shoulder socket

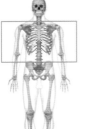

SHOULDER
AND UPPER ARM

The scapula and clavicle make up the shoulder girdle, which anchors the arm to the thorax. This is a very mobile attachment – the scapula "floats" on the ribcage, attached to it by muscles only (rather than by a true joint) that pull the scapula around on the underlying ribs, altering the position of the shoulder joint. The clavicle has joints – it articulates with the acromion of the scapula laterally (at the side) and the sternum at the other end – and helps to hold the shoulder out to the side while allowing the scapula to move around. The shoulder joint, the most mobile joint in the body, is a ball-and-socket joint, but the socket is small and shallow, allowing the ball-shaped head of the humerus to move freely.

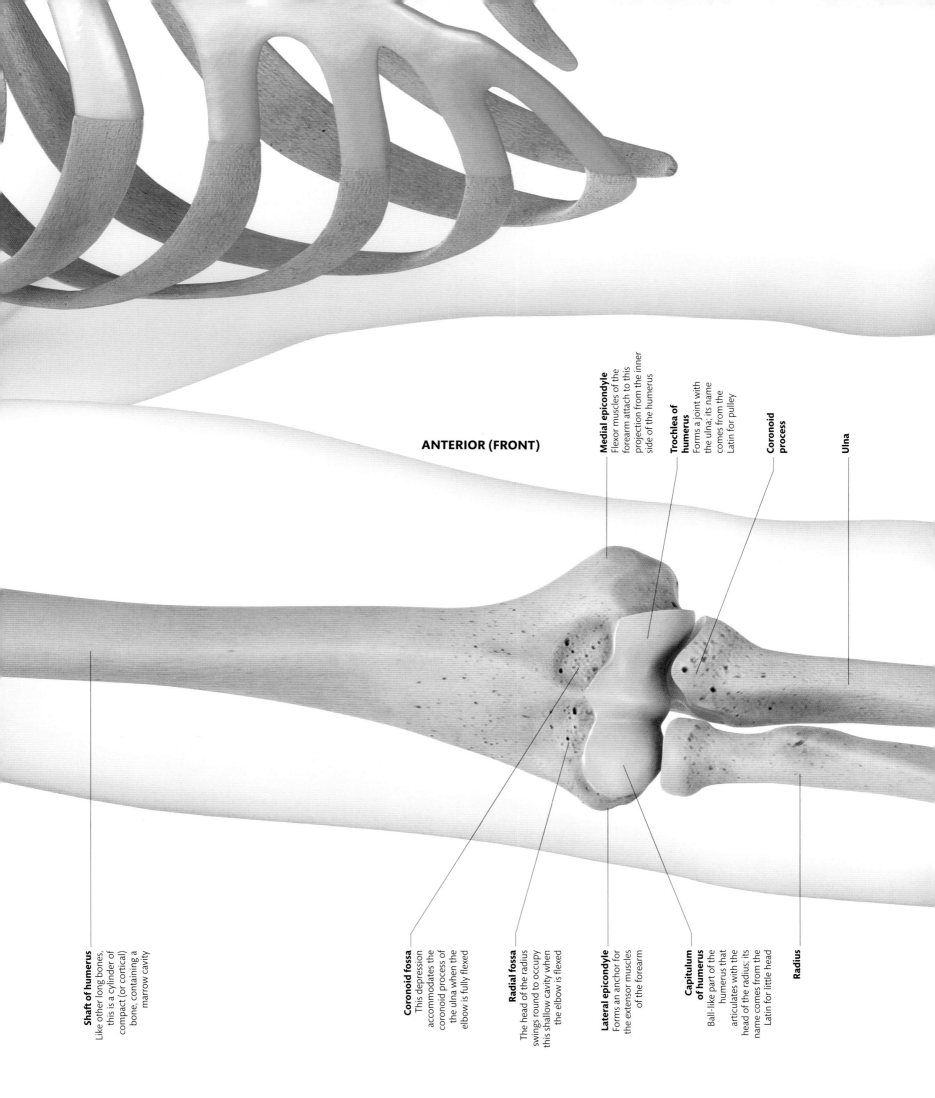

ANTERIOR (FRONT)

Shaft of humerus
Like other long bones, this is a cylinder of compact (or cortical) bone, containing a marrow cavity

Coronoid fossa
This depression accommodates the coronoid process of the ulna when the elbow is fully flexed

Radial fossa
The head of the radius swings round to occupy this shallow cavity when the elbow is flexed

Lateral epicondyle
Forms an anchor for the extensor muscles of the forearm

Capitulum of humerus
Ball-like part of the humerus that articulates with the head of the radius; its name comes from the Latin for little head

Medial epicondyle
Flexor muscles of the forearm attach to this projection from the inner side of the humerus

Trochlea of humerus
Forms a joint with the ulna; its name comes from the Latin for pulley

Coronoid process

Ulna

Radius

SHOULDER AND **UPPER ARM**

The back of the scapula is divided into two sections by its spine. The muscles that attach above this spine are called supraspinatus; those that attach below are called infraspinatus. They are part of the rotator cuff muscle group, which enables shoulder movements and stabilizes the shoulder joint. The spine of the scapula runs to the side and projects out above the shoulder joint to form the acromion, which can be easily felt on the top of the shoulder. The scapula rests in the position shown here when the arm is hanging at the side of the body. If the arm is abducted (raised to the side), the entire scapula rotates so that the glenoid cavity points upwards and the inferior angle moves outwards.

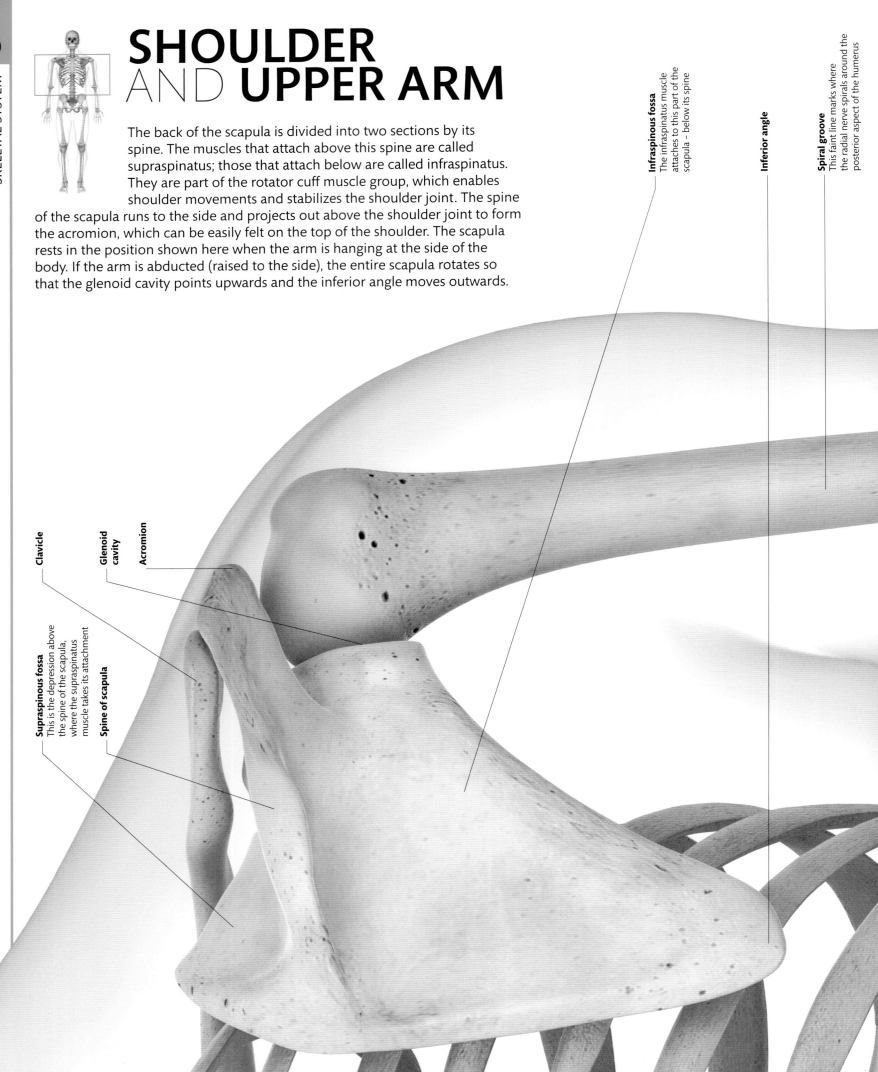

Infraspinous fossa
The infraspinatus muscle attaches to this part of the scapula – below its spine

Inferior angle

Spiral groove
This faint line marks where the radial nerve spirals around the posterior aspect of the humerus

Clavicle

Glenoid cavity

Acromion

Supraspinous fossa
This is the depression above the spine of the scapula, where the supraspinatus muscle takes its attachment

Spine of scapula

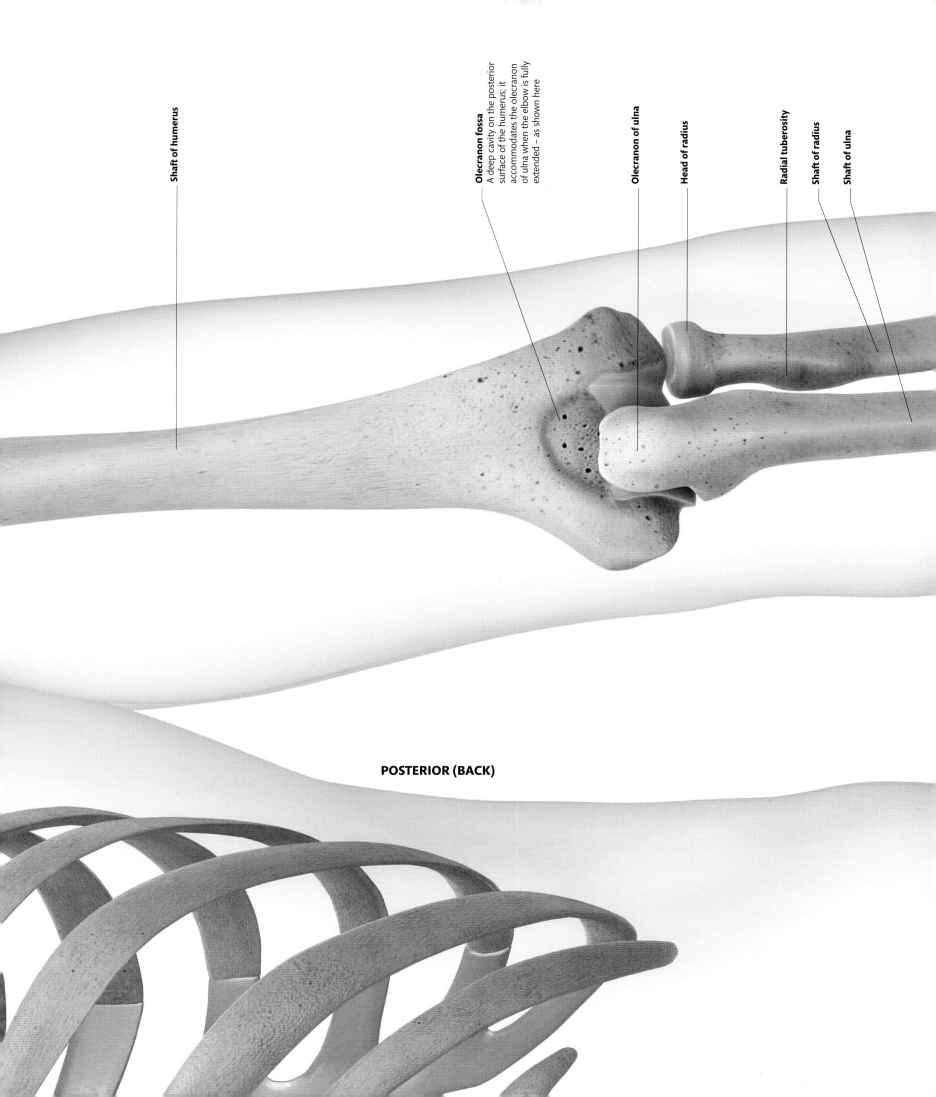

Shaft of humerus

Olecranon fossa
A deep cavity on the posterior surface of the humerus; it accommodates the olecranon of ulna when the elbow is fully extended – as shown here

Olecranon of ulna

Head of radius

Radial tuberosity

Shaft of radius

Shaft of ulna

POSTERIOR (BACK)

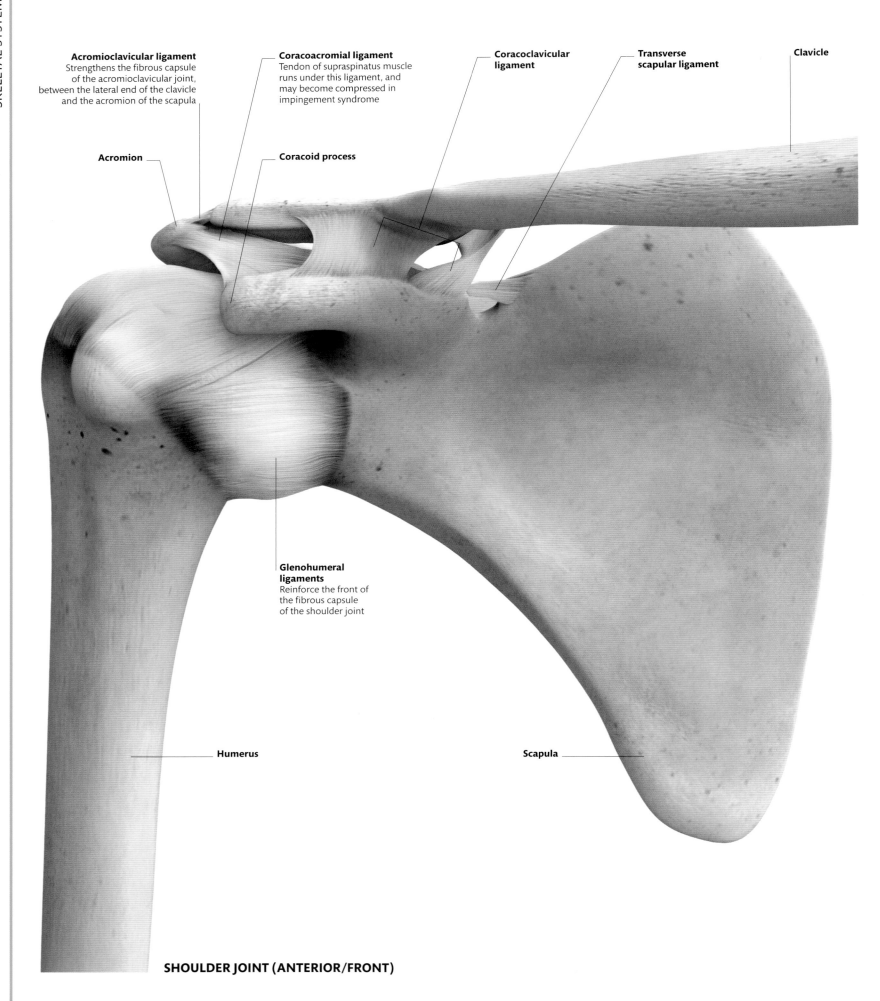

Acromioclavicular ligament
Strengthens the fibrous capsule
of the acromioclavicular joint,
between the lateral end of the clavicle
and the acromion of the scapula

Coracoacromial ligament
Tendon of supraspinatus muscle
runs under this ligament, and
may become compressed in
impingement syndrome

Coracoclavicular ligament

Transverse scapular ligament

Clavicle

Acromion

Coracoid process

Glenohumeral ligaments
Reinforce the front of
the fibrous capsule
of the shoulder joint

Humerus

Scapula

SHOULDER JOINT (ANTERIOR/FRONT)

SHOULDER AND UPPER ARM

In any joint, there is always a play-off between mobility and stability. The extremely mobile shoulder joint is therefore naturally unstable, and so it is not surprising that this is the most commonly dislocated joint in the body. The coracoacromial arch, formed by the acromion and coracoid process of the scapula with the strong coracoacromial ligament stretching between them, prevents upwards dislocation; when the head of the humerus dislocates, it usually does so in a downwards direction. The elbow joint is formed by the articulation of the humerus with the forearm bones: the trochlea articulates with the ulna, and the capitulum with the head of the radius. The elbow is a hinge joint, stabilized by collateral ligaments on each side.

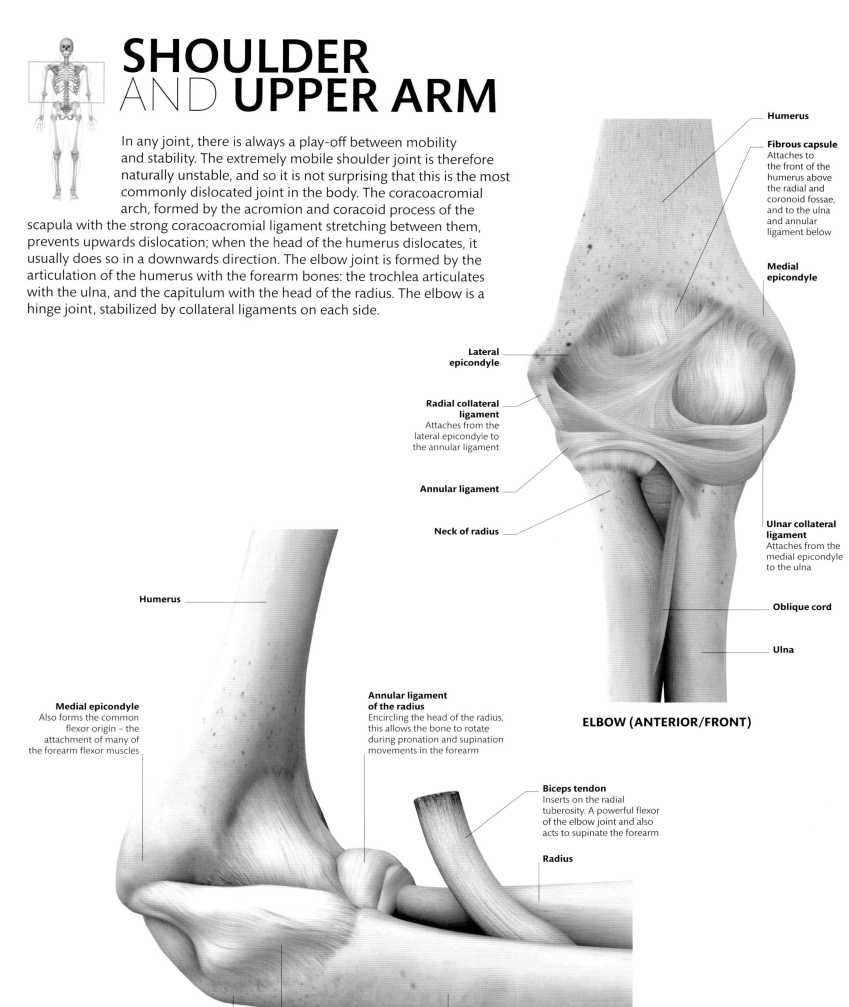

Humerus

Fibrous capsule
Attaches to the front of the humerus above the radial and coronoid fossae, and to the ulna and annular ligament below

Medial epicondyle

Lateral epicondyle

Radial collateral ligament
Attaches from the lateral epicondyle to the annular ligament

Annular ligament

Neck of radius

Ulnar collateral ligament
Attaches from the medial epicondyle to the ulna

Oblique cord

Ulna

ELBOW (ANTERIOR/FRONT)

Humerus

Medial epicondyle
Also forms the common flexor origin – the attachment of many of the forearm flexor muscles

Annular ligament of the radius
Encircling the head of the radius, this allows the bone to rotate during pronation and supination movements in the forearm

Biceps tendon
Inserts on the radial tuberosity. A powerful flexor of the elbow joint and also acts to supinate the forearm

Radius

Olecranon of ulna

Ulnar collateral ligament

Ulna

ELBOW (LATERAL/OUTER SIDE)

Medial epicondyle

Coronoid process
Forms anterior margin of the trochlear notch of the ulna, which accommodates the trochlea of the humerus

Radial notch of ulna
This concave surface articulates with the head of the radius, forming the proximal radioulnar joint

Tuberosity of ulna
Brachialis muscle attaches here

Interosseous border of radius
Sharp ridges on facing edges of the radius and ulna provide attachment for the forearm's interosseous membrane

Shaft of radius
Like the ulna, this is triangular in cross-section

Interosseous border of ulna

Shaft of ulna

Lunate
Crescent-shaped bone named after the Latin for moon

Head of ulna
Articulates with lower end of the radius, at the distal radioulnar joint

Styloid process of ulna
Where the ulnar collateral ligament attaches

Pisiform

Triquetral

Hamate
One of the carpal bones, along with the other bones between the radius and ulna

Lateral epicondyle

Capitulum of humerus

Head of radius
Bowl-shaped surface articulates with the capitulum of humerus

Trochlea of humerus

Radial tuberosity
Biceps tendon attaches here

Trapezoid
This four-sided bone's name means table-shaped in Greek

Scaphoid
Convex bone named after the Greek for boat-shaped

Styloid process of radius
The radial collateral ligament of the wrist attaches to this sharp point

Trapezium
Four-sided bone named after the Greek for table

Distal phalanx

Middle phalanx

Proximal phalanx

Fifth metacarpal

Capitate
Articulates with third and fourth metacarpals

Hamate
Articulates with fourth and fifth metacarpals

Triquetral
Latin for three-cornered

Pisiform
Latin for pea-shaped; articulates with the triquetral, and receives the tendon of the flexor carpi ulnaris muscle

Styloid process of ulna
Pointed projection taking its name from the Greek for pillar-shaped

Head of ulna

POSTERIOR (BACK)

Distal phalanx

Proximal phalanx

First metacarpal

Trapezoid
Articulates with second metacarpal of index finger

Trapezium
Articulates with first metacarpal of thumb

Scaphoid
The most commonly fractured wrist bone

Lunate
Articulates with scaphoid and radius to form wrist joint; this is the most commonly dislocated carpal (wrist) bone

Styloid process of radius

LOWER ARM AND HAND

The two forearm bones, the radius and ulna, are bound together by a flat sheet of ligament called the interosseous membrane, and by synovial joints between the ends of the two bones. Known as radioulnar joints, these joints allow the radius to move around the ulna. Hold your hand out in front of you, palm upwards. Now turn your hand so that the palm faces the ground. This movement is called pronation, and is achieved by bringing the radius to cross over the ulna. The movement that returns the palm to an upward-facing position is called supination. Since the forearm bones are bound together by ligaments, joints, and muscles, it is common for both bones to be involved in a serious forearm injury. Often, one bone is fractured and the other dislocated. The skeleton of the hand comprises the eight carpal bones (bones between the radius and ulna), five metacarpals, and fourteen phalanges.

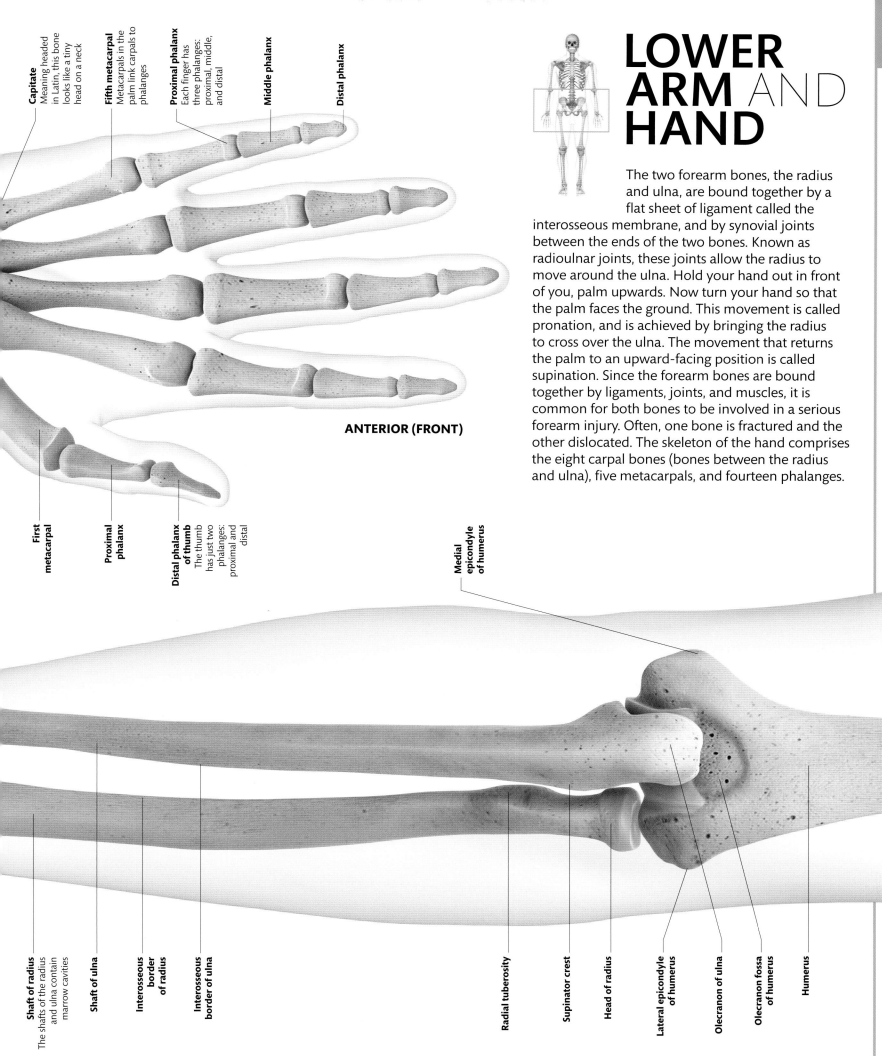

Capitate
Meaning headed in Latin, this bone looks like a tiny head on a neck

Fifth metacarpal
Metacarpals in the palm link carpals to phalanges

Proximal phalanx
Each finger has three phalanges; proximal, middle, and distal

Middle phalanx

Distal phalanx

ANTERIOR (FRONT)

First metacarpal

Proximal phalanx

Distal phalanx of thumb
The thumb has just two phalanges: proximal and distal

Medial epicondyle of humerus

Shaft of radius
The shafts of the radius and ulna contain marrow cavities

Shaft of ulna

Interosseous border of radius

Interosseous border of ulna

Radial tuberosity

Supinator crest

Head of radius

Lateral epicondyle of humerus

Olecranon of ulna

Olecranon fossa of humerus

Humerus

HAND AND WRIST JOINTS

The radius widens out at its distal (lower) end to form the wrist joint with the nearest two carpal bones, the lunate and scaphoid. This joint allows flexion, extension, adduction, and abduction (see pp.16–17). There are also synovial joints (see p.60) between the carpal bones in the wrist, which increase the range of motion during wrist flexion and extension. Synovial joints between metacarpals and phalanges allow us to spread or close our fingers, as well as flexing or extending the whole finger. Joints between the individual finger bones, or phalanges, enable fingers to bend and straighten. In common with many other primates, humans have opposable thumbs. The joints at the base of the thumb are shaped differently from those of the fingers. The joint between the metacarpal of the thumb and the wrist bones is especially mobile and allows the thumb to be brought across the palm of the hand so that the tip of the thumb can touch the other fingertips.

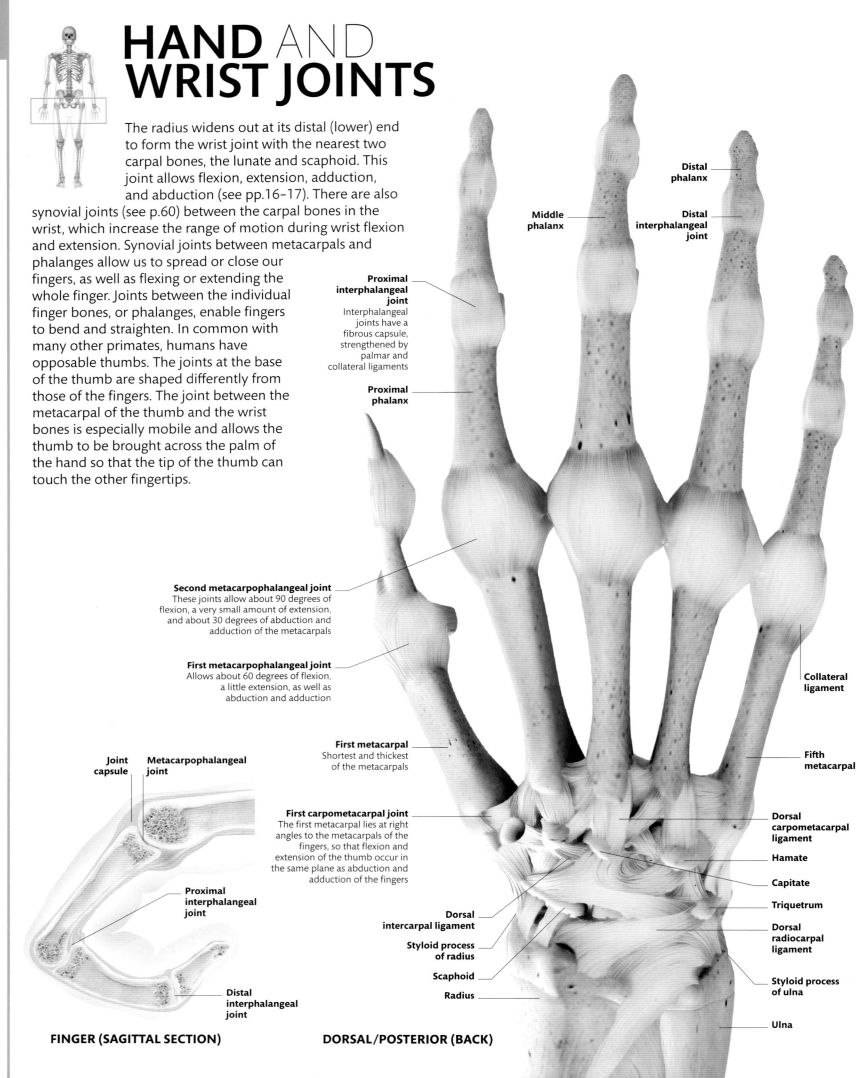

Distal phalanx

Middle phalanx

Distal interphalangeal joint

Proximal interphalangeal joint
Interphalangeal joints have a fibrous capsule, strengthened by palmar and collateral ligaments

Proximal phalanx

Second metacarpophalangeal joint
These joints allow about 90 degrees of flexion, a very small amount of extension, and about 30 degrees of abduction and adduction of the metacarpals

First metacarpophalangeal joint
Allows about 60 degrees of flexion, a little extension, as well as abduction and adduction

Collateral ligament

Fifth metacarpal

First metacarpal
Shortest and thickest of the metacarpals

First carpometacarpal joint
The first metacarpal lies at right angles to the metacarpals of the fingers, so that flexion and extension of the thumb occur in the same plane as abduction and adduction of the fingers

Dorsal carpometacarpal ligament

Hamate

Capitate

Triquetrum

Dorsal radiocarpal ligament

Dorsal intercarpal ligament

Styloid process of radius

Scaphoid

Radius

Styloid process of ulna

Ulna

Joint capsule

Metacarpophalangeal joint

Proximal interphalangeal joint

Distal interphalangeal joint

FINGER (SAGITTAL SECTION)

DORSAL/POSTERIOR (BACK)

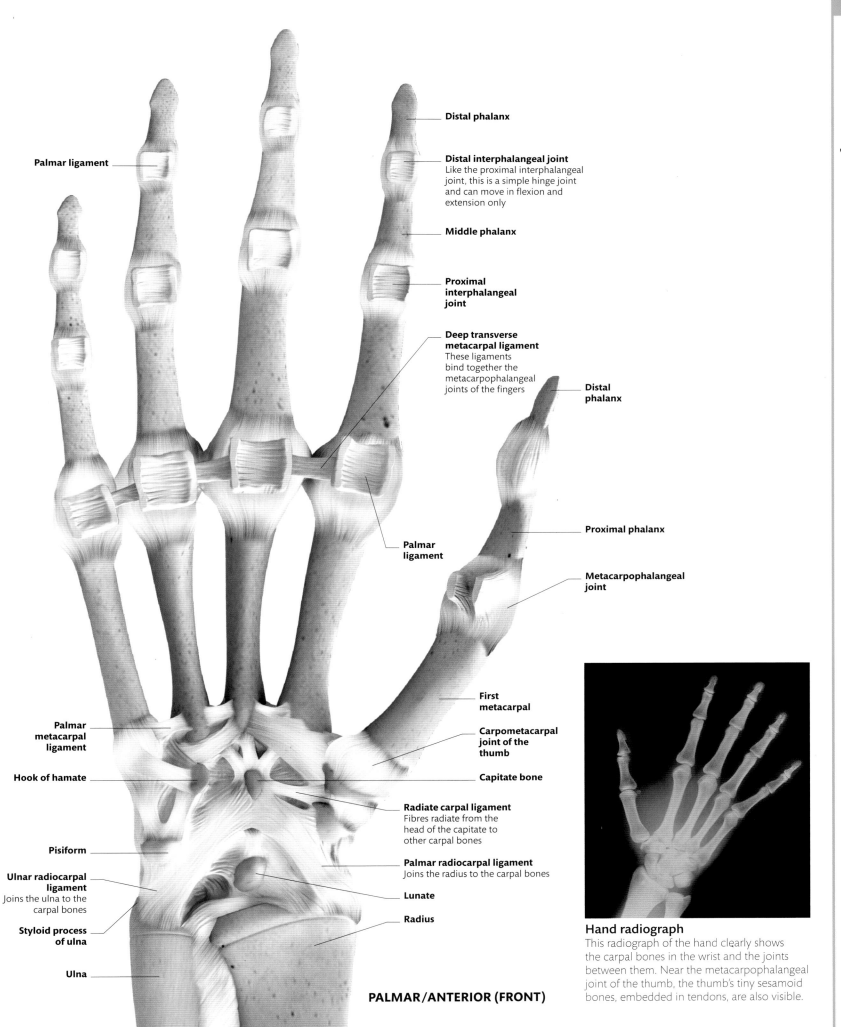

Palmar ligament

Distal phalanx

Distal interphalangeal joint
Like the proximal interphalangeal joint, this is a simple hinge joint and can move in flexion and extension only

Middle phalanx

Proximal interphalangeal joint

Deep transverse metacarpal ligament
These ligaments bind together the metacarpophalangeal joints of the fingers

Distal phalanx

Palmar ligament

Proximal phalanx

Metacarpophalangeal joint

First metacarpal

Palmar metacarpal ligament

Carpometacarpal joint of the thumb

Hook of hamate

Capitate bone

Radiate carpal ligament
Fibres radiate from the head of the capitate to other carpal bones

Pisiform

Palmar radiocarpal ligament
Joins the radius to the carpal bones

Ulnar radiocarpal ligament
Joins the ulna to the carpal bones

Lunate

Styloid process of ulna

Radius

Ulna

PALMAR/ANTERIOR (FRONT)

Hand radiograph

This radiograph of the hand clearly shows the carpal bones in the wrist and the joints between them. Near the metacarpophalangeal joint of the thumb, the thumb's tiny sesamoid bones, embedded in tendons, are also visible.

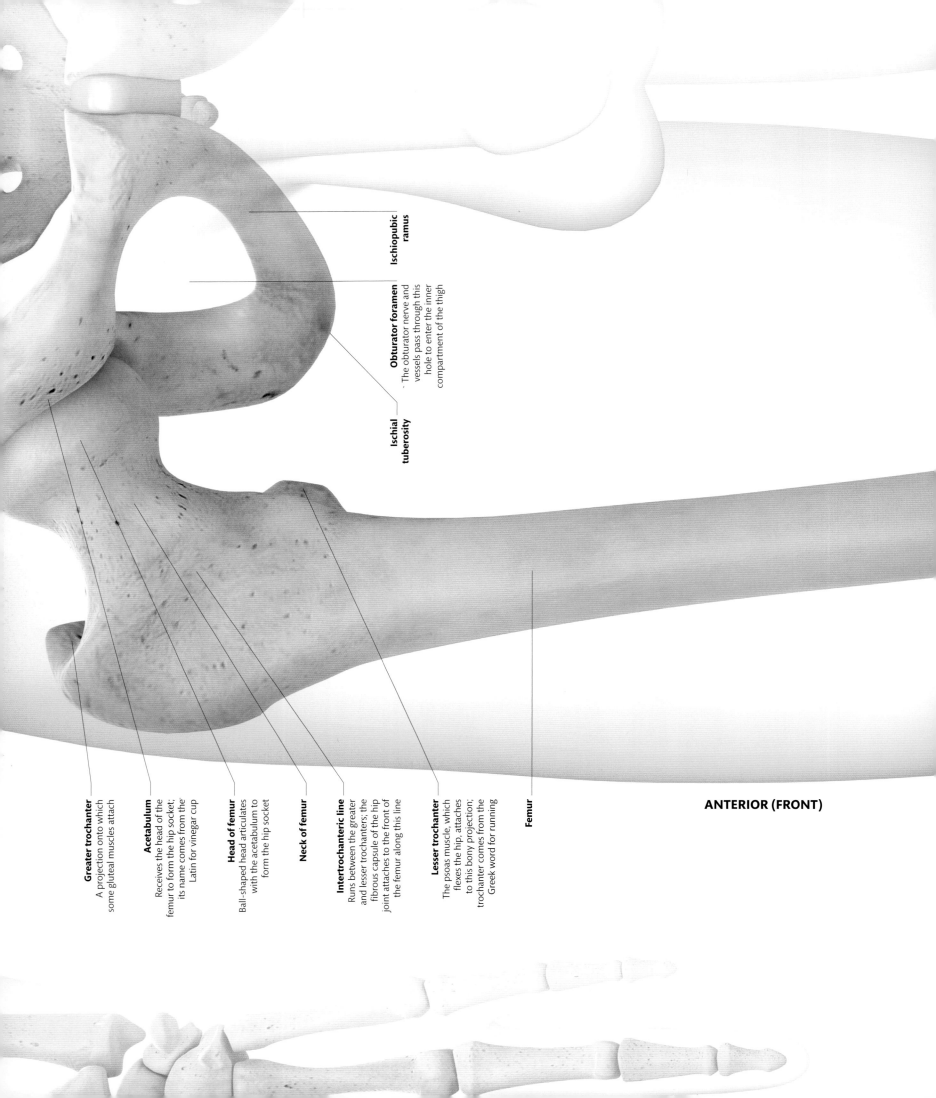

Ischiopubic ramus

Obturator foramen
The obturator nerve and vessels pass through this hole to enter the inner compartment of the thigh

Ischial tuberosity

Greater trochanter
A projection onto which some gluteal muscles attach

Acetabulum
Receives the head of the femur to form the hip socket; its name comes from the Latin for vinegar cup

Head of femur
Ball-shaped head articulates with the acetabulum to form the hip socket

Neck of femur

Intertrochanteric line
Runs between the greater and lesser trochanters; the fibrous capsule of the hip joint attaches to the front of the femur along this line

Lesser trochanter
The psoas muscle, which flexes the hip, attaches to this bony projection; trochanter comes from the Greek word for running

Femur

ANTERIOR (FRONT)

Shaft of femur
This is not vertical, but angled inwards slightly, to bring the knees under the body

Adductor tubercle
The point at which the tendon of adductor magnus attaches to the femur

Medial epicondyle

Patella
The technical name for the kneecap comes from the Latin for small dish

Medial condyle

Tibia

Base of patella

Patellar surface of the femur

Lateral epicondyle
The term epicondyle (meaning close to the condyle) describes a projecting part of bone near a joint that provides a point of attachment for muscles

Lateral condyle of the femur
Condyle comes from the Greek word for knuckle; the term describes parts of the ends of bones that form joints

Apex of patella

HIP AND THIGH

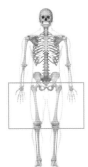

The leg or, to be anatomically precise, the lower limb, is attached to the spine by the pelvic bones. This is a much more stable arrangement than that of the shoulder girdle, which anchors the arm, because the legs and pelvis must bear our body weight as we stand or move around. The sacroiliac joint provides a strong attachment between the ilium of the pelvis and the sacrum, and the hip joint is a much deeper and more stable ball-and-socket joint than that in the shoulder. The neck of the femur joins the head at an obtuse angle. A slightly raised diagonal line on the front of the neck (the intertrochanteric line) shows where the fibrous capsule of the hip joint attaches to the bone.

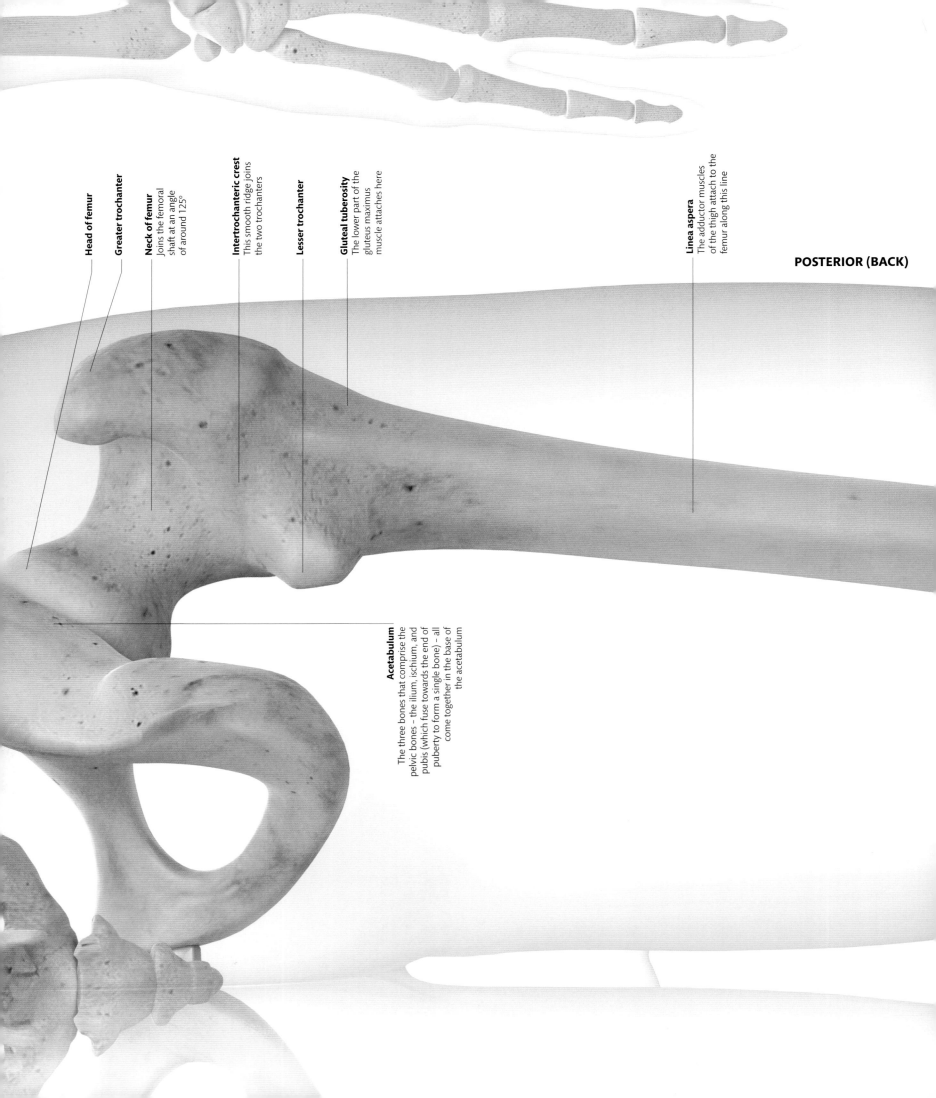

Head of femur

Greater trochanter

Neck of femur
Joins the femoral
shaft at an angle
of around 125°

Intertrochanteric crest
This smooth ridge joins
the two trochanters

Lesser trochanter

Gluteal tuberosity
The lower part of the
gluteus maximus
muscle attaches here

Linea aspera
The adductor muscles
of the thigh attach to the
femur along this line

POSTERIOR (BACK)

Acetabulum
The three bones that comprise the
pelvic bones – the ilium, ischium, and
pubis (which fuse towards the end of
puberty to form a single bone) – all
come together in the base of
the acetabulum

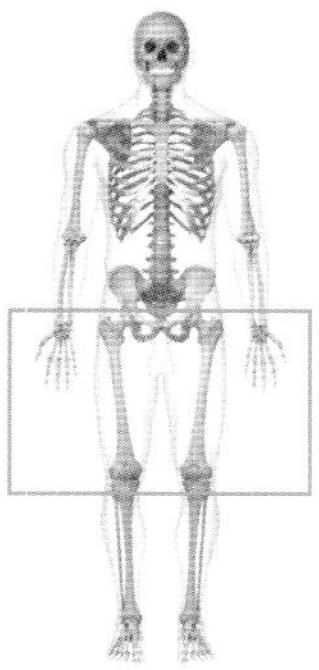

HIP AND THIGH

The shaft of the femur (thighbone) is cylindrical, with a marrow cavity. The linea aspera runs down along the back of the femoral shaft. This line is where the inner thigh's adductor muscles attach to the femur. Parts of the quadriceps muscle also wrap right around the back of the femur to attach to the linea aspera. At the bottom – or distal – end, towards the knee, the femur widens to form the knee joint with the tibia and the patella. From the back, the distal end of the femur has a distinct double-knuckle shape, with two condyles (rounded projections) that articulate with the tibia.

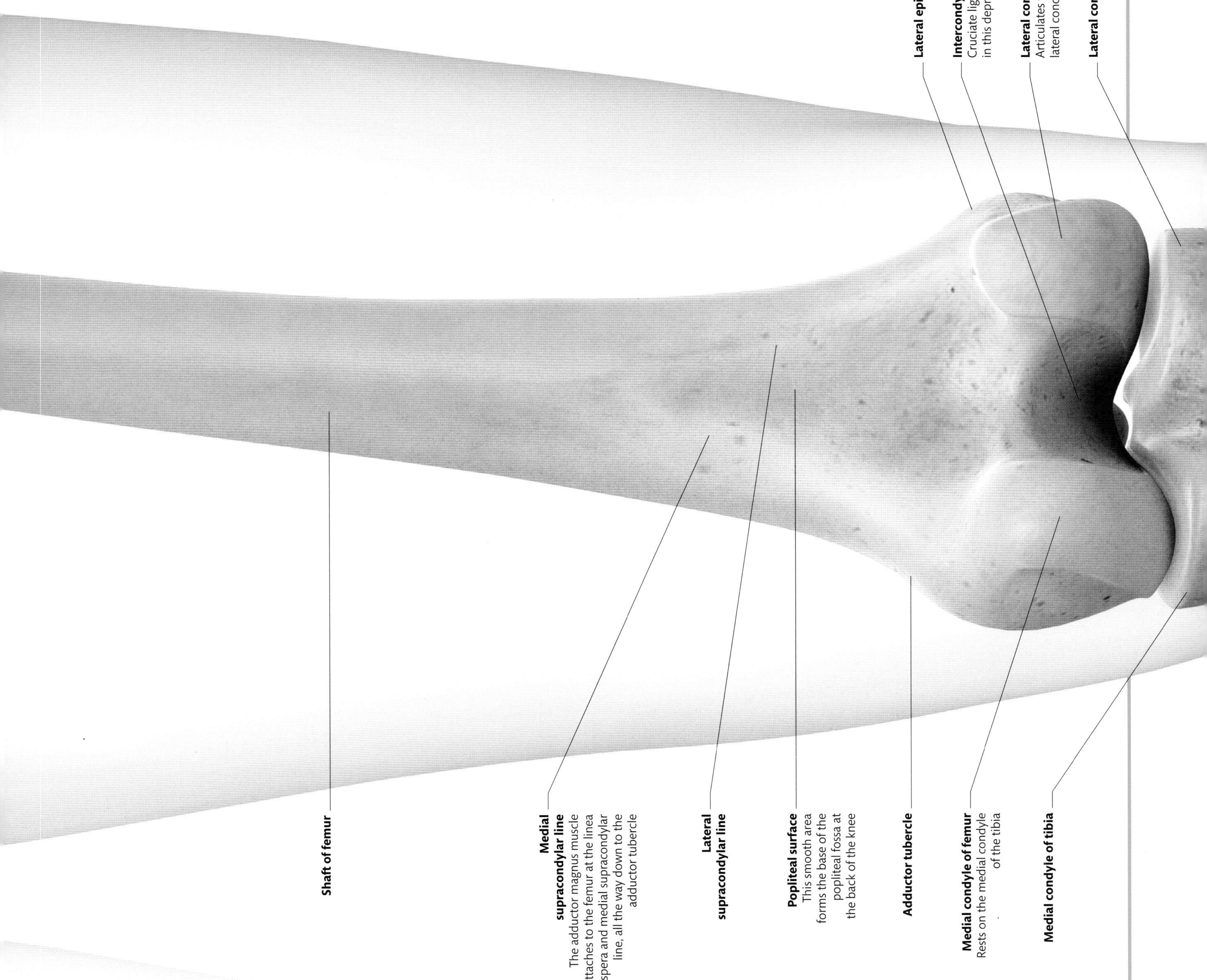

Lateral epicondyle

Intercondylar fossa
Cruciate ligaments attach to the femur in this depression between the condyles

Lateral condyle of femur
Articulates with the slightly concave lateral condyle of the tibia

Lateral condyle of tibia

Shaft of femur

Medial supracondylar line
The adductor magnus muscle attaches to the femur at the linea aspera and medial supracondylar line, all the way down to the adductor tubercle

Lateral supracondylar line

Popliteal surface
This smooth area forms the base of the popliteal fossa at the back of the knee

Adductor tubercle

Medial condyle of femur
Rests on the medial condyle of the tibia

Medial condyle of tibia

HIP AND KNEE

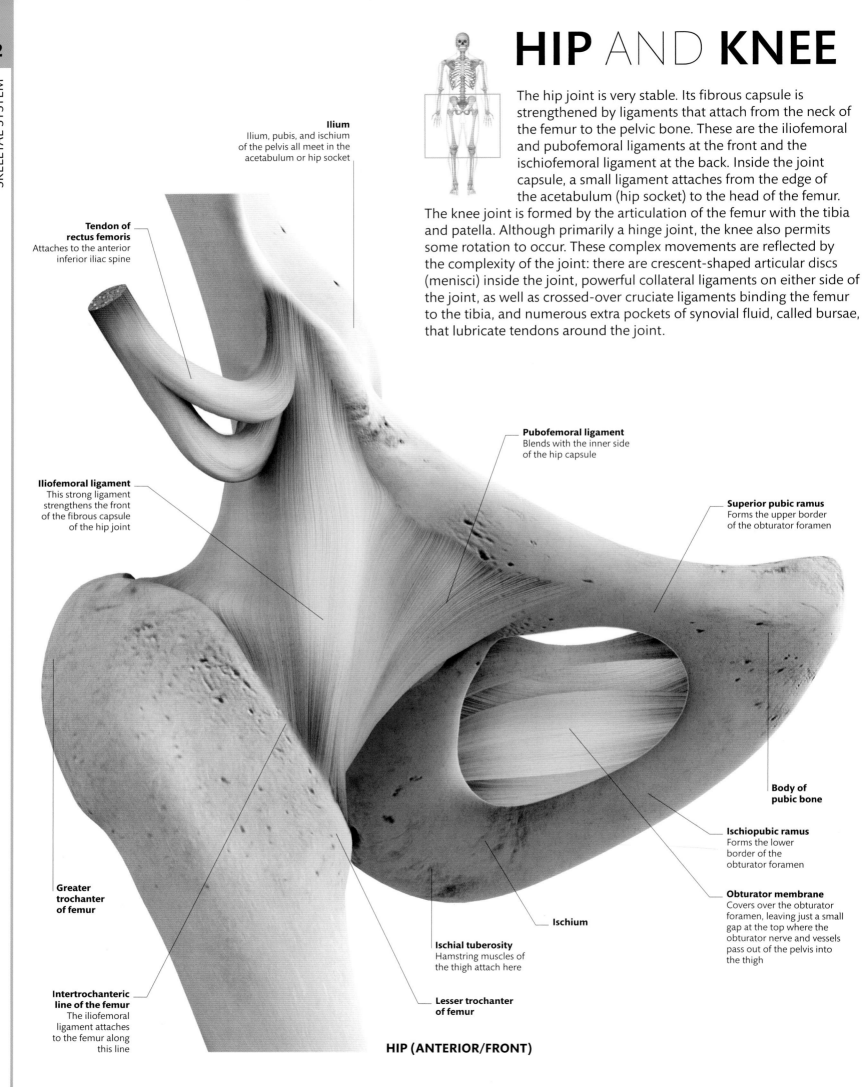

The hip joint is very stable. Its fibrous capsule is strengthened by ligaments that attach from the neck of the femur to the pelvic bone. These are the iliofemoral and pubofemoral ligaments at the front and the ischiofemoral ligament at the back. Inside the joint capsule, a small ligament attaches from the edge of the acetabulum (hip socket) to the head of the femur. The knee joint is formed by the articulation of the femur with the tibia and patella. Although primarily a hinge joint, the knee also permits some rotation to occur. These complex movements are reflected by the complexity of the joint: there are crescent-shaped articular discs (menisci) inside the joint, powerful collateral ligaments on either side of the joint, as well as crossed-over cruciate ligaments binding the femur to the tibia, and numerous extra pockets of synovial fluid, called bursae, that lubricate tendons around the joint.

Ilium
Ilium, pubis, and ischium of the pelvis all meet in the acetabulum or hip socket

Tendon of rectus femoris
Attaches to the anterior inferior iliac spine

Iliofemoral ligament
This strong ligament strengthens the front of the fibrous capsule of the hip joint

Pubofemoral ligament
Blends with the inner side of the hip capsule

Superior pubic ramus
Forms the upper border of the obturator foramen

Body of pubic bone

Ischiopubic ramus
Forms the lower border of the obturator foramen

Obturator membrane
Covers over the obturator foramen, leaving just a small gap at the top where the obturator nerve and vessels pass out of the pelvis into the thigh

Greater trochanter of femur

Ischium

Intertrochanteric line of the femur
The iliofemoral ligament attaches to the femur along this line

Ischial tuberosity
Hamstring muscles of the thigh attach here

Lesser trochanter of femur

HIP (ANTERIOR/FRONT)

SYNOVIAL JOINTS

The majority of the body's 320 or so joints, including those in the finger, knee, and shoulder, are free-moving synovial joints. The joint surfaces are lined with smooth hyaline cartilage to reduce friction, and contain lubricating synovial fluid.

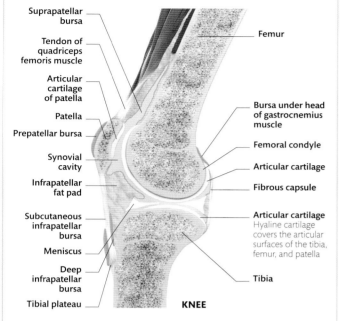

Suprapatellar bursa
Tendon of quadriceps femoris muscle
Articular cartilage of patella
Patella
Prepatellar bursa
Synovial cavity
Infrapatellar fat pad
Subcutaneous infrapatellar bursa
Meniscus
Deep infrapatellar bursa
Tibial plateau
Femur
Bursa under head of gastrocnemius muscle
Femoral condyle
Articular cartilage
Fibrous capsule
Articular cartilage
Hyaline cartilage covers the articular surfaces of the tibia, femur, and patella
Tibia

KNEE

Complex joint

A complex synovial joint, such as the knee, has articular discs or menisci inside the synovial cavity. The knee is also a compound hinge joint, as it involves more than two bones. Its complex anatomy allows it to move in flexion and extension, but some sliding and axial rotation of the femur on the tibia also occurs.

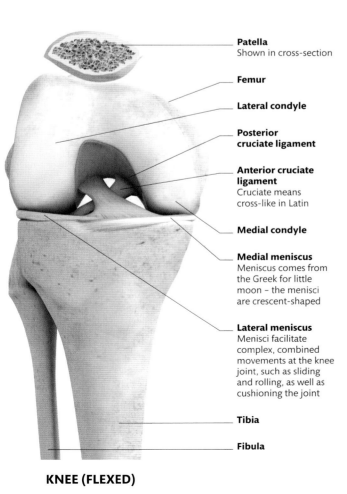

Patella
Shown in cross-section

Femur

Lateral condyle

Posterior cruciate ligament

Anterior cruciate ligament
Cruciate means cross-like in Latin

Medial condyle

Medial meniscus
Meniscus comes from the Greek for little moon – the menisci are crescent-shaped

Lateral meniscus
Menisci facilitate complex, combined movements at the knee joint, such as sliding and rolling, as well as cushioning the joint

Tibia

Fibula

KNEE (FLEXED)

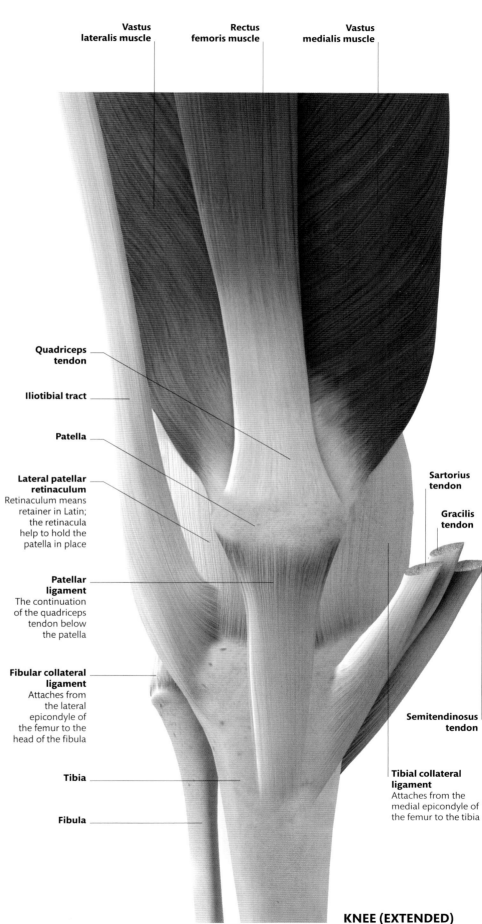

Vastus lateralis muscle
Rectus femoris muscle
Vastus medialis muscle

Quadriceps tendon

Iliotibial tract

Patella

Lateral patellar retinaculum
Retinaculum means retainer in Latin; the retinacula help to hold the patella in place

Patellar ligament
The continuation of the quadriceps tendon below the patella

Fibular collateral ligament
Attaches from the lateral epicondyle of the femur to the head of the fibula

Tibia

Fibula

Sartorius tendon

Gracilis tendon

Semitendinosus tendon

Tibial collateral ligament
Attaches from the medial epicondyle of the femur to the tibia

KNEE (EXTENDED)

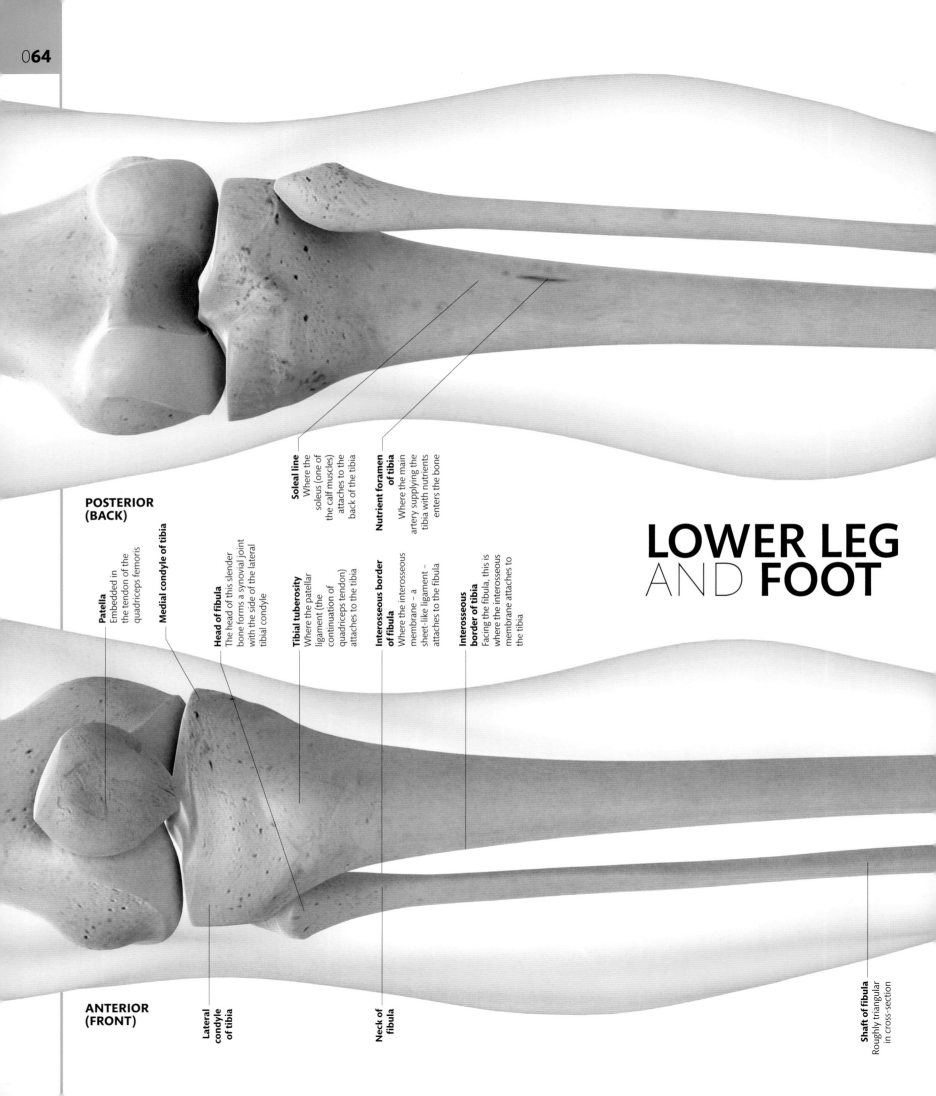

LOWER LEG AND FOOT

POSTERIOR (BACK)

ANTERIOR (FRONT)

Soleal line
Where the soleus (one of the calf muscles) attaches to the back of the tibia

Nutrient foramen of tibia
Where the main artery supplying the tibia with nutrients enters the bone

Patella
Embedded in the tendon of the quadriceps femoris

Medial condyle of tibia

Head of fibula
The head of this slender bone forms a synovial joint with the side of the lateral tibial condyle

Tibial tuberosity
Where the patellar ligament (the continuation of quadriceps tendon) attaches to the tibia

Interosseous border of fibula
Where the interosseous membrane – a sheet-like ligament – attaches to the fibula

Interosseous border of tibia
Facing the fibula, this is where the interosseous membrane attaches to the tibia

Lateral condyle of tibia

Neck of fibula

Shaft of fibula
Roughly triangular in cross-section

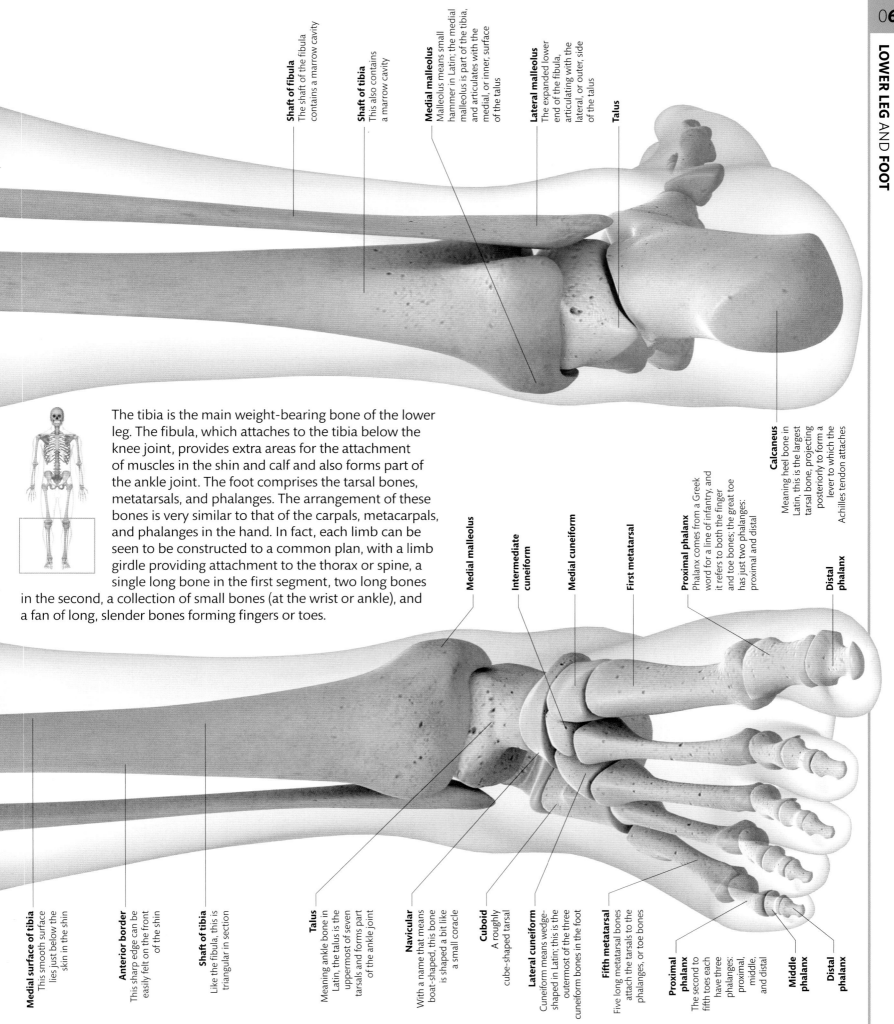

Shaft of fibula
The shaft of the fibula contains a marrow cavity

Shaft of tibia
This also contains a marrow cavity

Medial malleolus
Malleolus means small hammer in Latin; the medial malleolus is part of the tibia, and articulates with the medial, or inner, surface of the talus

Lateral malleolus
The expanded lower end of the fibula, articulating with the lateral, or outer, side of the talus

Talus

The tibia is the main weight-bearing bone of the lower leg. The fibula, which attaches to the tibia below the knee joint, provides extra areas for the attachment of muscles in the shin and calf and also forms part of the ankle joint. The foot comprises the tarsal bones, metatarsals, and phalanges. The arrangement of these bones is very similar to that of the carpals, metacarpals, and phalanges in the hand. In fact, each limb can be seen to be constructed to a common plan, with a limb girdle providing attachment to the thorax or spine, a single long bone in the first segment, two long bones in the second, a collection of small bones (at the wrist or ankle), and a fan of long, slender bones forming fingers or toes.

Calcaneus
Meaning heel bone in Latin, this is the largest tarsal bone, projecting posteriorly to form a lever to which the Achilles tendon attaches

Medial malleolus

Intermediate cuneiform

Medial cuneiform

First metatarsal

Proximal phalanx
Phalanx comes from a Greek word for a line of infantry, and it refers to both the finger and toe bones; the great toe has just two phalanges: proximal and distal

Distal phalanx

Medial surface of tibia
This smooth surface lies just below the skin in the shin

Anterior border
This sharp edge can be easily felt on the front of the shin

Shaft of tibia
Like the fibula, this is triangular in section

Talus
Meaning ankle bone in Latin, the talus is the uppermost of seven tarsals and forms part of the ankle joint

Navicular
With a name that means boat-shaped, this bone is shaped a bit like a small coracle

Cuboid
A roughly cube-shaped tarsal

Lateral cuneiform
Cuneiform means wedge-shaped in Latin; this is the outermost of the three cuneiform bones in the foot

Fifth metatarsal
Five long metatarsal bones attach the tarsals to the phalanges, or toe bones

Proximal phalanx
The second to fifth toes each have three phalanges: proximal, middle, and distal

Middle phalanx

Distal phalanx

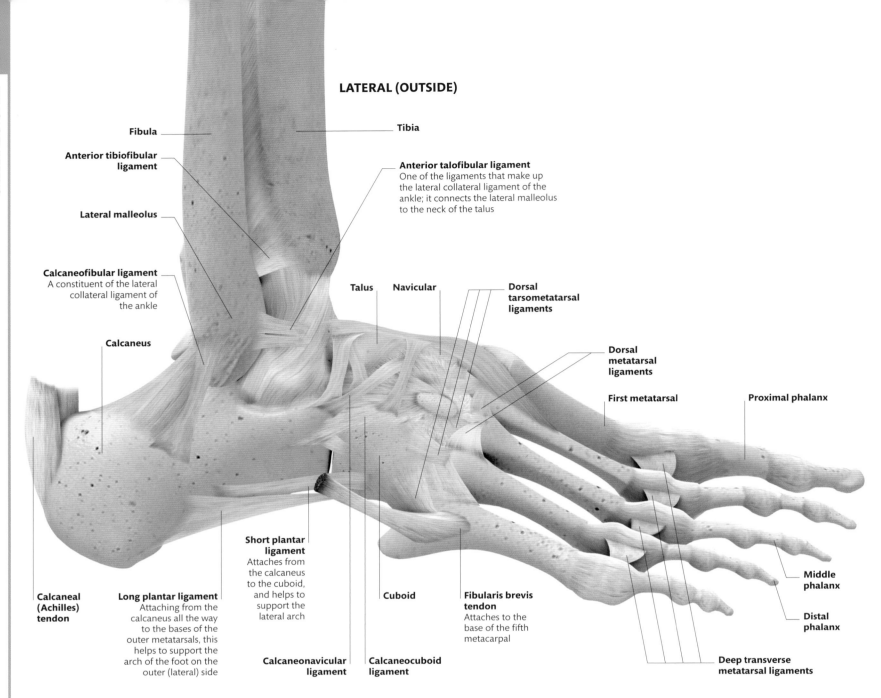

LATERAL (OUTSIDE)

Fibula

Tibia

Anterior tibiofibular ligament

Anterior talofibular ligament
One of the ligaments that make up the lateral collateral ligament of the ankle; it connects the lateral malleolus to the neck of the talus

Lateral malleolus

Calcaneofibular ligament
A constituent of the lateral collateral ligament of the ankle

Talus

Navicular

Dorsal tarsometatarsal ligaments

Calcaneus

Dorsal metatarsal ligaments

First metatarsal

Proximal phalanx

Short plantar ligament
Attaches from the calcaneus to the cuboid, and helps to support the lateral arch

Calcaneal (Achilles) tendon

Long plantar ligament
Attaching from the calcaneus all the way to the bases of the outer metatarsals, this helps to support the arch of the foot on the outer (lateral) side

Cuboid

Fibularis brevis tendon
Attaches to the base of the fifth metacarpal

Middle phalanx

Distal phalanx

Calcaneonavicular ligament

Calcaneocuboid ligament

Deep transverse metatarsal ligaments

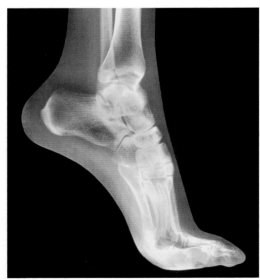

X-ray on tiptoe
This radiograph shows the foot in action. The calf muscles are pulling up on the lever of the calcaneus to flex the ankle down (plantarflex), while the metatarsophalangeal joints are extended.

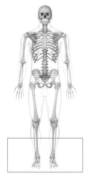

FOOT AND ANKLE

The ankle joint is a simple hinge joint. The lower ends of the tibia and fibula are firmly bound together by ligaments, forming a strong fibrous joint, and making a spanner shape that neatly sits around the nut of the talus. The joint is stabilized by strong collateral ligaments on either side. The talus forms synovial joints (see p.61) with the calcaneus beneath it, and the navicular bone in front of it. Level with the joint between the talus and the navicular is a joint between the calcaneus and the cuboid. These joints together allow the foot to be angled inwards or outwards – these movements are called inversion and eversion respectively. The skeleton of the foot is a sprung structure, with the bones forming arches, held together by ligaments and also supported by tendons.

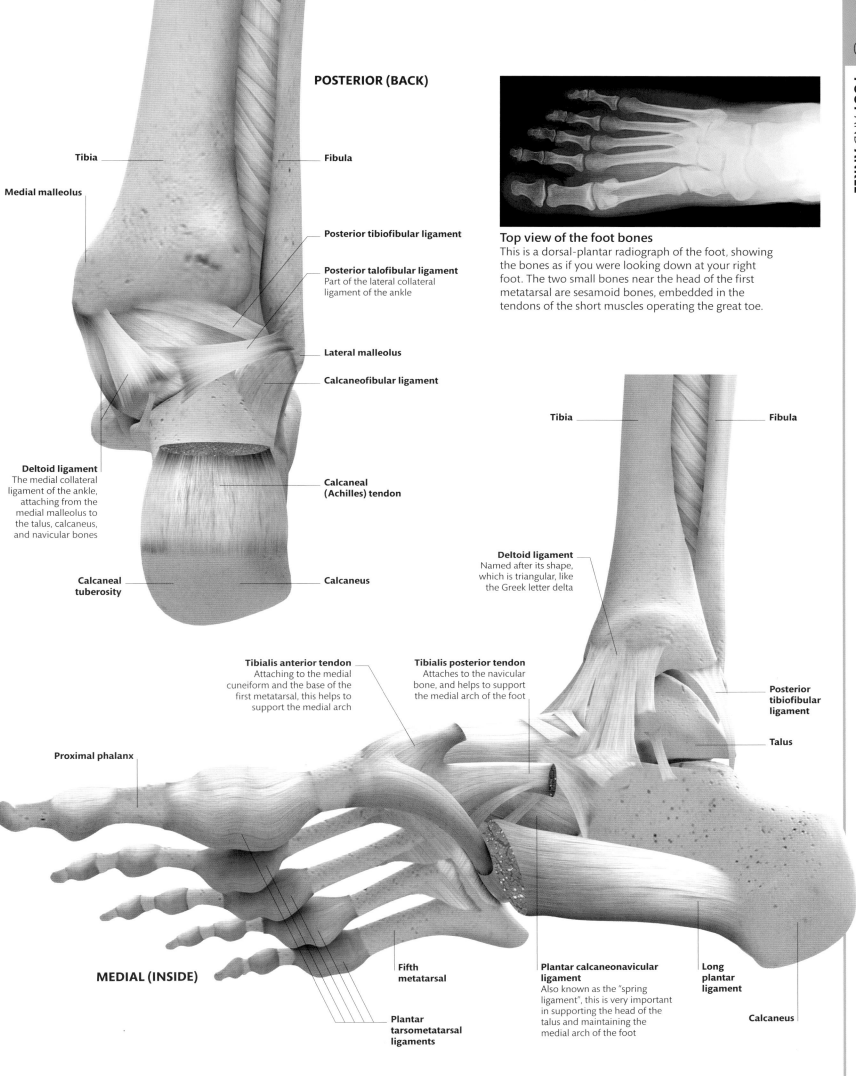

POSTERIOR (BACK)

Tibia

Fibula

Medial malleolus

Posterior tibiofibular ligament

Posterior talofibular ligament
Part of the lateral collateral
ligament of the ankle

Lateral malleolus

Calcaneofibular ligament

Deltoid ligament
The medial collateral
ligament of the ankle,
attaching from the
medial malleolus to
the talus, calcaneus,
and navicular bones

Calcaneal
(Achilles) tendon

Calcaneal
tuberosity

Calcaneus

Top view of the foot bones

This is a dorsal-plantar radiograph of the foot, showing
the bones as if you were looking down at your right
foot. The two small bones near the head of the first
metatarsal are sesamoid bones, embedded in the
tendons of the short muscles operating the great toe.

Tibia

Fibula

Deltoid ligament
Named after its shape,
which is triangular, like
the Greek letter delta

Posterior
tibiofibular
ligament

Talus

Tibialis anterior tendon
Attaching to the medial
cuneiform and the base of the
first metatarsal, this helps to
support the medial arch

Tibialis posterior tendon
Attaches to the navicular
bone, and helps to support
the medial arch of the foot

Proximal phalanx

MEDIAL (INSIDE)

Fifth
metatarsal

Plantar
tarsometatarsal
ligaments

**Plantar calcaneonavicular
ligament**
Also known as the "spring
ligament", this is very important
in supporting the head of the
talus and maintaining the
medial arch of the foot

Long
plantar
ligament

Calcaneus

MUSCULAR SYSTEM OVERVIEW

There are three types of muscle in the body: skeletal, smooth, and cardiac. The main role of skeletal muscles is to generate movement. A muscle's movement, or "action" is produced when it contracts. The force it generates depends on the shape of the muscle. For instance, long, thin muscles contract a lot but exert low forces. Muscles attach to the skeleton by means of tendons, aponeuroses, and connective tissue called fascia. While muscles are well supplied with blood vessels and appear reddish, tendons have a sparse vascular supply and look white. The muscles in our body are located at varied depths. The deep layer sits closest to the bone, while the superficial one lies beneath the skin.

SKELETAL MUSCLE STRUCTURE

Skeletal muscle includes familiar muscles such as biceps or quadriceps. It is built up of parallel bundles of muscle fibres, which are conglomerations of many cells. These muscles are supplied by somatic motor nerves, which are part of the peripheral nervous system and are generally under conscious control.

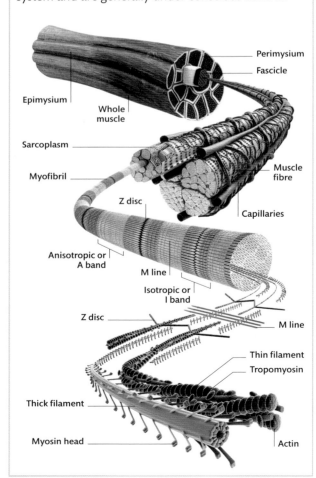

- Perimysium
- Fascicle
- Epimysium
- Whole muscle
- Sarcoplasm
- Muscle fibre
- Myofibril
- Z disc
- Capillaries
- Anisotropic or A band
- M line
- Isotropic or I band
- Z disc
- M line
- Thin filament
- Tropomyosin
- Thick filament
- Myosin head
- Actin

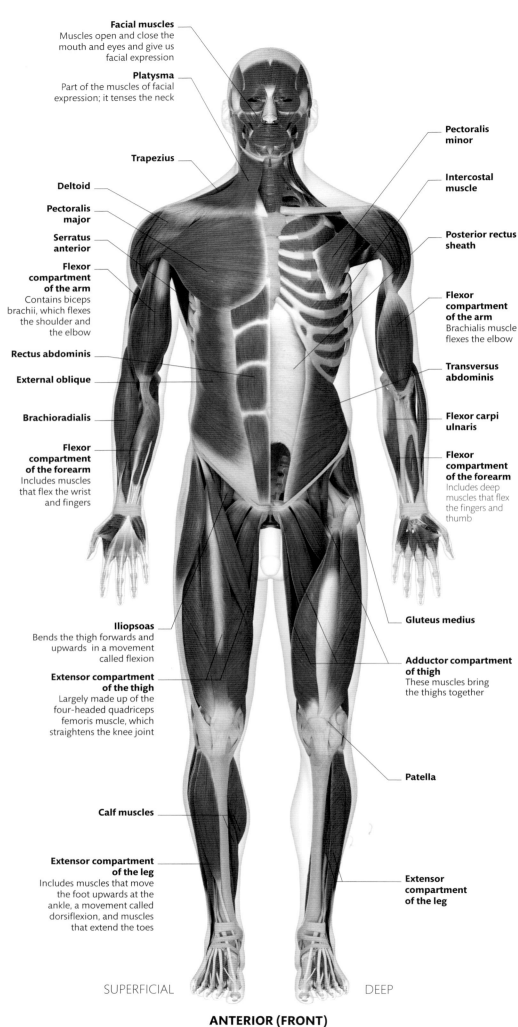

Facial muscles
Muscles open and close the mouth and eyes and give us facial expression

Platysma
Part of the muscles of facial expression; it tenses the neck

Trapezius

Deltoid

Pectoralis major

Serratus anterior

Flexor compartment of the arm
Contains biceps brachii, which flexes the shoulder and the elbow

Rectus abdominis

External oblique

Brachioradialis

Flexor compartment of the forearm
Includes muscles that flex the wrist and fingers

Iliopsoas
Bends the thigh forwards and upwards in a movement called flexion

Extensor compartment of the thigh
Largely made up of the four-headed quadriceps femoris muscle, which straightens the knee joint

Calf muscles

Extensor compartment of the leg
Includes muscles that move the foot upwards at the ankle, a movement called dorsiflexion, and muscles that extend the toes

Pectoralis minor

Intercostal muscle

Posterior rectus sheath

Flexor compartment of the arm
Brachialis muscle flexes the elbow

Transversus abdominis

Flexor carpi ulnaris

Flexor compartment of the forearm
Includes deep muscles that flex the fingers and thumb

Gluteus medius

Adductor compartment of thigh
These muscles bring the thighs together

Patella

Extensor compartment of the leg

SUPERFICIAL   DEEP

ANTERIOR (FRONT)

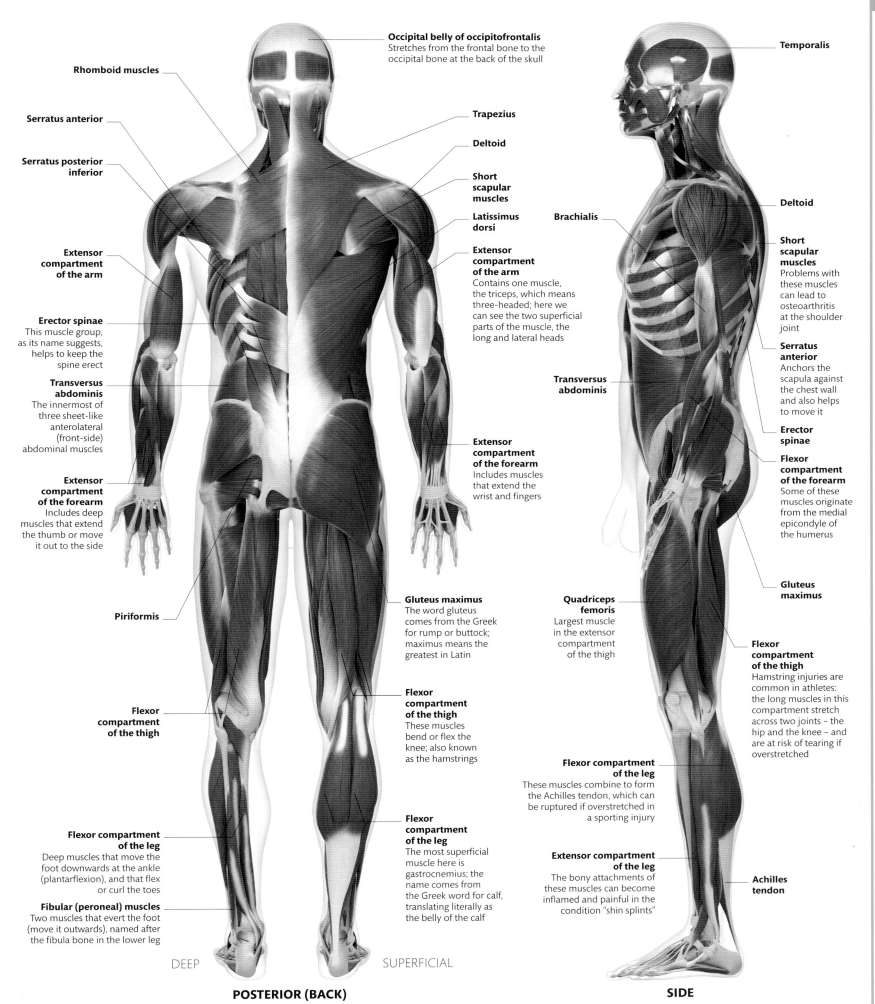

Rhomboid muscles

Serratus anterior

Serratus posterior inferior

Extensor compartment of the arm

Erector spinae
This muscle group, as its name suggests, helps to keep the spine erect

Transversus abdominis
The innermost of three sheet-like anterolateral (front-side) abdominal muscles

Extensor compartment of the forearm
Includes deep muscles that extend the thumb or move it out to the side

Piriformis

Flexor compartment of the thigh

Flexor compartment of the leg
Deep muscles that move the foot downwards at the ankle (plantarflexion), and that flex or curl the toes

Fibular (peroneal) muscles
Two muscles that evert the foot (move it outwards), named after the fibula bone in the lower leg

Occipital belly of occipitofrontalis
Stretches from the frontal bone to the occipital bone at the back of the skull

Trapezius

Deltoid

Short scapular muscles

Latissimus dorsi

Extensor compartment of the arm
Contains one muscle, the triceps, which means three-headed; here we can see the two superficial parts of the muscle, the long and lateral heads

Extensor compartment of the forearm
Includes muscles that extend the wrist and fingers

Gluteus maximus
The word gluteus comes from the Greek for rump or buttock; maximus means the greatest in Latin

Flexor compartment of the thigh
These muscles bend or flex the knee; also known as the hamstrings

Flexor compartment of the leg
The most superficial muscle here is gastrocnemius; the name comes from the Greek word for calf, translating literally as the belly of the calf

DEEP   SUPERFICIAL

POSTERIOR (BACK)

Temporalis

Brachialis

Deltoid

Short scapular muscles
Problems with these muscles can lead to osteoarthritis at the shoulder joint

Serratus anterior
Anchors the scapula against the chest wall and also helps to move it

Transversus abdominis

Erector spinae

Flexor compartment of the forearm
Some of these muscles originate from the medial epicondyle of the humerus

Gluteus maximus

Quadriceps femoris
Largest muscle in the extensor compartment of the thigh

Flexor compartment of the thigh
Hamstring injuries are common in athletes: the long muscles in this compartment stretch across two joints – the hip and the knee – and are at risk of tearing if overstretched

Flexor compartment of the leg
These muscles combine to form the Achilles tendon, which can be ruptured if overstretched in a sporting injury

Extensor compartment of the leg
The bony attachments of these muscles can become inflamed and painful in the condition "shin splints"

Achilles tendon

SIDE

HEAD
AND NECK

The muscles of the face have very important functions. They open and close the apertures in our faces – our eyes, noses, and mouths. But they also play an extremely important role in communication, and this is why these muscles are often known, collectively, as "the muscles of facial expression". These muscles are attached to bone at one end and skin at the other. It is these muscles that allow us to raise our eyebrows in surprise, frown, or knit our brows in concentration, to scrunch up our noses in distaste, to smile gently or to grin widely, and to pout. As we age, and our skin forms creases and wrinkles, these reflect the expressions we have used throughout our lives. The wrinkles and creases lie perpendicular to the direction of the underlying muscle fibres.

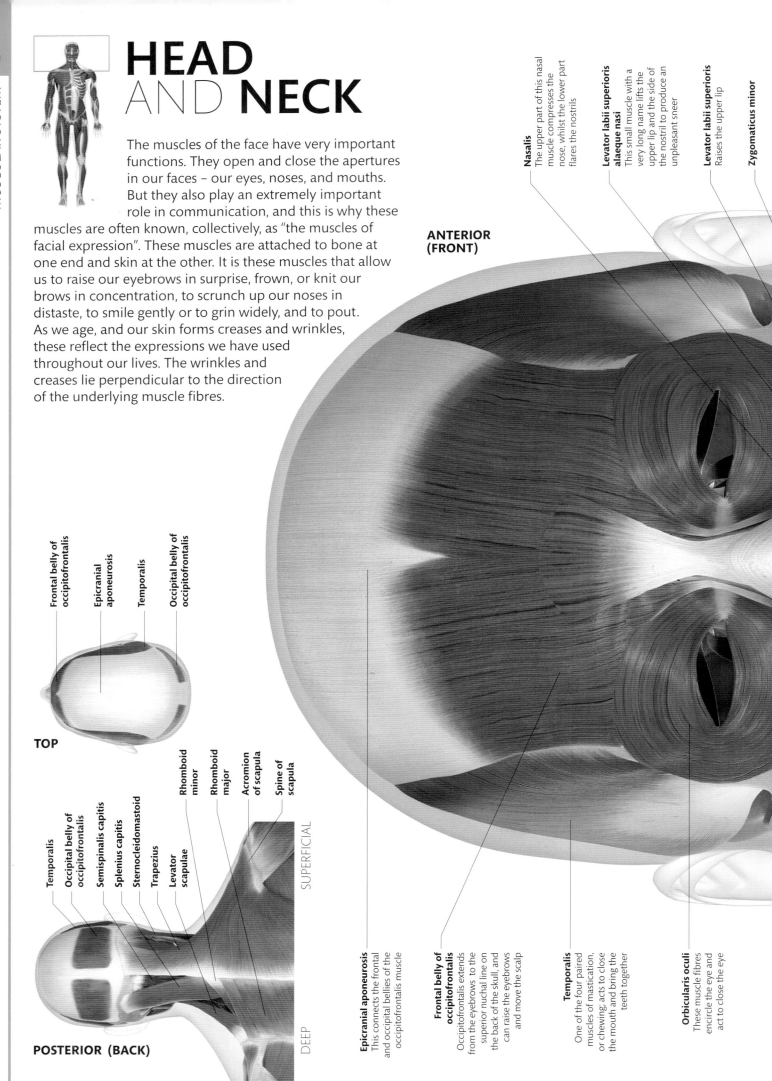

ANTERIOR (FRONT)

TOP

POSTERIOR (BACK)

SUPERFICIAL

DEEP

Frontal belly of occipitofrontalis

Epicranial aponeurosis

Temporalis

Occipital belly of occipitofrontalis

Temporalis

Occipital belly of occipitofrontalis

Semispinalis capitis

Splenius capitis

Sternocleidomastoid

Trapezius

Levator scapulae

Rhomboid minor

Rhomboid major

Acromion of scapula

Spine of scapula

Nasalis
The upper part of this nasal muscle compresses the nose, whilst the lower part flares the nostrils

Levator labii superioris alaeque nasi
This small muscle with a very long name lifts the upper lip and the side of the nostril to produce an unpleasant sneer

Levator labii superioris
Raises the upper lip

Zygomaticus minor

Zygomaticus major
Both the zygomaticus major and minor attach from the zygomatic arch (cheek bone) to the side of the upper lip, and are used in smiling

Epicranial aponeurosis
This connects the frontal and occipital bellies of the occipitofrontalis muscle

Frontal belly of occipitofrontalis
Occipitofrontalis extends from the eyebrows to the superior nuchal line on the back of the skull, and can raise the eyebrows and move the scalp

Temporalis
One of the four paired muscles of mastication, or chewing; acts to close the mouth and bring the teeth together

Orbicularis oculi
These muscle fibres encircle the eye and act to close the eye

Cartilage of the external nose

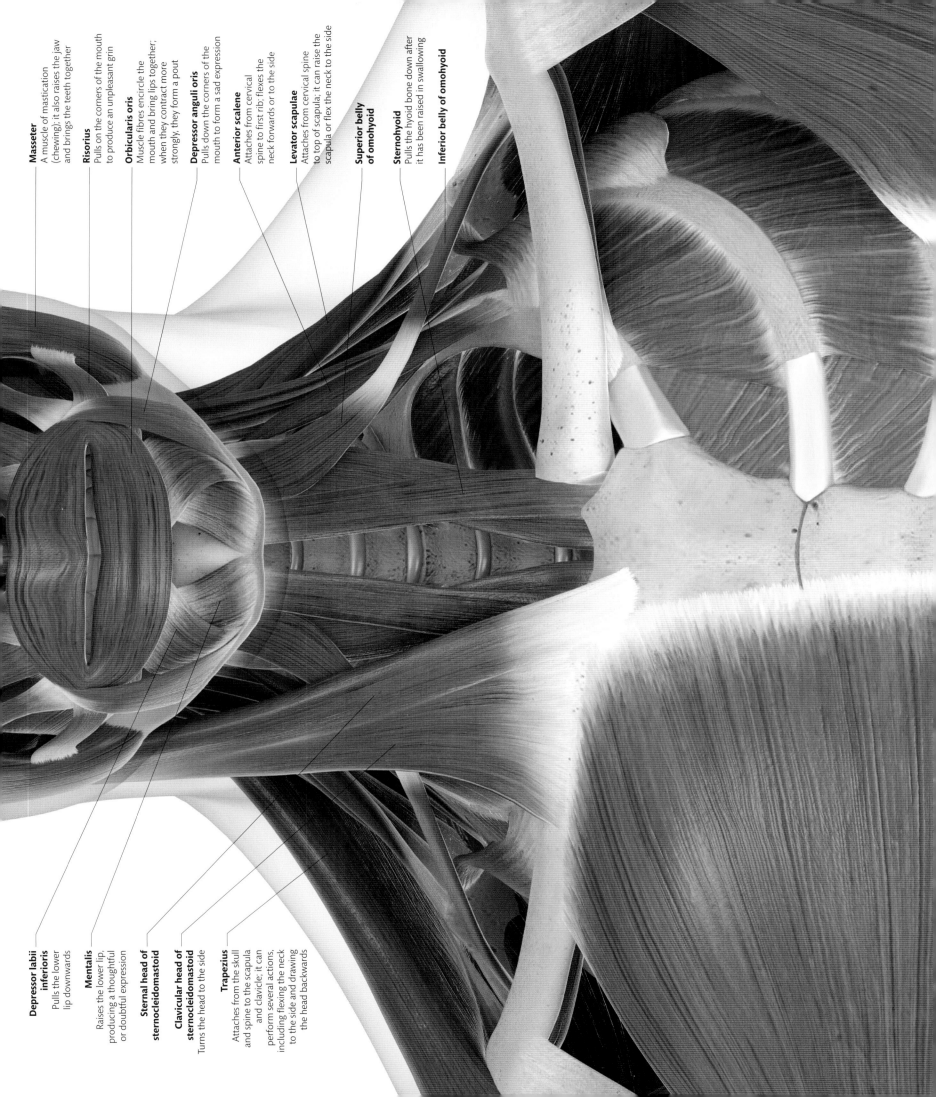

Masseter
A muscle of mastication (chewing); it also raises the jaw and brings the teeth together

Risorius
Pulls on the corners of the mouth to produce an unpleasant grin

Orbicularis oris
Muscle fibres encircle the mouth and bring lips together; when they contract more strongly, they form a pout

Depressor anguli oris
Pulls down the corners of the mouth to form a sad expression

Anterior scalene
Attaches from cervical spine to first rib; flexes the neck forwards or to the side

Levator scapulae
Attaches from cervical spine to top of scapula; it can raise the scapula or flex the neck to the side

Superior belly of omohyoid

Sternohyoid
Pulls the hyoid bone down after it has been raised in swallowing

Inferior belly of omohyoid

Depressor labii inferioris
Pulls the lower lip downwards

Mentalis
Raises the lower lip, producing a thoughtful or doubtful expression

Sternal head of sternocleidomastoid

Clavicular head of sternocleidomastoid
Turns the head to the side

Trapezius
Attaches from the skull and spine to the scapula and clavicle; it can perform several actions, including flexing the neck to the side and drawing the head backwards

072

HEAD AND NECK

The muscles of mastication (chewing) attach from the skull to the mandible (jawbone), operating to open and shut the mouth, and to grind the teeth together to crush the food we eat. In this side view, we can see the two largest muscles of mastication, the temporalis and masseter muscles. Two smaller muscles attach to the inner surface of the mandible. Human jaws don't just open and close, they also move from side to side, and these four muscles act in concert to produce complex chewing movements. In this view, we can also see how the frontal bellies (fleshy central parts) of the occipitofrontalis muscle are connected to occipital bellies at the back of the head by a thin, flat tendon, or aponeurosis. This makes the entire scalp movable on the skull.

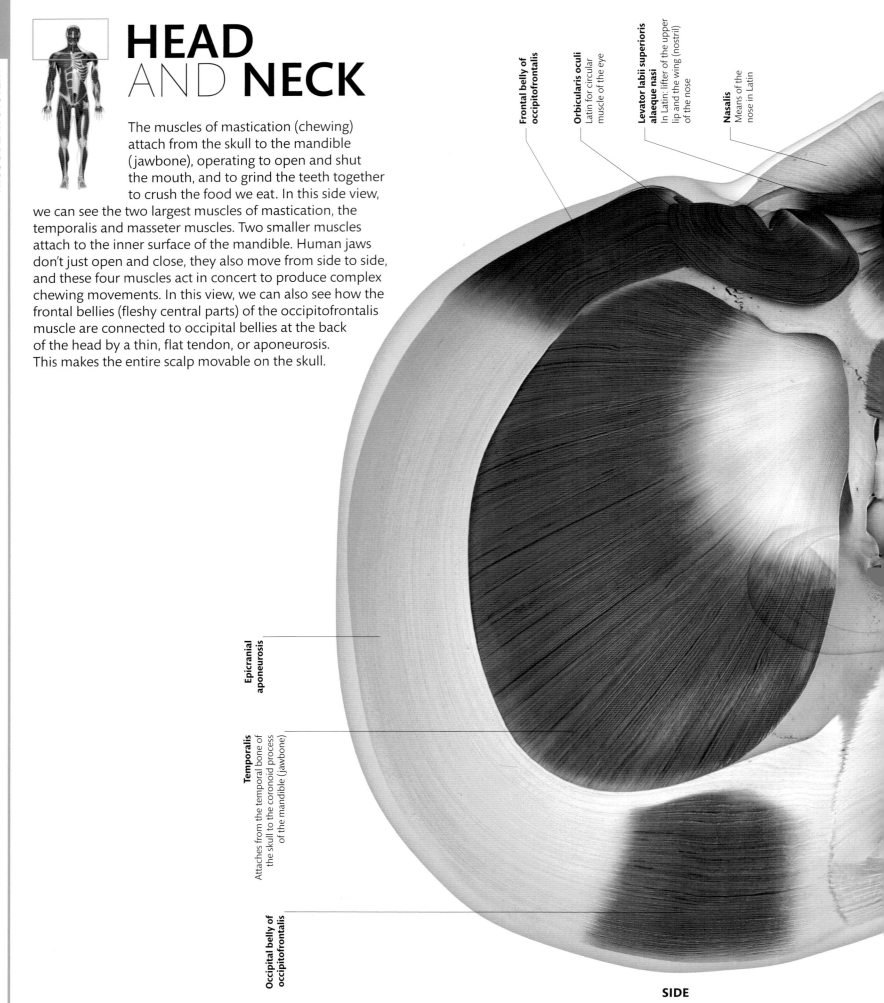

Frontal belly of occipitofrontalis

Orbicularis oculi
Latin for circular muscle of the eye

Levator labii superioris alaeque nasi
In Latin: lifter of the upper lip and the wing (nostril) of the nose

Nasalis
Means of the nose in Latin

Epicranial aponeurosis

Temporalis
Attaches from the temporal bone of the skull to the coronoid process of the mandible (jawbone)

Occipital belly of occipitofrontalis

SIDE

Levator labii superioris
Literally, lifter of the upper lip

Zygomaticus major
Attaches from the zygomatic arch (cheek bone)

Orbicularis oris
Latin for circular muscle of the mouth

Risorius
From the Latin for laughter

Depressor labii inferioris
The depressor of the lower lip

Mentalis
This means of the chin in Latin

Depressor anguli oris
Literally, the depressor of the corner of the mouth

Masseter
From the Greek for chewer

Anterior belly of digastric
Digastric means two-bellied

Posterior belly of digastric
The digastric pulls the mandible (lower jawbone) down to open the mouth, and pulls the hyoid bone up in swallowing

Thyrohyoid
Attaches from the hyoid bone to the thyroid cartilage of the larynx

Superior belly of omohyoid
Omo comes from the Greek for shoulder; this muscle is named after its attachments – from the hyoid bone to the shoulder blade

Sternohyoid
Attaches from the sternum to the hyoid bone

Sternothyroid
Attaches from the sternum to the thyroid cartilage

Inferior belly of omohyoid

Splenius capitis
Named after the Latin for bandage of the head, this muscle draws the head backwards

Sternocleidomastoid
Turns the head to the side

Superior constrictor of pharynx

Trapezius

Levator scapulae
In Latin this means lifter of the shoulder blade

Middle scalene

Anterior scalene
The scalene muscles are shaped like scalene triangles (where each side is a different length)

Posterior scalene

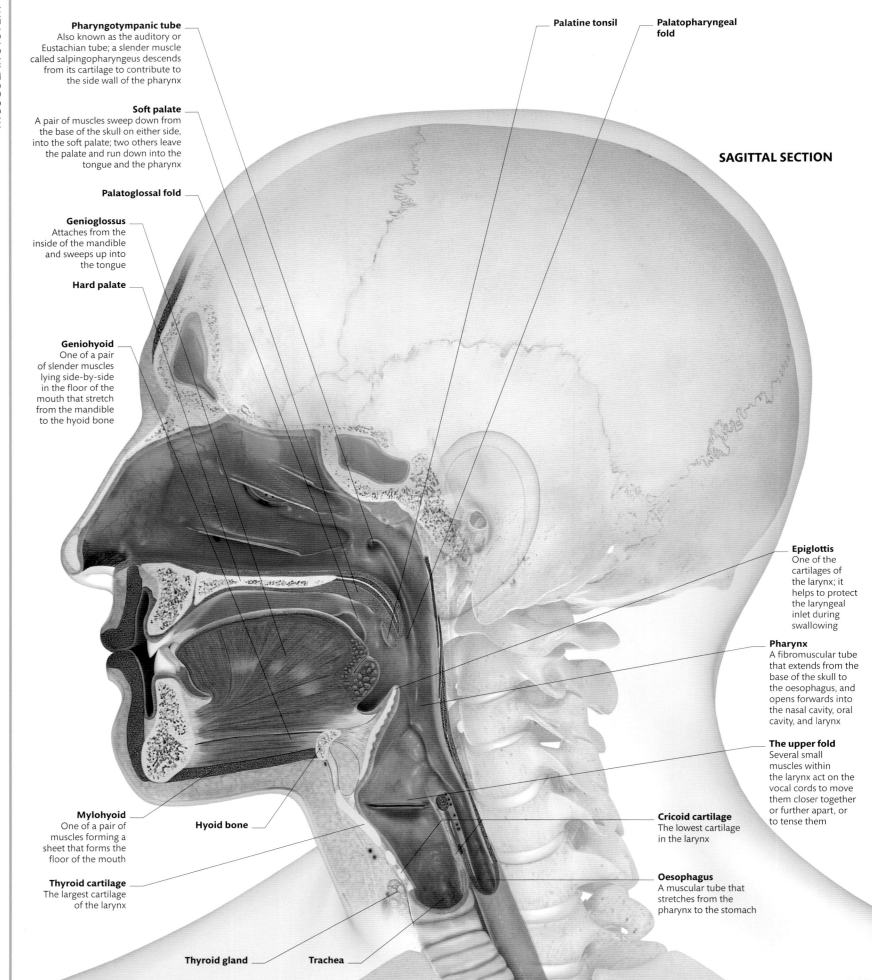

SAGITTAL SECTION

Pharyngotympanic tube
Also known as the auditory or Eustachian tube; a slender muscle called salpingopharyngeus descends from its cartilage to contribute to the side wall of the pharynx

Soft palate
A pair of muscles sweep down from the base of the skull on either side, into the soft palate; two others leave the palate and run down into the tongue and the pharynx

Palatoglossal fold

Genioglossus
Attaches from the inside of the mandible and sweeps up into the tongue

Hard palate

Geniohyoid
One of a pair of slender muscles lying side-by-side in the floor of the mouth that stretch from the mandible to the hyoid bone

Palatine tonsil

Palatopharyngeal fold

Epiglottis
One of the cartilages of the larynx; it helps to protect the laryngeal inlet during swallowing

Pharynx
A fibromuscular tube that extends from the base of the skull to the oesophagus, and opens forwards into the nasal cavity, oral cavity, and larynx

The upper fold
Several small muscles within the larynx act on the vocal cords to move them closer together or further apart, or to tense them

Mylohyoid
One of a pair of muscles forming a sheet that forms the floor of the mouth

Hyoid bone

Cricoid cartilage
The lowest cartilage in the larynx

Thyroid cartilage
The largest cartilage of the larynx

Oesophagus
A muscular tube that stretches from the pharynx to the stomach

Thyroid gland

Trachea

HEAD AND NECK

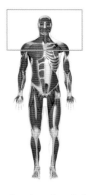

In the section through the head (opposite), we see the soft palate, tongue, pharynx, and larynx, all of which contain muscles. The soft palate comprises five pairs of muscles. When relaxed, it hangs down at the back of the mouth but, during swallowing, it thickens and is drawn upwards to block off the airway. The tongue is a great mass of muscle, covered in mucosa. Some of its muscles arise from the hyoid bone and the mandible, and anchor it to these bones and move it around. Other muscle fibres are entirely within the tongue and change its shape. The pharyngeal muscles are important in swallowing, and the laryngeal muscles control the vocal cords. The muscles that move the eye can be seen on p.122.

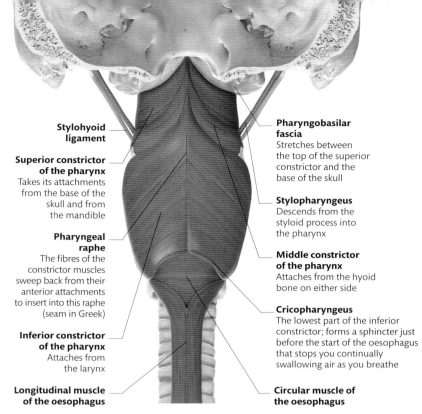

Stylohyoid ligament

Superior constrictor of the pharynx
Takes its attachments from the base of the skull and from the mandible

Pharyngeal raphe
The fibres of the constrictor muscles sweep back from their anterior attachments to insert into this raphe (seam in Greek)

Inferior constrictor of the pharynx
Attaches from the larynx

Longitudinal muscle of the oesophagus

Pharyngobasilar fascia
Stretches between the top of the superior constrictor and the base of the skull

Stylopharyngeus
Descends from the styloid process into the pharynx

Middle constrictor of the pharynx
Attaches from the hyoid bone on either side

Cricopharyngeus
The lowest part of the inferior constrictor; forms a sphincter just before the start of the oesophagus that stops you continually swallowing air as you breathe

Circular muscle of the oesophagus

PHARYNX POSTERIOR (BACK)

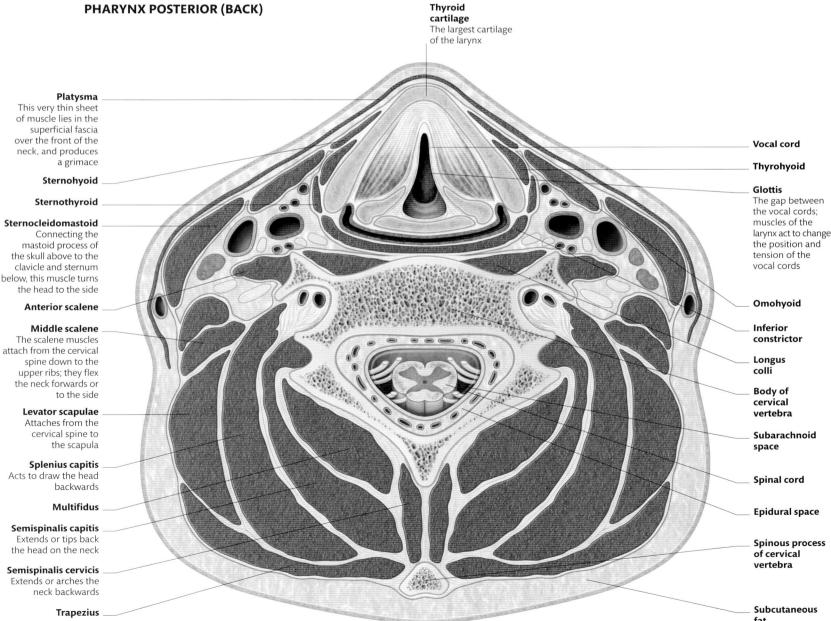

Thyroid cartilage
The largest cartilage of the larynx

Platysma
This very thin sheet of muscle lies in the superficial fascia over the front of the neck, and produces a grimace

Sternohyoid

Sternothyroid

Sternocleidomastoid
Connecting the mastoid process of the skull above to the clavicle and sternum below, this muscle turns the head to the side

Anterior scalene

Middle scalene
The scalene muscles attach from the cervical spine down to the upper ribs; they flex the neck forwards or to the side

Levator scapulae
Attaches from the cervical spine to the scapula

Splenius capitis
Acts to draw the head backwards

Multifidus

Semispinalis capitis
Extends or tips back the head on the neck

Semispinalis cervicis
Extends or arches the neck backwards

Trapezius

Vocal cord

Thyrohyoid

Glottis
The gap between the vocal cords; muscles of the larynx act to change the position and tension of the vocal cords

Omohyoid

Inferior constrictor

Longus colli

Body of cervical vertebra

Subarachnoid space

Spinal cord

Epidural space

Spinous process of cervical vertebra

Subcutaneous fat

TRANSVERSE SECTION OF THE NECK AT THE VOCAL CORDS

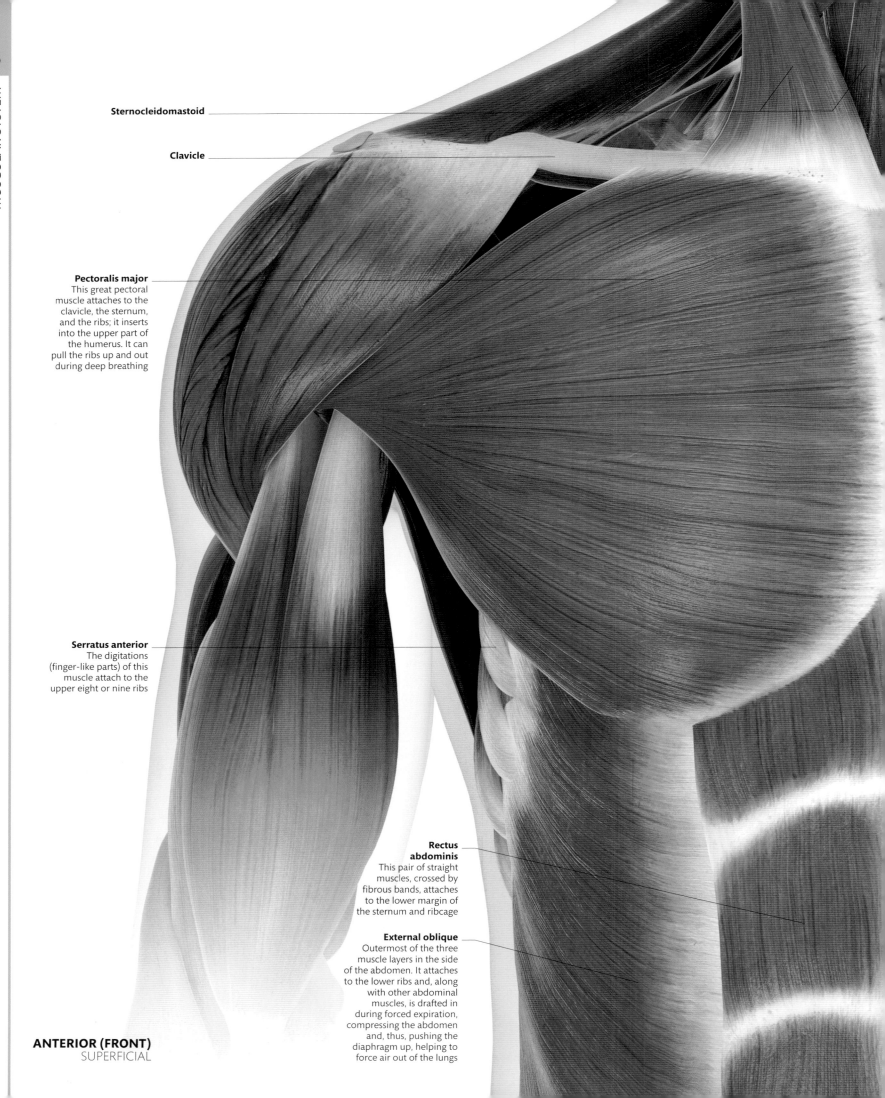

Sternocleidomastoid

Clavicle

Pectoralis major
This great pectoral
muscle attaches to the
clavicle, the sternum,
and the ribs; it inserts
into the upper part of
the humerus. It can
pull the ribs up and out
during deep breathing

Serratus anterior
The digitations
(finger-like parts) of this
muscle attach to the
upper eight or nine ribs

**Rectus
abdominis**
This pair of straight
muscles, crossed by
fibrous bands, attaches
to the lower margin of
the sternum and ribcage

External oblique
Outermost of the three
muscle layers in the side
of the abdomen. It attaches
to the lower ribs and, along
with other abdominal
muscles, is drafted in
during forced expiration,
compressing the abdomen
and, thus, pushing the
diaphragm up, helping to
force air out of the lungs

ANTERIOR (FRONT)
SUPERFICIAL

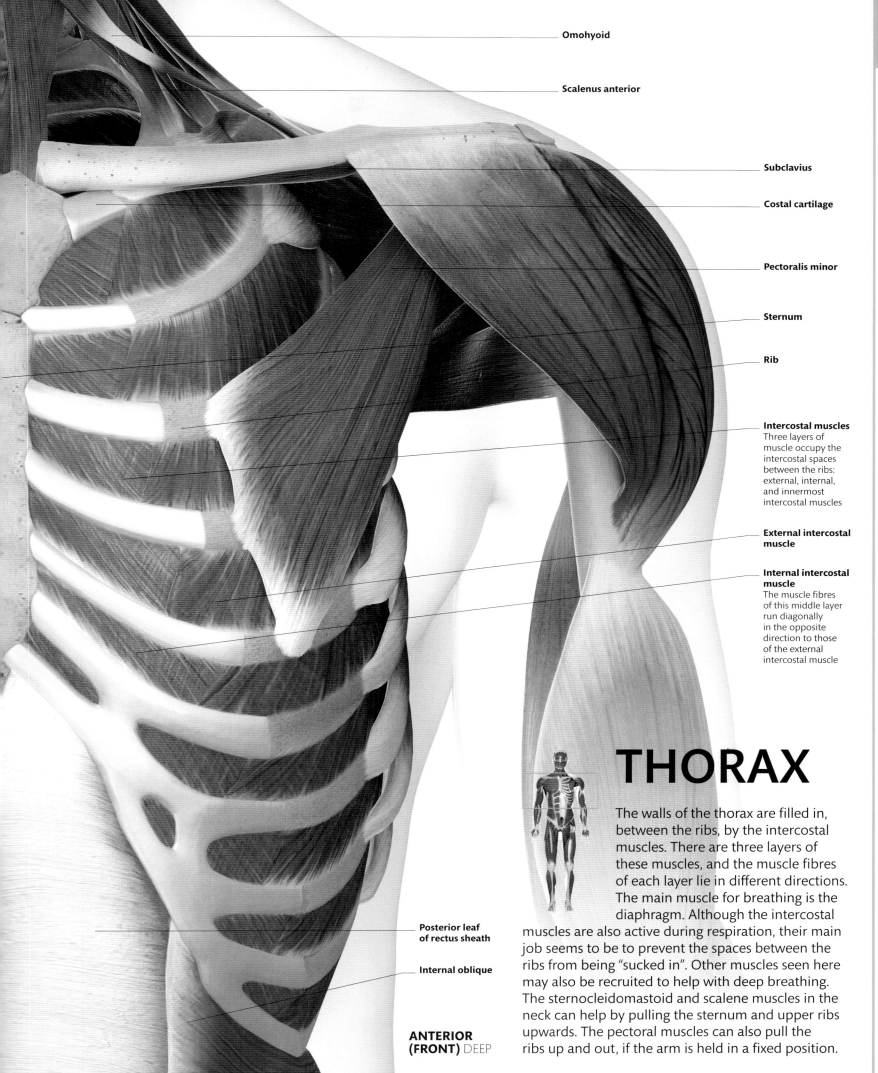

Omohyoid

Scalenus anterior

Subclavius

Costal cartilage

Pectoralis minor

Sternum

Rib

Intercostal muscles
Three layers of
muscle occupy the
intercostal spaces
between the ribs:
external, internal,
and innermost
intercostal muscles

**External intercostal
muscle**

**Internal intercostal
muscle**
The muscle fibres
of this middle layer
run diagonally
in the opposite
direction to those
of the external
intercostal muscle

Posterior leaf
of rectus sheath

Internal oblique

**ANTERIOR
(FRONT)** DEEP

THORAX

The walls of the thorax are filled in,
between the ribs, by the intercostal
muscles. There are three layers of
these muscles, and the muscle fibres
of each layer lie in different directions.
The main muscle for breathing is the
diaphragm. Although the intercostal
muscles are also active during respiration, their main
job seems to be to prevent the spaces between the
ribs from being "sucked in". Other muscles seen here
may also be recruited to help with deep breathing.
The sternocleidomastoid and scalene muscles in the
neck can help by pulling the sternum and upper ribs
upwards. The pectoral muscles can also pull the
ribs up and out, if the arm is held in a fixed position.

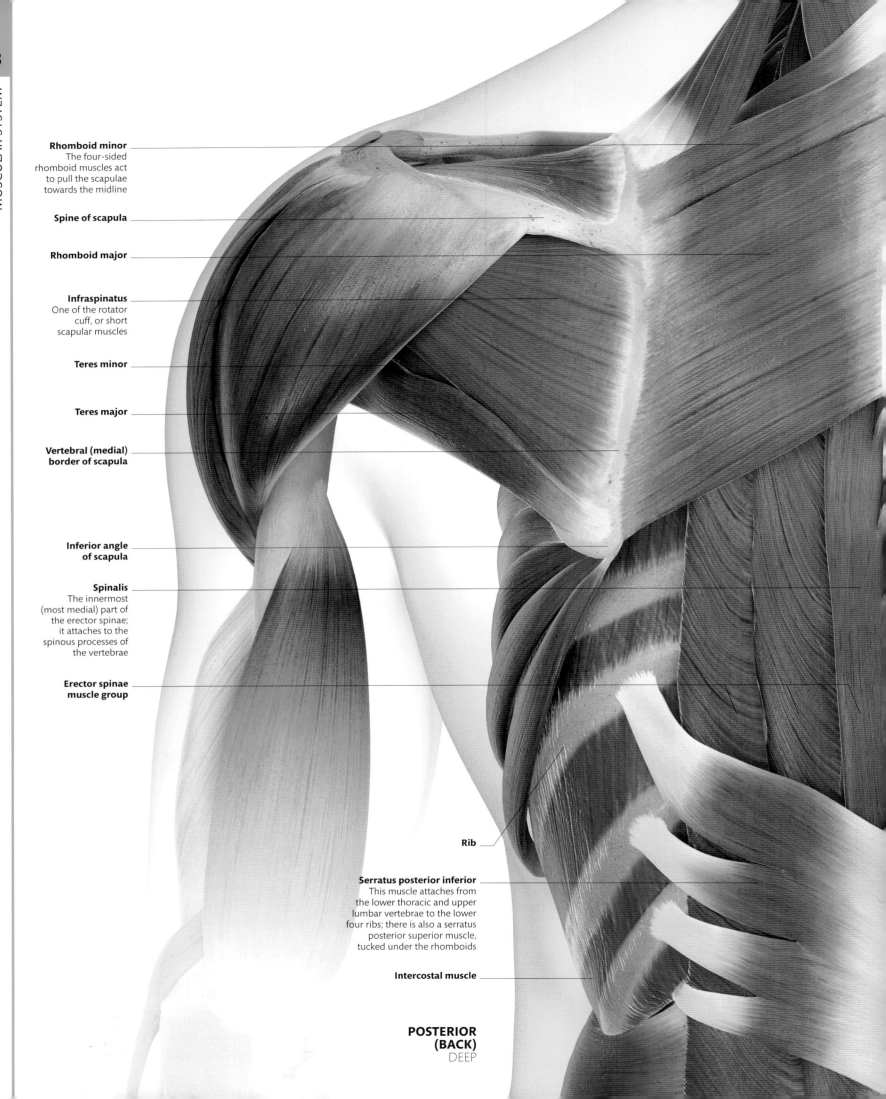

Rhomboid minor
The four-sided rhomboid muscles act to pull the scapulae towards the midline

Spine of scapula

Rhomboid major

Infraspinatus
One of the rotator cuff, or short scapular muscles

Teres minor

Teres major

Vertebral (medial) border of scapula

Inferior angle of scapula

Spinalis
The innermost (most medial) part of the erector spinae; it attaches to the spinous processes of the vertebrae

Erector spinae muscle group

Rib

Serratus posterior inferior
This muscle attaches from the lower thoracic and upper lumbar vertebrae to the lower four ribs; there is also a serratus posterior superior muscle, tucked under the rhomboids

Intercostal muscle

POSTERIOR (BACK)
DEEP

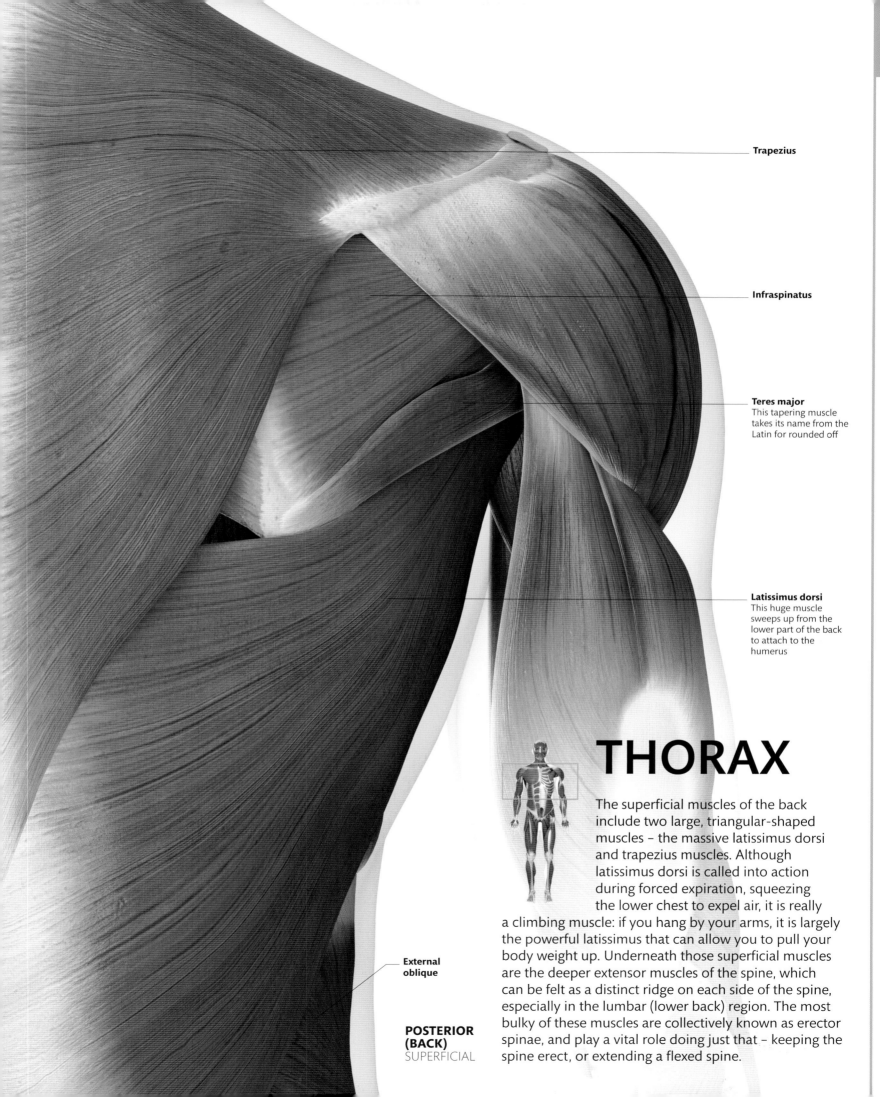

Trapezius

Infraspinatus

Teres major
This tapering muscle takes its name from the Latin for rounded off

Latissimus dorsi
This huge muscle sweeps up from the lower part of the back to attach to the humerus

External oblique

POSTERIOR (BACK)
SUPERFICIAL

THORAX

The superficial muscles of the back include two large, triangular-shaped muscles – the massive latissimus dorsi and trapezius muscles. Although latissimus dorsi is called into action during forced expiration, squeezing the lower chest to expel air, it is really a climbing muscle: if you hang by your arms, it is largely the powerful latissimus that can allow you to pull your body weight up. Underneath those superficial muscles are the deeper extensor muscles of the spine, which can be felt as a distinct ridge on each side of the spine, especially in the lumbar (lower back) region. The most bulky of these muscles are collectively known as erector spinae, and play a vital role doing just that – keeping the spine erect, or extending a flexed spine.

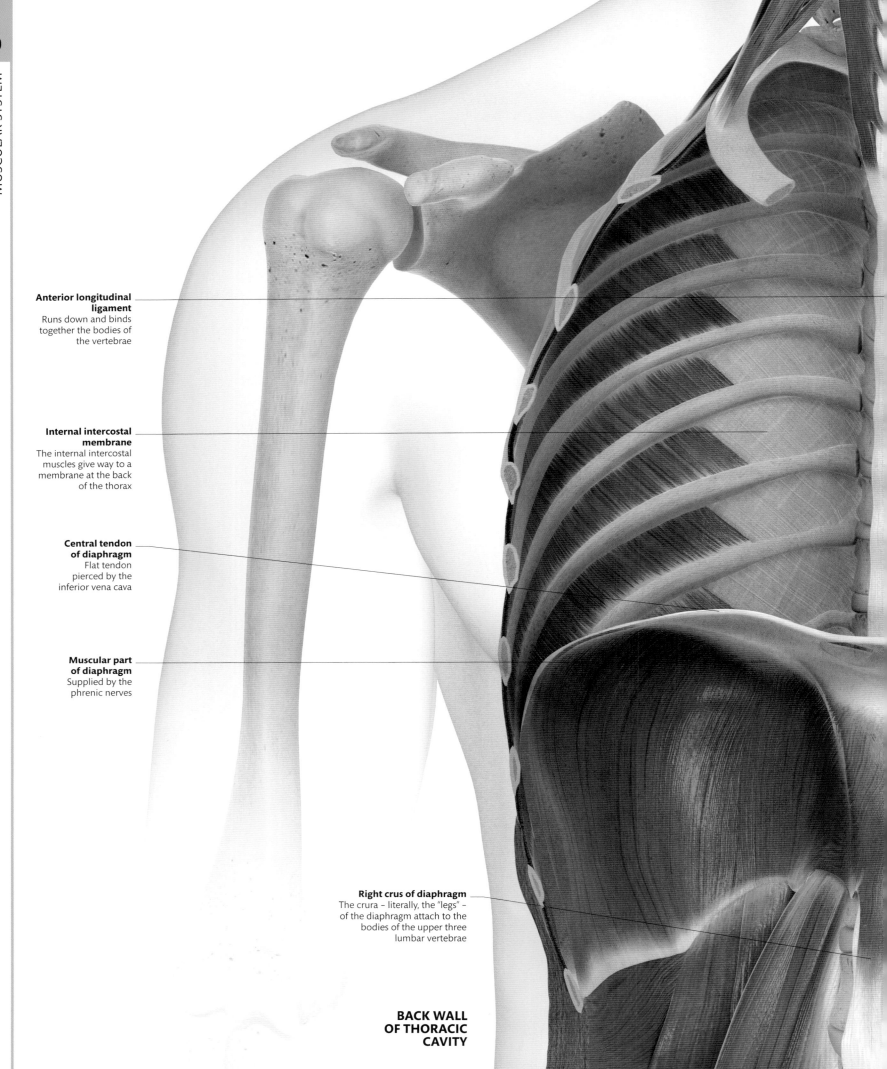

Anterior longitudinal ligament
Runs down and binds together the bodies of the vertebrae

Internal intercostal membrane
The internal intercostal muscles give way to a membrane at the back of the thorax

Central tendon of diaphragm
Flat tendon pierced by the inferior vena cava

Muscular part of diaphragm
Supplied by the phrenic nerves

Right crus of diaphragm
The crura – literally, the "legs" – of the diaphragm attach to the bodies of the upper three lumbar vertebrae

BACK WALL OF THORACIC CAVITY

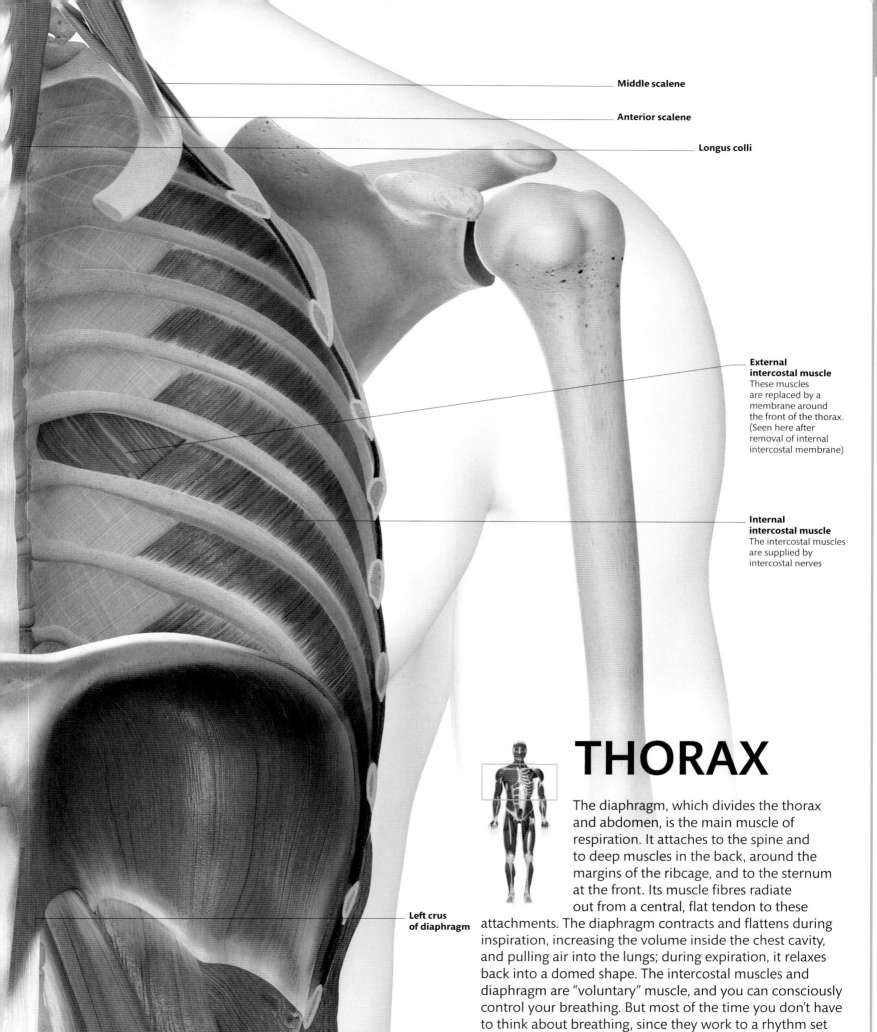

Middle scalene

Anterior scalene

Longus colli

External intercostal muscle
These muscles are replaced by a membrane around the front of the thorax. (Seen here after removal of internal intercostal membrane)

Internal intercostal muscle
The intercostal muscles are supplied by intercostal nerves

Left crus of diaphragm

THORAX

The diaphragm, which divides the thorax and abdomen, is the main muscle of respiration. It attaches to the spine and to deep muscles in the back, around the margins of the ribcage, and to the sternum at the front. Its muscle fibres radiate out from a central, flat tendon to these attachments. The diaphragm contracts and flattens during inspiration, increasing the volume inside the chest cavity, and pulling air into the lungs; during expiration, it relaxes back into a domed shape. The intercostal muscles and diaphragm are "voluntary" muscle, and you can consciously control your breathing. But most of the time you don't have to think about breathing, since they work to a rhythm set by the brainstem, producing about 12 to 20 breaths per minute in an adult.

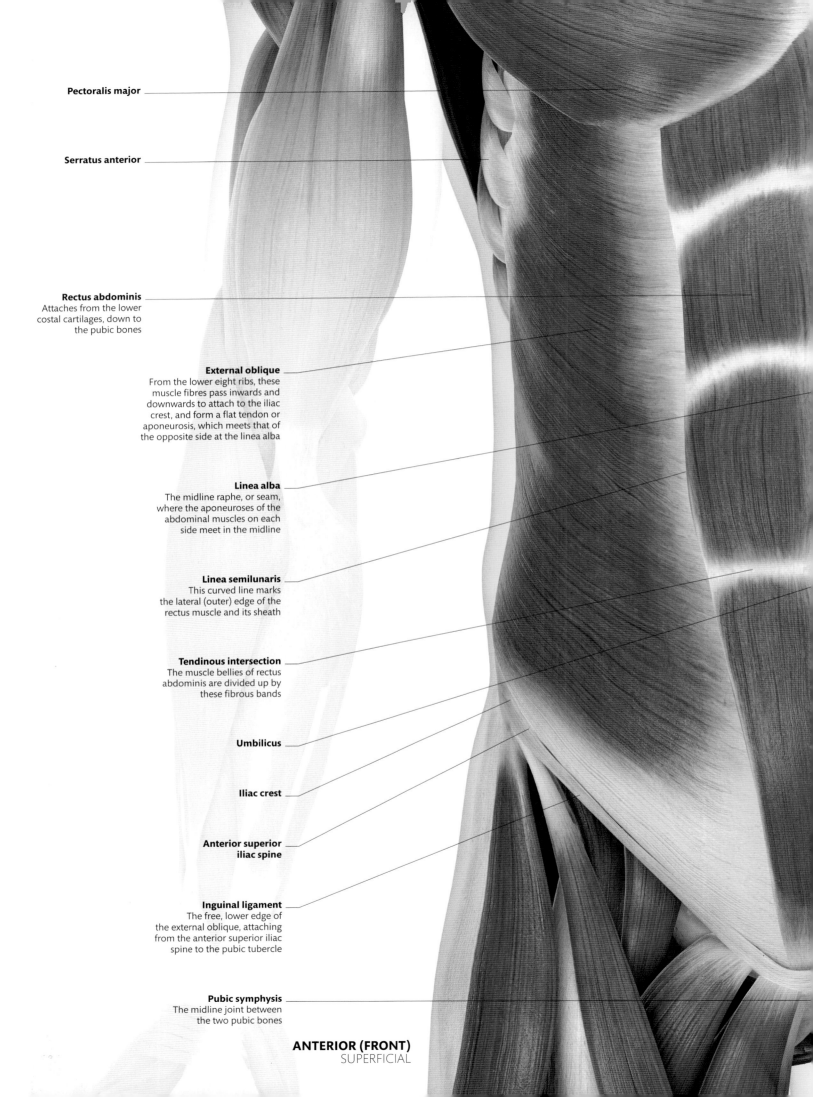

Pectoralis major

Serratus anterior

Rectus abdominis
Attaches from the lower
costal cartilages, down to
the pubic bones

External oblique
From the lower eight ribs, these
muscle fibres pass inwards and
downwards to attach to the iliac
crest, and form a flat tendon or
aponeurosis, which meets that of
the opposite side at the linea alba

Linea alba
The midline raphe, or seam,
where the aponeuroses of the
abdominal muscles on each
side meet in the midline

Linea semilunaris
This curved line marks
the lateral (outer) edge of the
rectus muscle and its sheath

Tendinous intersection
The muscle bellies of rectus
abdominis are divided up by
these fibrous bands

Umbilicus

Iliac crest

**Anterior superior
iliac spine**

Inguinal ligament
The free, lower edge of
the external oblique, attaching
from the anterior superior iliac
spine to the pubic tubercle

Pubic symphysis
The midline joint between
the two pubic bones

ANTERIOR (FRONT)
SUPERFICIAL

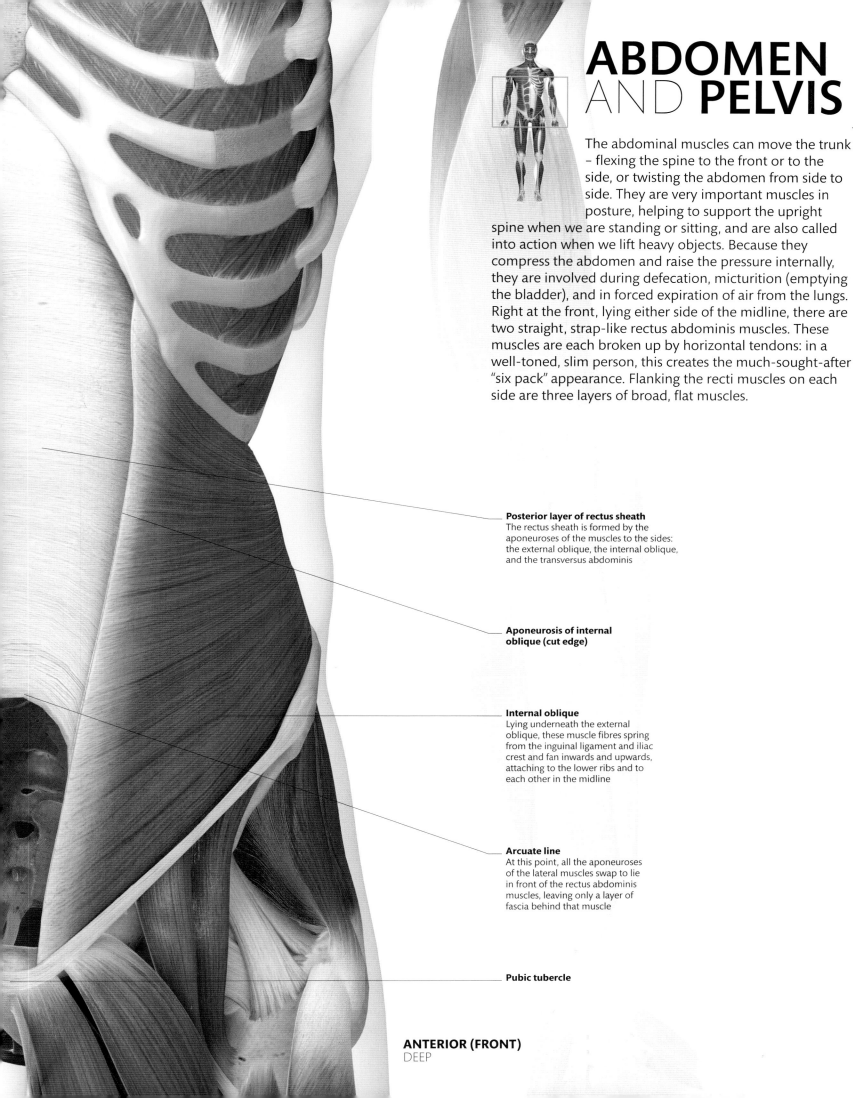

ABDOMEN AND PELVIS

The abdominal muscles can move the trunk – flexing the spine to the front or to the side, or twisting the abdomen from side to side. They are very important muscles in posture, helping to support the upright spine when we are standing or sitting, and are also called into action when we lift heavy objects. Because they compress the abdomen and raise the pressure internally, they are involved during defecation, micturition (emptying the bladder), and in forced expiration of air from the lungs. Right at the front, lying either side of the midline, there are two straight, strap-like rectus abdominis muscles. These muscles are each broken up by horizontal tendons: in a well-toned, slim person, this creates the much-sought-after "six pack" appearance. Flanking the recti muscles on each side are three layers of broad, flat muscles.

Posterior layer of rectus sheath
The rectus sheath is formed by the aponeuroses of the muscles to the sides: the external oblique, the internal oblique, and the transversus abdominis

Aponeurosis of internal oblique (cut edge)

Internal oblique
Lying underneath the external oblique, these muscle fibres spring from the inguinal ligament and iliac crest and fan inwards and upwards, attaching to the lower ribs and to each other in the midline

Arcuate line
At this point, all the aponeuroses of the lateral muscles swap to lie in front of the rectus abdominis muscles, leaving only a layer of fascia behind that muscle

Pubic tubercle

ANTERIOR (FRONT)
DEEP

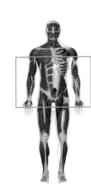

ABDOMEN AND PELVIS

The most superficial muscle of the lower back is the incredibly broad latissimus dorsi. Underneath this, lying along the spine on each side, there is a large bulk of muscle that forms two ridges in the lumbar region in a well-toned person. This muscle mass is collectively known as the erector spinae, and its name suggests its importance in keeping the spine upright. When the spine is flexed forwards, the erector spinae can pull it back into an upright position, and even take it further, into extension. The muscle can be divided up into three main strips on each side: iliocostalis, longissimus, and spinalis. Most of the muscle bulk of the buttock comes down to just one muscle: the fleshy gluteus maximus, which extends the hip joint. Hidden beneath the gluteus maximus are a range of smaller muscles that also move the hip.

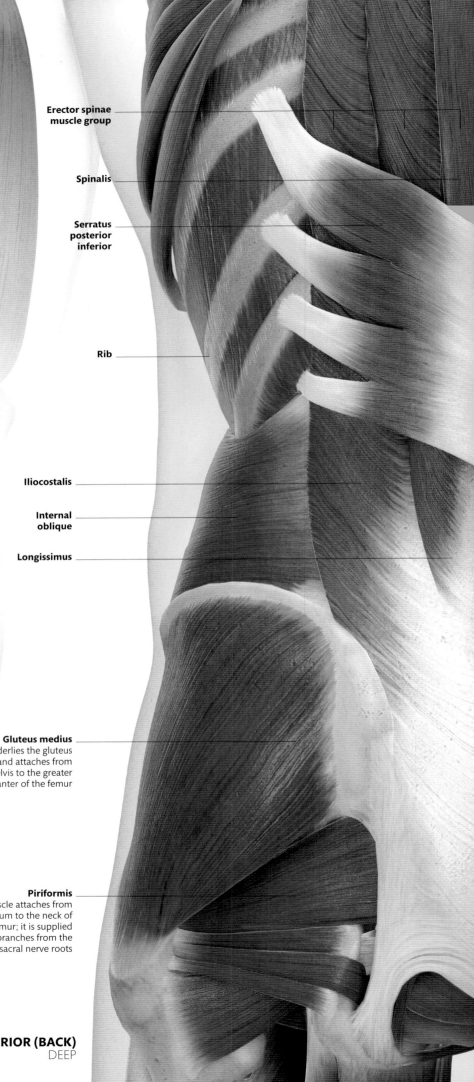

Erector spinae muscle group

Spinalis

Serratus posterior inferior

Rib

Iliocostalis

Internal oblique

Longissimus

Gluteus medius
Underlies the gluteus maximus, and attaches from the pelvis to the greater trochanter of the femur

Piriformis
This muscle attaches from the sacrum to the neck of the femur; it is supplied by branches from the sacral nerve roots

POSTERIOR (BACK)
DEEP

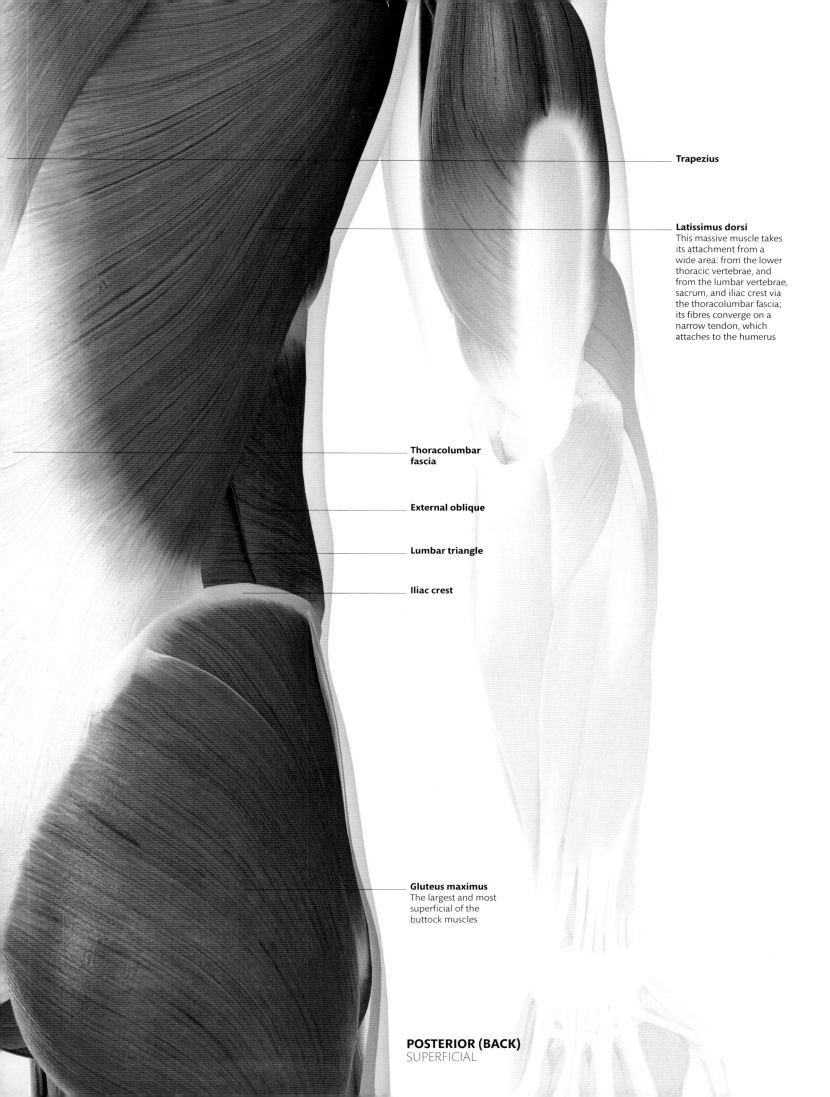

Trapezius

Latissimus dorsi
This massive muscle takes
its attachment from a
wide area: from the lower
thoracic vertebrae, and
from the lumbar vertebrae,
sacrum, and iliac crest via
the thoracolumbar fascia;
its fibres converge on a
narrow tendon, which
attaches to the humerus

**Thoracolumbar
fascia**

External oblique

Lumbar triangle

Iliac crest

Gluteus maximus
The largest and most
superficial of the
buttock muscles

POSTERIOR (BACK)
SUPERFICIAL

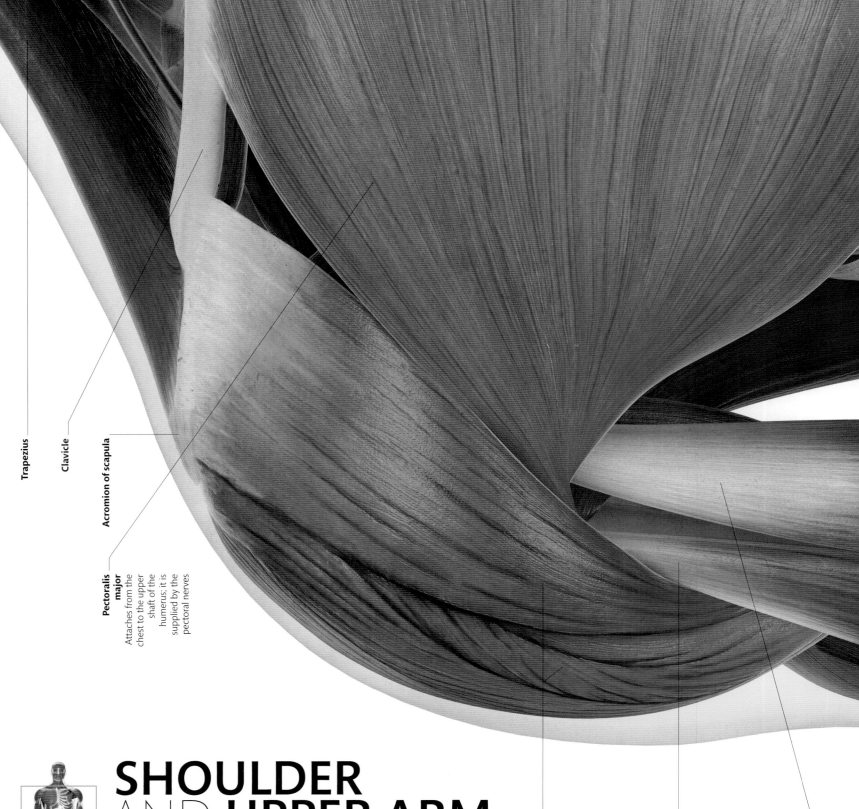

Trapezius

Clavicle

Acromion of scapula

Pectoralis major
Attaches from the chest to the upper shaft of the humerus; it is supplied by the pectoral nerves

SHOULDER
AND **UPPER ARM**

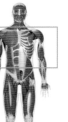

The triangular deltoid muscle lies over the shoulder. Acting as a whole, this muscle raises the arm to the side (abduction), but the fibres of the deltoid attaching to the front of the clavicle can also move the arm forwards. The pectoralis major muscle can also act on the shoulder joint, flexing the arm forwards or pulling it in to the side of the chest (adduction). The biceps brachii muscle forms much of the muscle bulk on the front of the arm. The biceps tendon inserts on the radius, and also has an aponeurosis (flat tendon) that fans out over the forearm muscles. The biceps is a powerful flexor of the elbow, and can also rotate the radius to position the lower arm so the palm faces upwards (supination).

Deltoid
This powerful muscle attaches from the clavicle, acromion, and spine of the scapula to the deltoid tuberosity on the side of the humerus

Long head of biceps
This tendon disappears under the deltoid sooner than the short head, so it appears to be the shorter of the two, but it runs right over the head of the humerus to attach to the scapula above the glenoid fossa

Short head of biceps
Attaches to the coracoid process of the scapula

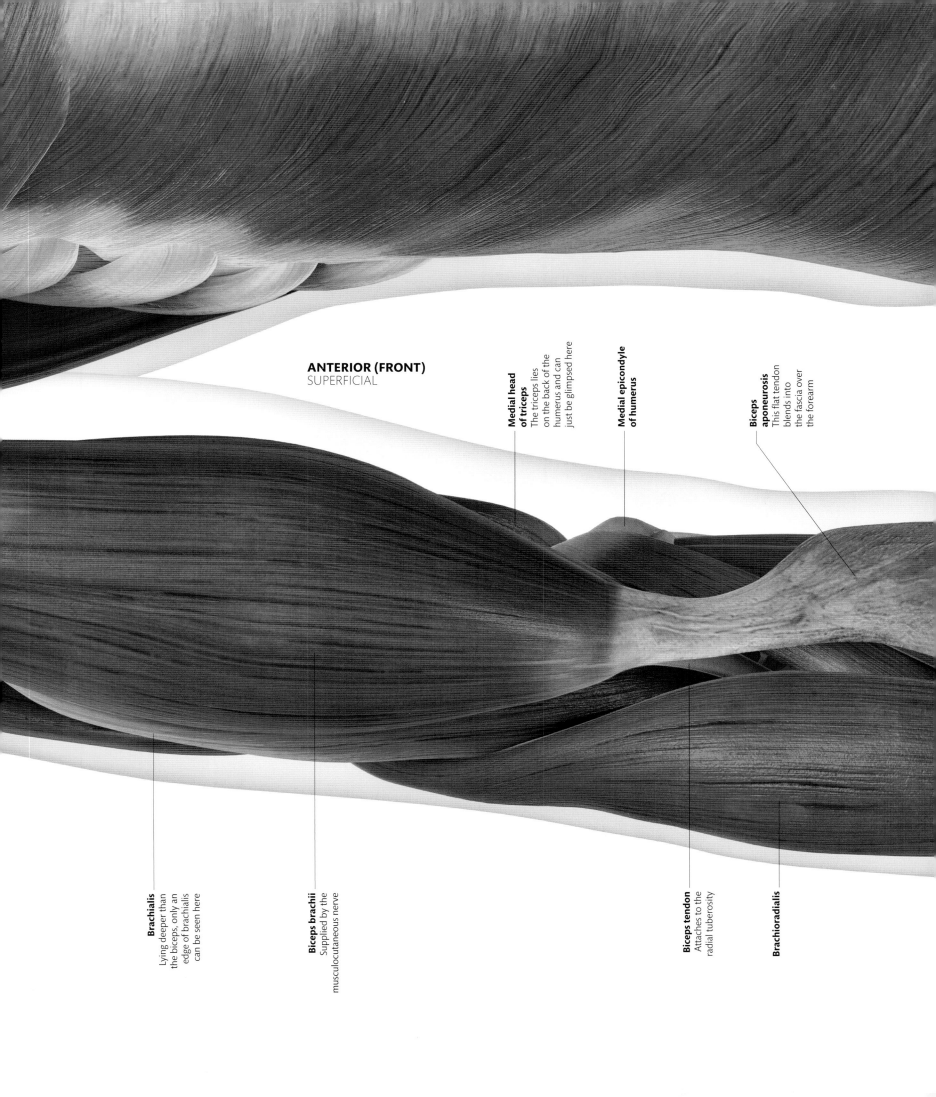

ANTERIOR (FRONT)
SUPERFICIAL

**Medial head
of triceps**
The triceps lies
on the back of the
humerus and can
just be glimpsed here

**Medial epicondyle
of humerus**

**Biceps
aponeurosis**
This flat tendon
blends into
the fascia over
the forearm

Brachialis
Lying deeper than
the biceps, only an
edge of brachialis
can be seen here

Biceps brachii
Supplied by the
musculocutaneous nerve

Biceps tendon
Attaches to the
radial tuberosity

Brachioradialis

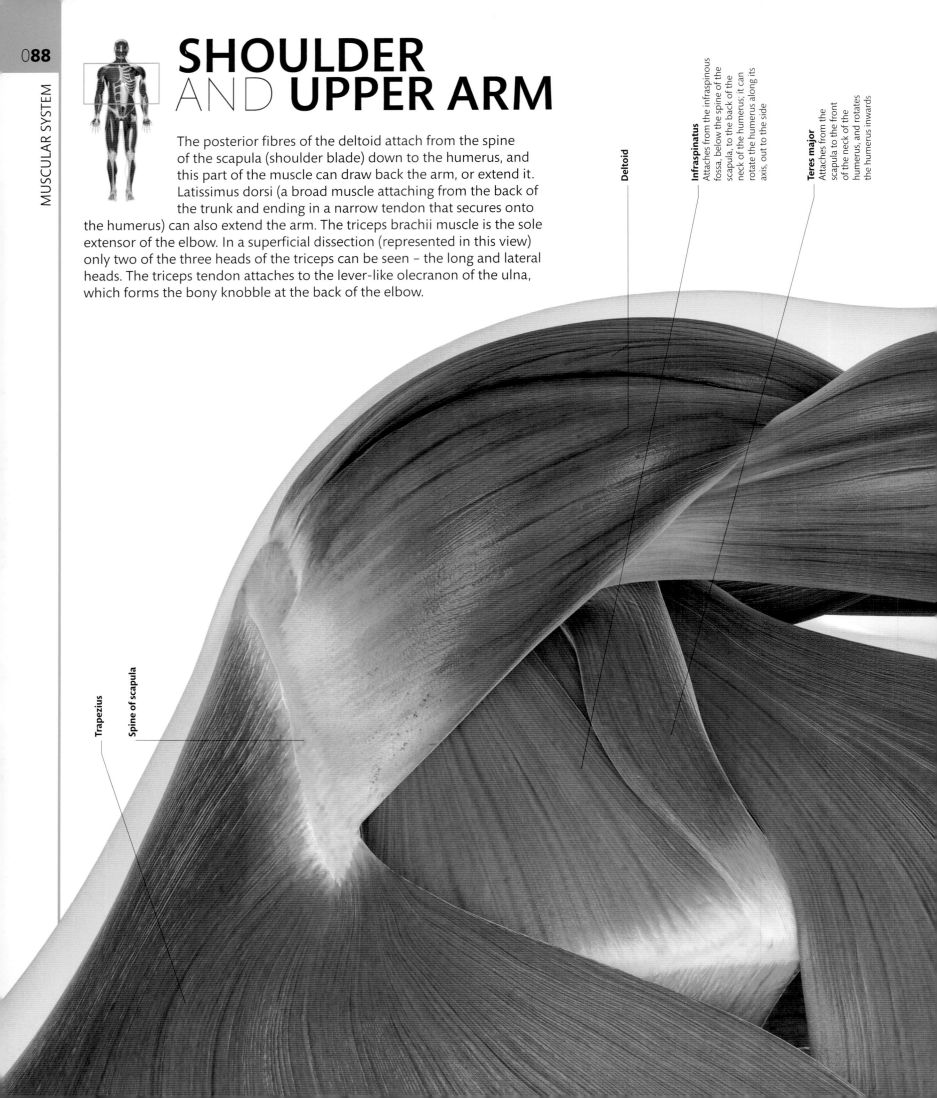

SHOULDER
AND UPPER ARM

The posterior fibres of the deltoid attach from the spine of the scapula (shoulder blade) down to the humerus, and this part of the muscle can draw back the arm, or extend it. Latissimus dorsi (a broad muscle attaching from the back of the trunk and ending in a narrow tendon that secures onto the humerus) can also extend the arm. The triceps brachii muscle is the sole extensor of the elbow. In a superficial dissection (represented in this view) only two of the three heads of the triceps can be seen – the long and lateral heads. The triceps tendon attaches to the lever-like olecranon of the ulna, which forms the bony knobble at the back of the elbow.

Deltoid

Infraspinatus
Attaches from the infraspinous fossa, below the spine of the scapula, to the back of the neck of the humerus; it can rotate the humerus along its axis, out to the side

Teres major
Attaches from the scapula to the front of the neck of the humerus, and rotates the humerus inwards

Trapezius

Spine of scapula

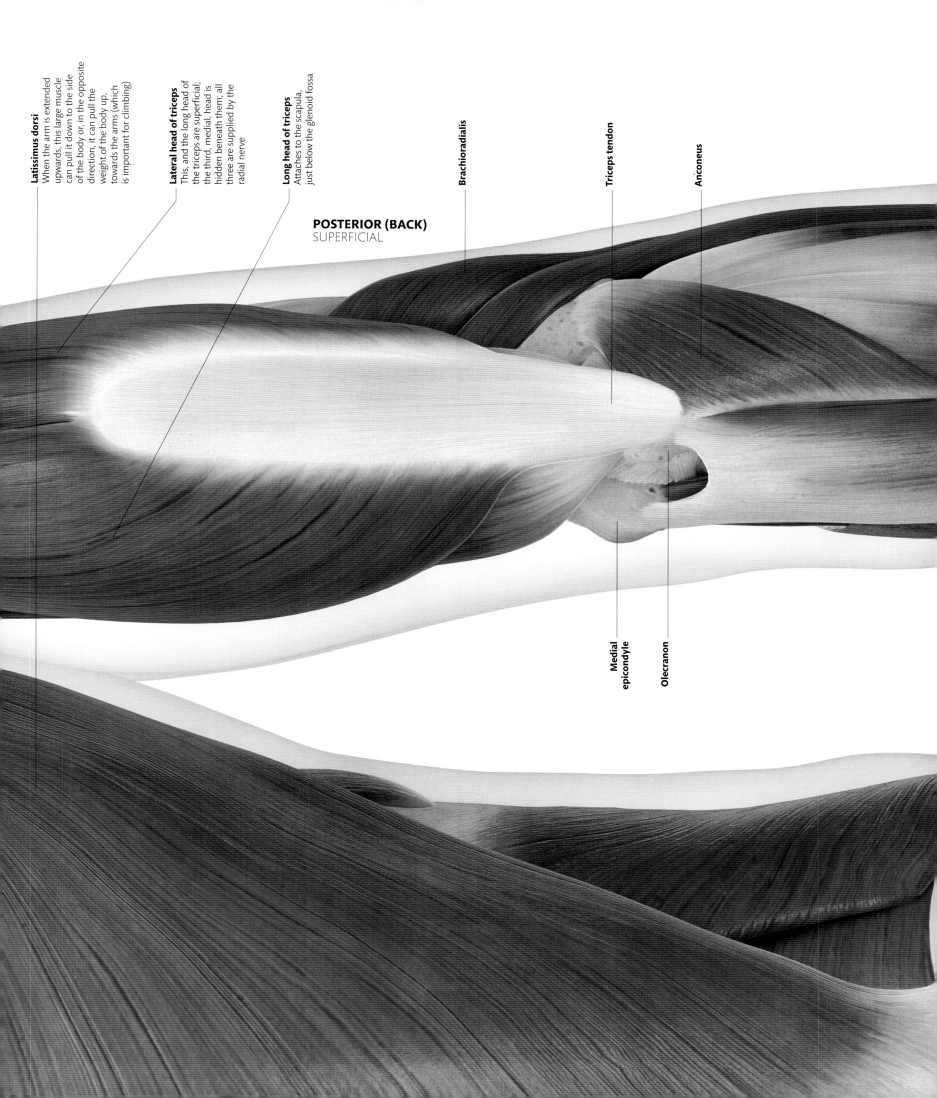

Latissimus dorsi
When the arm is extended upwards, this large muscle can pull it down to the side of the body or, in the opposite direction, it can pull the body up, towards the arms (which is important for climbing)

Lateral head of triceps
This, and the long head of the triceps are superficial; the third, medial, head is hidden beneath them; all three are supplied by the radial nerve

Long head of triceps
Attaches to the scapula, just below the glenoid fossa

POSTERIOR (BACK)
SUPERFICIAL

Brachioradialis

Triceps tendon

Anconeus

Medial epicondyle

Olecranon

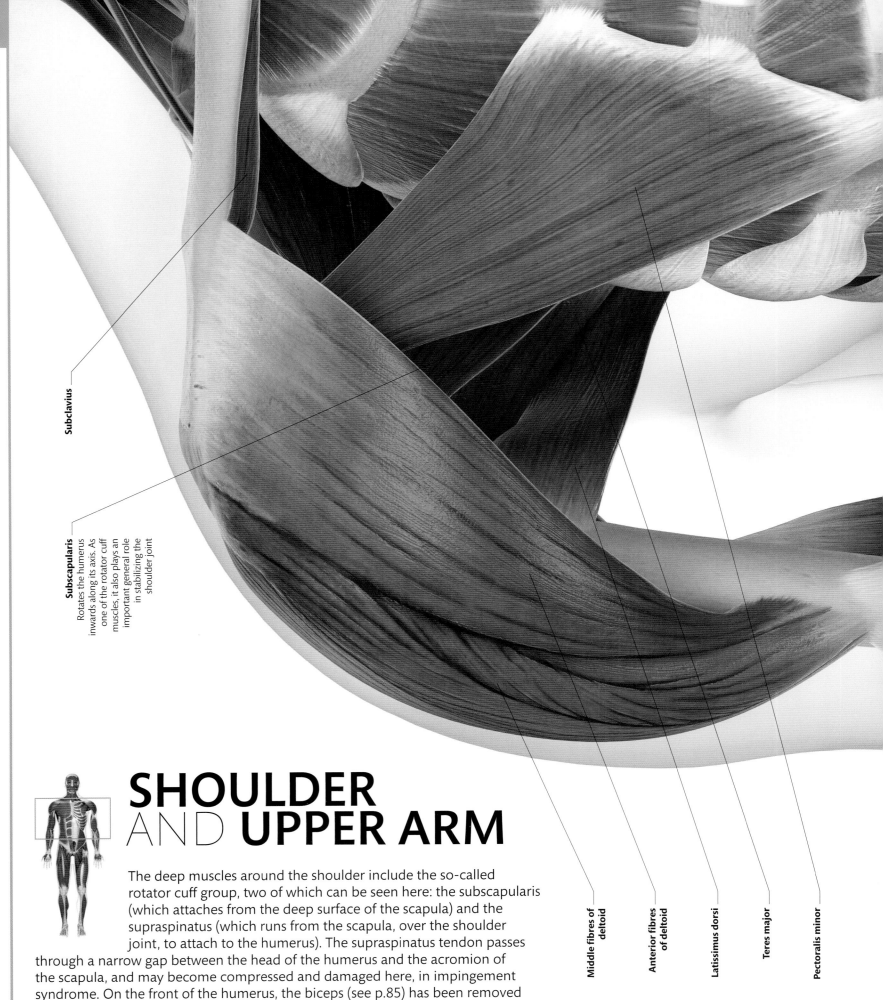

Subclavius

Subscapularis
Rotates the humerus inwards along its axis. As one of the rotator cuff muscles, it also plays an important general role in stabilizing the shoulder joint

Middle fibres of deltoid

Anterior fibres of deltoid

Latissimus dorsi

Teres major

Pectoralis minor

SHOULDER
AND UPPER ARM

The deep muscles around the shoulder include the so-called rotator cuff group, two of which can be seen here: the subscapularis (which attaches from the deep surface of the scapula) and the supraspinatus (which runs from the scapula, over the shoulder joint, to attach to the humerus). The supraspinatus tendon passes through a narrow gap between the head of the humerus and the acromion of the scapula, and may become compressed and damaged here, in impingement syndrome. On the front of the humerus, the biceps (see p.85) has been removed to reveal brachialis, which runs from the lower humerus down to the ulna. Like the biceps, brachialis is a flexor of the elbow.

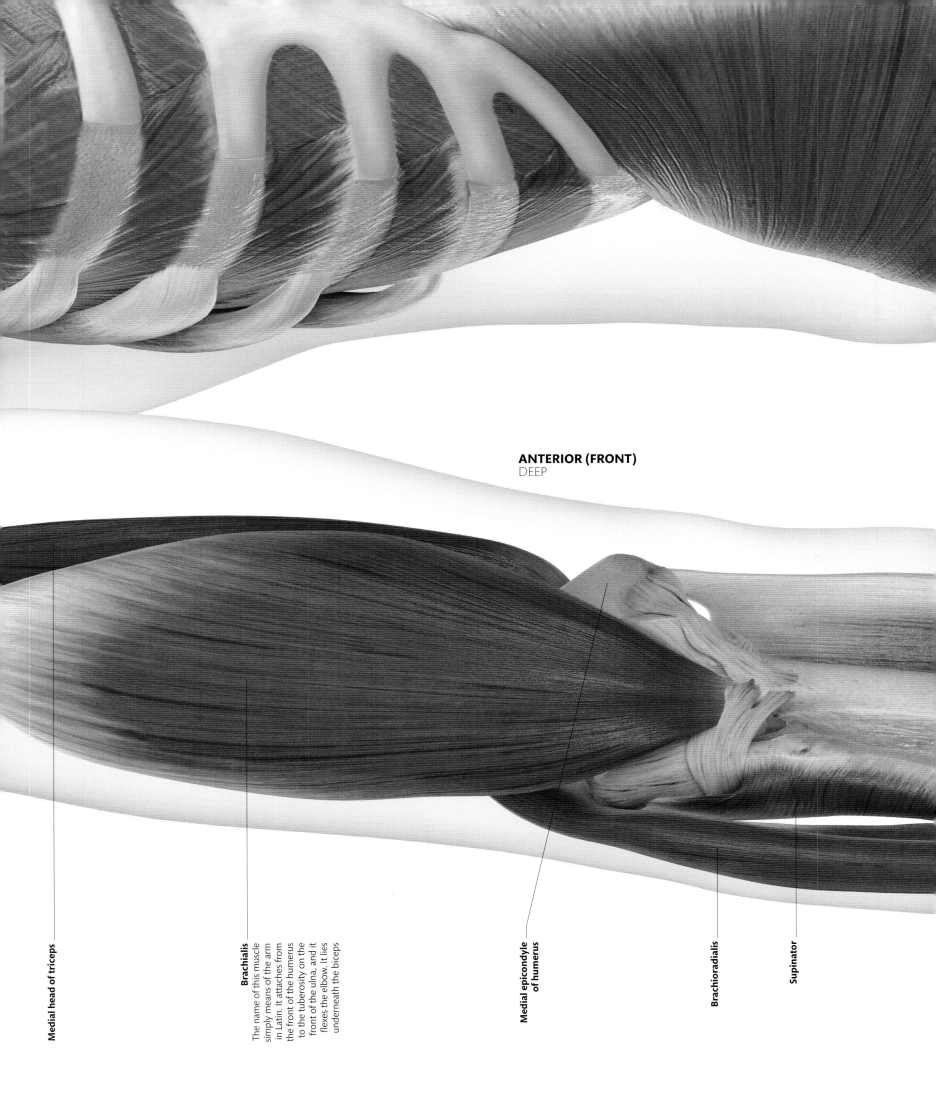

ANTERIOR (FRONT)
DEEP

Medial head of triceps

Brachialis

The name of this muscle simply means of the arm in Latin. It attaches from the front of the humerus to the tuberosity on the front of the ulna, and it flexes the elbow. It lies underneath the biceps

Medial epicondyle of humerus

Brachioradialis

Supinator

SHOULDER
AND **UPPER ARM**

More of the rotator cuff muscles – the supraspinatus, infraspinatus, and teres minor – can be seen from the back. As well as moving the shoulder joint in various directions, including rotation, these muscles are important in helping to stabilize the shoulder joint: they hug the head of the humerus into its socket during movements at the shoulder. On the back of the arm, a deeper view reveals the third, medial head of the triceps, which attaches from the back of the humerus. It joins with the lateral and long heads to form the triceps tendon, attaching to the olecranon. Most of the forearm muscles take their attachment from the epicondyles of the humerus, just above the elbow, but the brachioradialis and extensor carpi radialis longus have higher origins from the side of the humerus, as shown here.

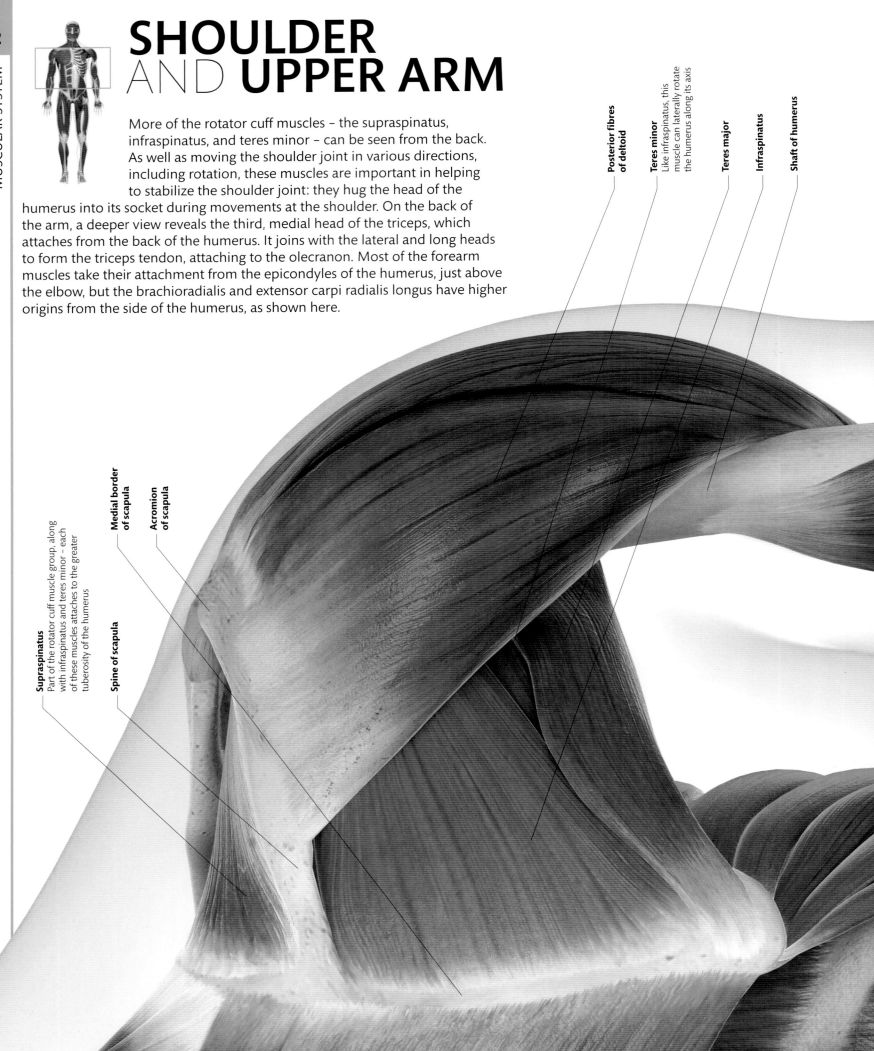

Posterior fibres of deltoid

Teres minor
Like infraspinatus, this muscle can laterally rotate the humerus along its axis

Teres major

Infraspinatus

Shaft of humerus

Supraspinatus
Part of the rotator cuff muscle group, along with infraspinatus and teres minor – each of these muscles attaches to the greater tuberosity of the humerus

Medial border of scapula

Acromion of scapula

Spine of scapula

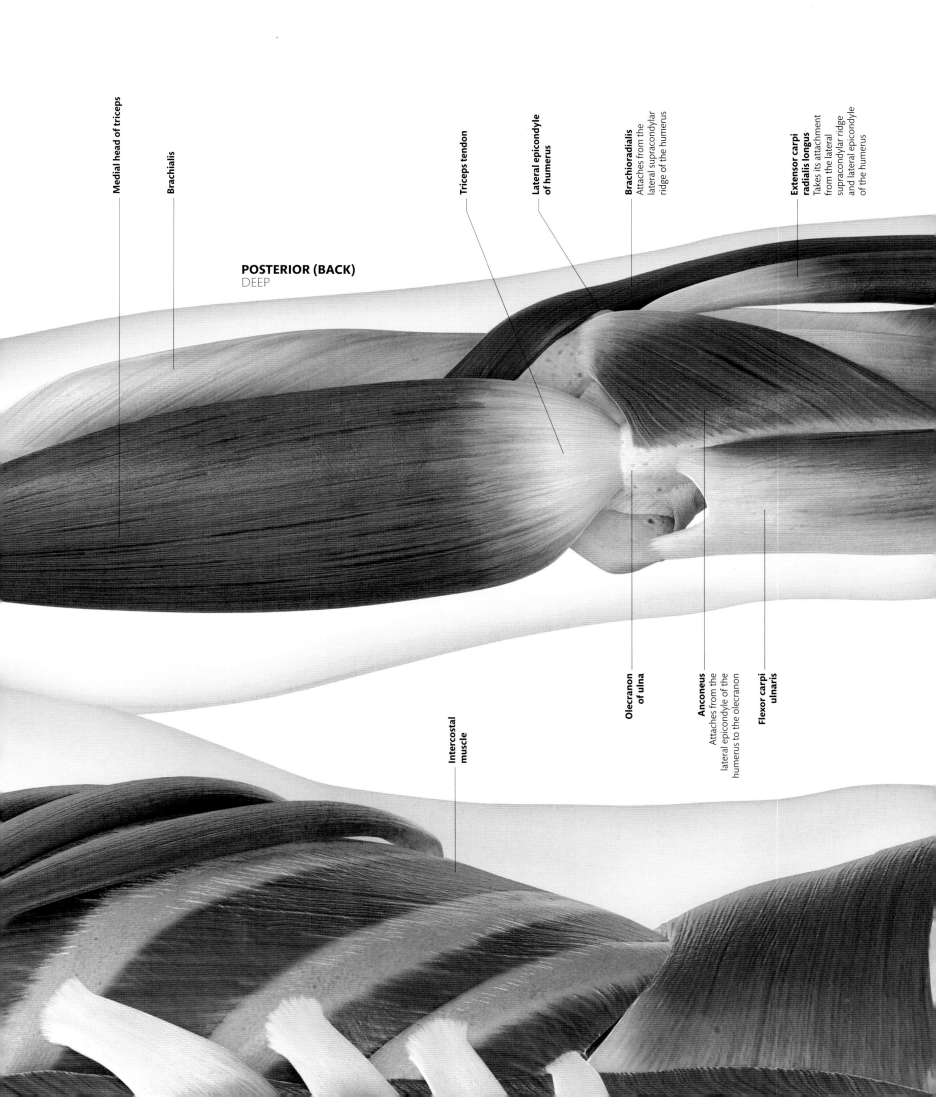

Medial head of triceps

Brachialis

Triceps tendon

Lateral epicondyle
of humerus

Brachioradialis
Attaches from the
lateral supracondylar
ridge of the humerus

**Extensor carpi
radialis longus**
Takes its attachment
from the lateral
supracondylar ridge
and lateral epicondyle
of the humerus

POSTERIOR (BACK)
DEEP

Olecranon
of ulna

Anconeus
Attaches from the
lateral epicondyle of the
humerus to the olecranon

**Flexor carpi
ulnaris**

Intercostal
muscle

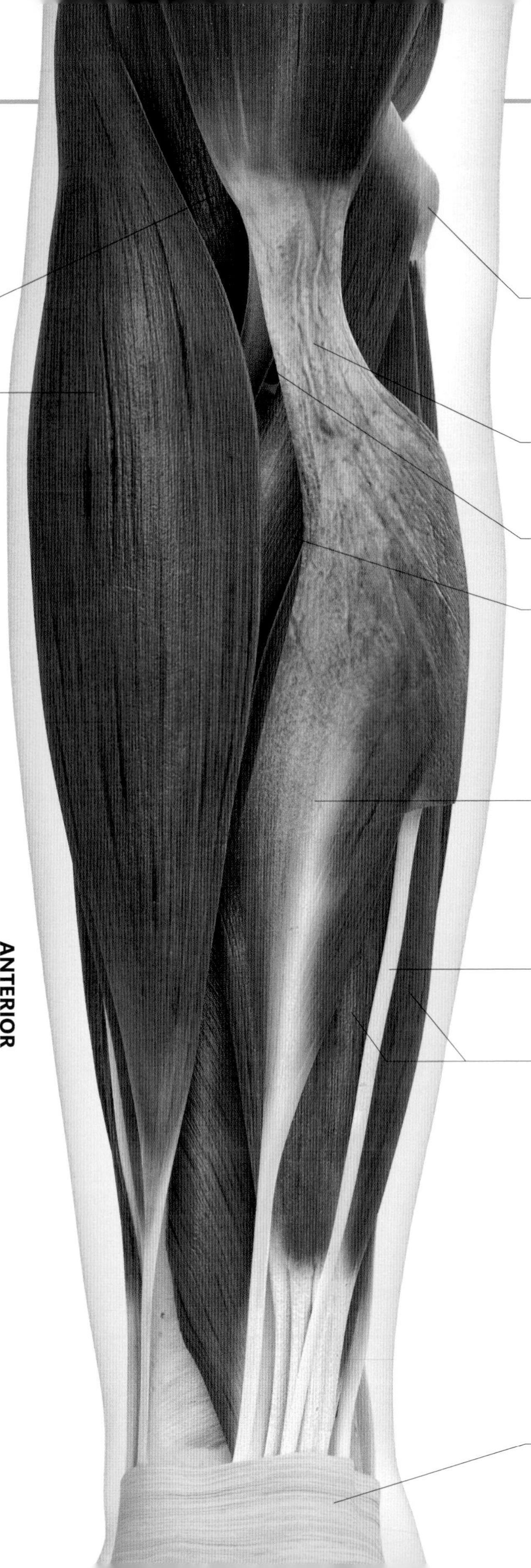

Medial epicondyle of humerus
Also called the common flexor origin; many of the superficial flexor muscles attach from this point

Biceps aponeurosis

Biceps tendon

Pronator teres
Attaches from the humerus and ulna down to the outer edge of the radius; it pronates the forearm, rotating the lower end of the radius around the ulna

Flexor carpi radialis
Radial extensor of the wrist; it arises from the medial epicondyle of the femur and secures onto the base of the second metacarpal; it flexes the wrist and abducts the hand

Palmaris longus tendon

Flexor digitorum superficialis
Takes its attachment from the humerus, ulna, and radius, and splits into four tendons that run across the wrist into the hand, to flex the fingers

Flexor retinaculum
This fibrous band keeps the flexor tendons close to the wrist and stops them bow-stringing outwards

Brachialis

Brachioradialis
Runs along the outer edge of the forearm and attaches to the end of the radius; it flexes and stabilizes the elbow

ANTERIOR (FRONT)
SUPERFICIAL

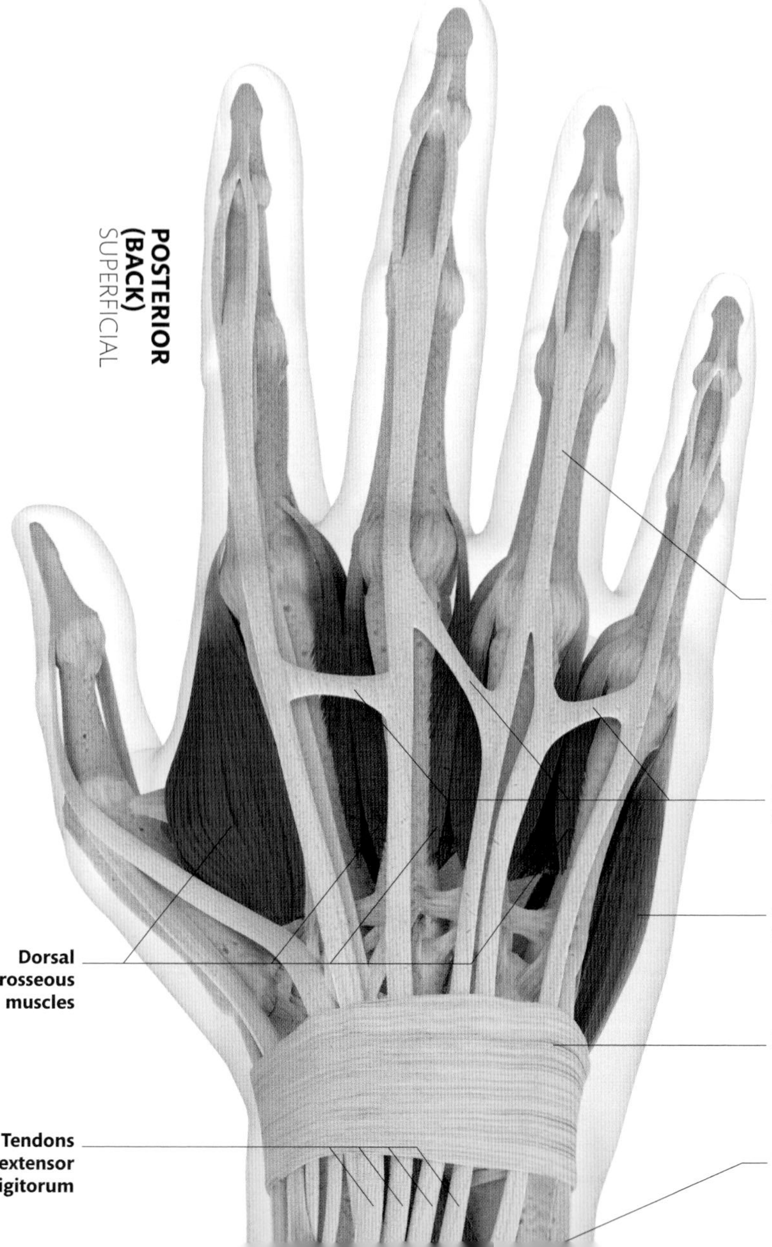

Extensor expansion

Intertendinous connections

Abductor digiti minimi

Extensor retinaculum
This fibrous band keeps the extensor tendons close to the wrist

Ulna

POSTERIOR (BACK)
SUPERFICIAL

Dorsal interosseous muscles

Tendons of extensor digitorum

LOWER ARM AND HAND

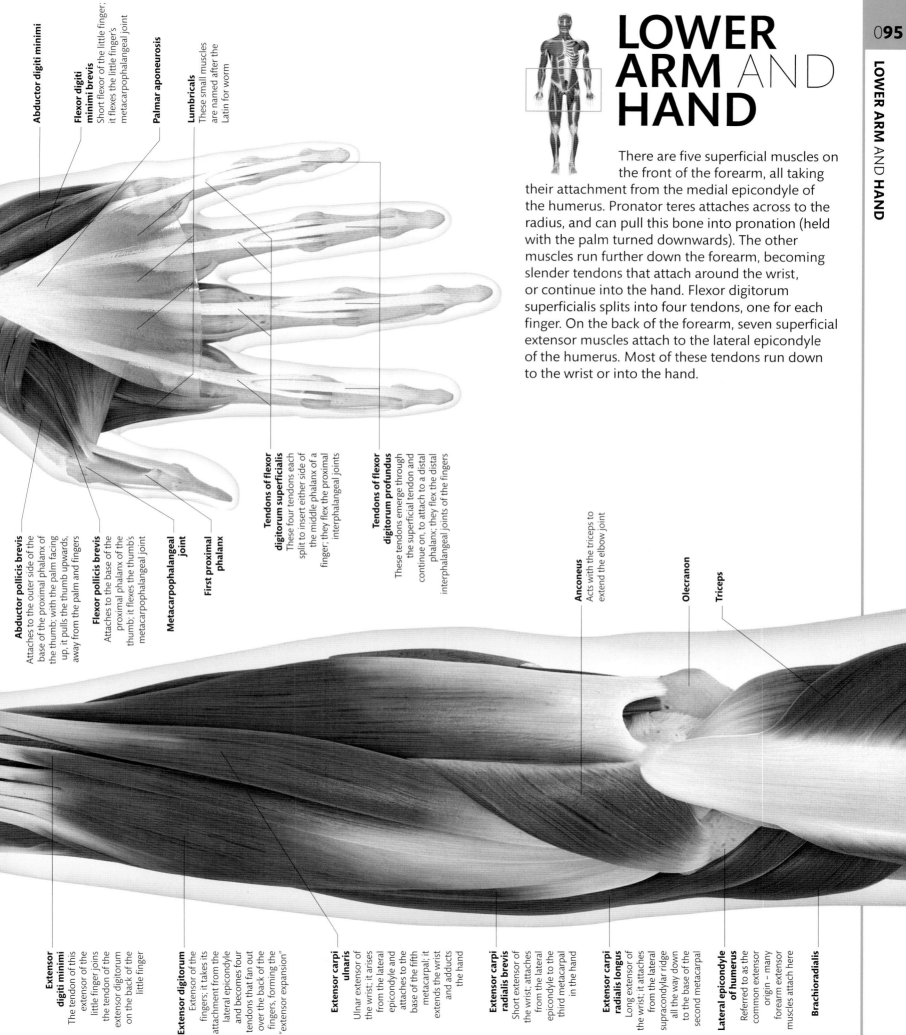

There are five superficial muscles on the front of the forearm, all taking their attachment from the medial epicondyle of the humerus. Pronator teres attaches across to the radius, and can pull this bone into pronation (held with the palm turned downwards). The other muscles run further down the forearm, becoming slender tendons that attach around the wrist, or continue into the hand. Flexor digitorum superficialis splits into four tendons, one for each finger. On the back of the forearm, seven superficial extensor muscles attach to the lateral epicondyle of the humerus. Most of these tendons run down to the wrist or into the hand.

Abductor digiti minimi

Flexor digiti minimi brevis
Short flexor of the little finger; it flexes the little finger's metacarpophalangeal joint

Palmar aponeurosis

Lumbricals
These small muscles are named after the Latin for worm

Abductor pollicis brevis
Attaches to the outer side of the base of the proximal phalanx of the thumb; with the palm facing up, it pulls the thumb upwards, away from the palm and fingers

Flexor pollicis brevis
Attaches to the base of the proximal phalanx of the thumb; it flexes the thumb's metacarpophalangeal joint

Metacarpophalangeal joint

First proximal phalanx

Tendons of flexor digitorum superficialis
These four tendons each split to insert either side of the middle phalanx of a finger; they flex the proximal interphalangeal joints

Tendons of flexor digitorum profundus
These tendons emerge through the superficial tendon and continue on, to attach to a distal phalanx; they flex the distal interphalangeal joints of the fingers

Anconeus
Acts with the triceps to extend the elbow joint

Olecranon

Triceps

Extensor digiti minimi
The tendon of this extensor of the little finger joins the tendon of the extensor digitorum on the back of the little finger

Extensor digitorum
Extensor of the fingers; it takes its attachment from the lateral epicondyle and becomes four tendons that fan out over the back of the fingers, forming the "extensor expansion"

Extensor carpi ulnaris
Ulnar extensor of the wrist; it arises from the lateral epicondyle and attaches to the base of the fifth metacarpal; it extends the wrist and adducts the hand

Extensor carpi radialis brevis
Short extensor of the wrist; it attaches from the lateral epicondyle to the third metacarpal in the hand

Extensor carpi radialis longus
Long extensor of the wrist; it attaches from the lateral supracondylar ridge all the way down to the base of the second metacarpal

Lateral epicondyle of humerus
Referred to as the common extensor origin – many forearm extensor muscles attach here

Brachioradialis

Brachialis

Medial epicondyle of humerus
Also known as the common flexor origin

ANTERIOR (FRONT)
DEEP

Flexor carpi ulnaris

Flexor retinaculum

Supinator

Brachioradialis

Flexor pollicis longus
This long flexor of the thumb arises from the radius and interosseous membrane; its tendon runs into the thumb to attach to the base of the distal phalanx

POSTERIOR (BACK)
DEEP

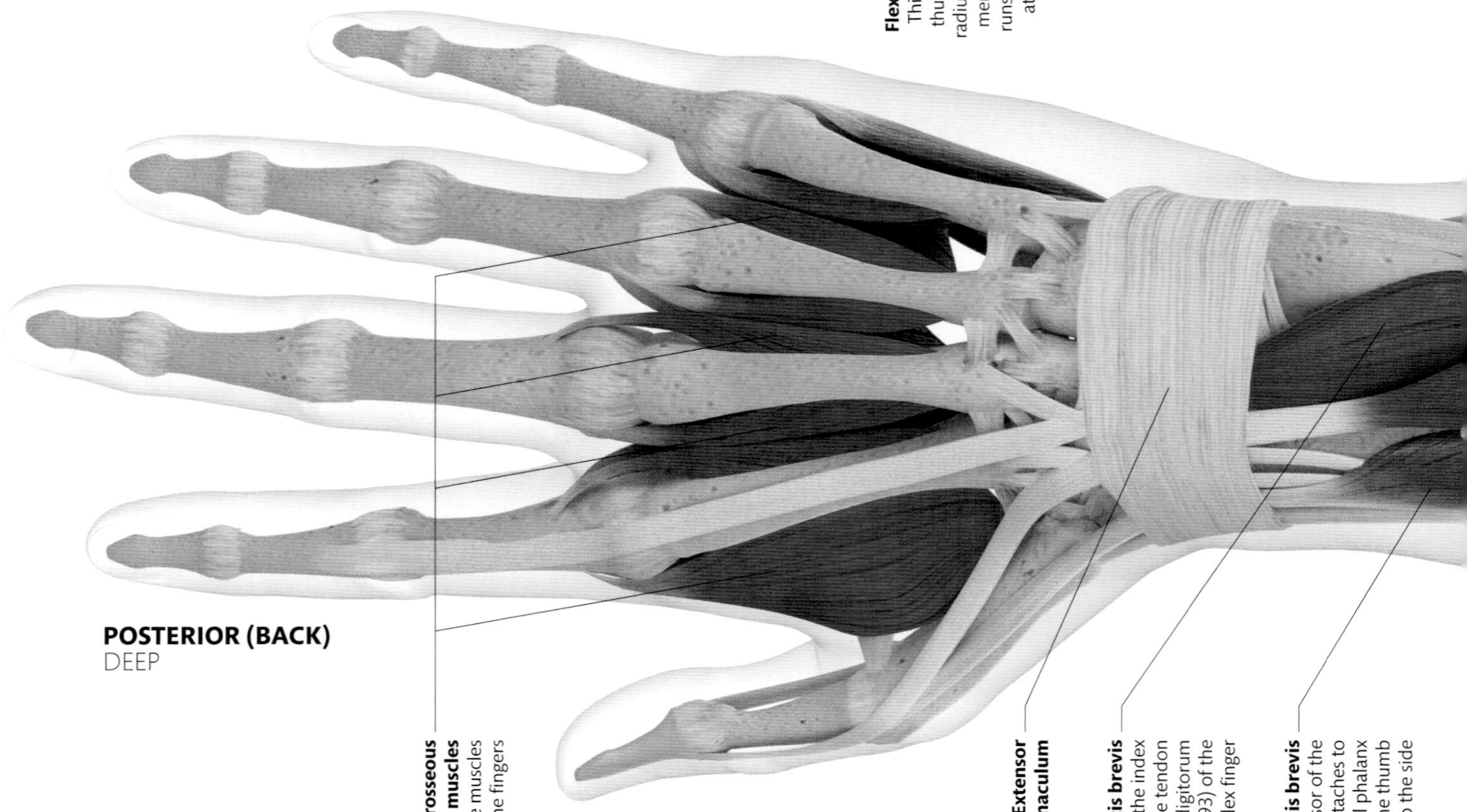

Dorsal interosseous muscles
These muscles spread the fingers

Extensor retinaculum

Extensor pollicis brevis
Extensor of the index finger; it joins the tendon of extensor digitorum (see pp.92–93) of the index finger

Extensor pollicis brevis
Short extensor of the thumb; it attaches to the proximal phalanx and pulls the thumb out to the side

LOWER ARM AND HAND

Stripping away the superficial muscles on the front of the forearm reveals a deeper layer attaching to the radius and ulna, and to the interosseous membrane between the bones. The long, quill-like flexor of the thumb (flexor pollicis longus) can be seen clearly. Deep muscles on the back of the forearm include the long extensors of the thumb and index finger and the supinator, which pulls on the radius to rotate the pronated arm (held with palm facing downwards) into supination (with palm facing up). In the hand, a deep dissection reveals the interosseous muscles that act on the metacarpophalangeal joints in order to either spread or close the fingers.

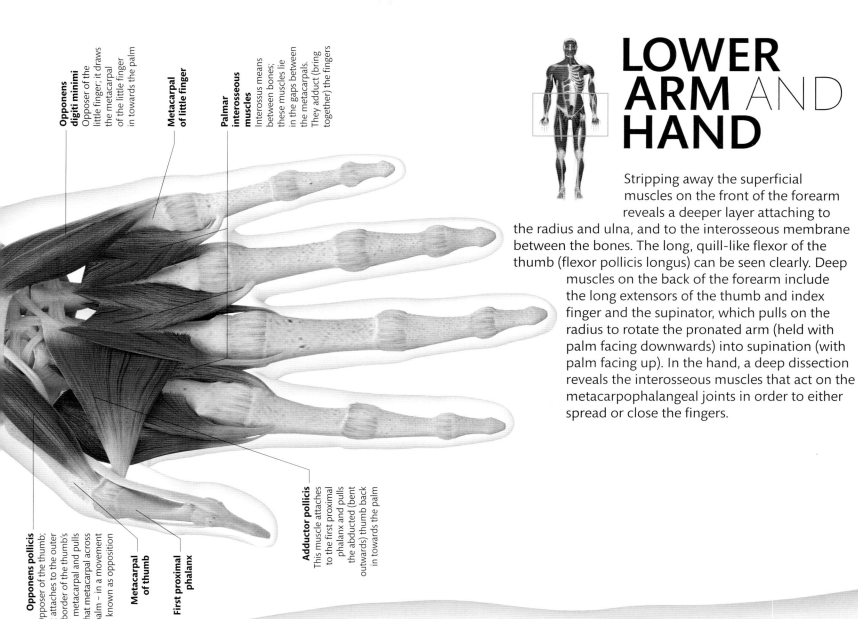

Opponens digiti minimi
Opposer of the little finger; it draws the metacarpal of the little finger in towards the palm

Metacarpal of little finger

Palmar interosseous muscles
Interossus means between bones; these muscles lie in the gaps between the metacarpals. They adduct (bring together) the fingers

Opponens pollicis
Opposer of the thumb; it attaches to the outer border of the thumb's metacarpal and pulls that metacarpal across the palm – in a movement known as opposition

Metacarpal of thumb

First proximal phalanx

Adductor pollicis
This muscle attaches to the first proximal phalanx and pulls the abducted (bent outwards) thumb back in towards the palm

Extensor pollicis longus
Long extensor of the thumb; it attaches onto the thumb's distal phalanx

Abductor pollicis longus
Long abductor of the thumb; it attaches to the base of the thumb's metacarpal

Extensor carpi ulnaris

Supinator
Arises from the lateral epicondyle of the humerus and wraps around the radius; it pulls the pronated forearm back into supination

Extensor carpi radialis brevis

Extensor carpi radialis longus

Anconeus

Triceps

Inguinal ligament

Adductor longus
Attaches from the pubis to the middle third of the linea aspera, a ridge on the back of the femur

Gracilis
This long, thin muscle attaches from the pubis down to the inner (medial) surface of the tibia, and adducts the thigh

Iliopsoas

Pubic symphysis

Pectineus
This muscle attaches from the pubic bone to the femur, and flexes and adducts the hip

Tensor fasciae latae
Tensor of the deep fascia; it attaches from the iliac crest on top of the pelvis and inserts into the iliotibial tract. It helps to steady the thigh while standing upright

Sartorius
Named after the Latin for tailor, this muscle flexes, abducts, and laterally rotates the hip while flexing the knee – producing a cross-legged position, apparently the traditional posture of tailors

Iliotibial tract
A thickening of the deep fascia over the outer (lateral) thigh, reaching from the iliac crest to the tibia

Vastus medialis
Another large head of the quadriceps femoris

Quadriceps tendon
The four heads of quadriceps femoris come together in one tendon at the knee

Pre-patellar bursa

Patellar ligament
The continuation of quadriceps tendon below the patella

Rectus femoris
The part of the quadriceps that can flex the hip as well as extend the knee

Vastus lateralis
The name of this part of the quadriceps reflects its impressive size

ANTERIOR (FRONT)
SUPERFICIAL

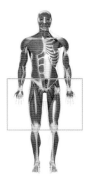

HIP AND THIGH

Most of the muscle bulk on the front of the leg is the four-headed quadriceps femoris. Three of its heads can be seen in a superficial dissection of the thigh: the rectus femoris, vastus lateralis, and vastus medialis. The quadriceps extends the knee, but it can also flex the hip, since the rectus femoris part has an attachment from the pelvis, above the hip joint. The patella is embedded in the quadriceps tendon; this may protect the tendon from wear and tear, but it also helps to give the quadriceps good leverage in extending the knee. The part of the tendon below the patella is usually called the patellar ligament. Tapping this with a tendon hammer produces a reflex contraction in the quadriceps – the "knee jerk".

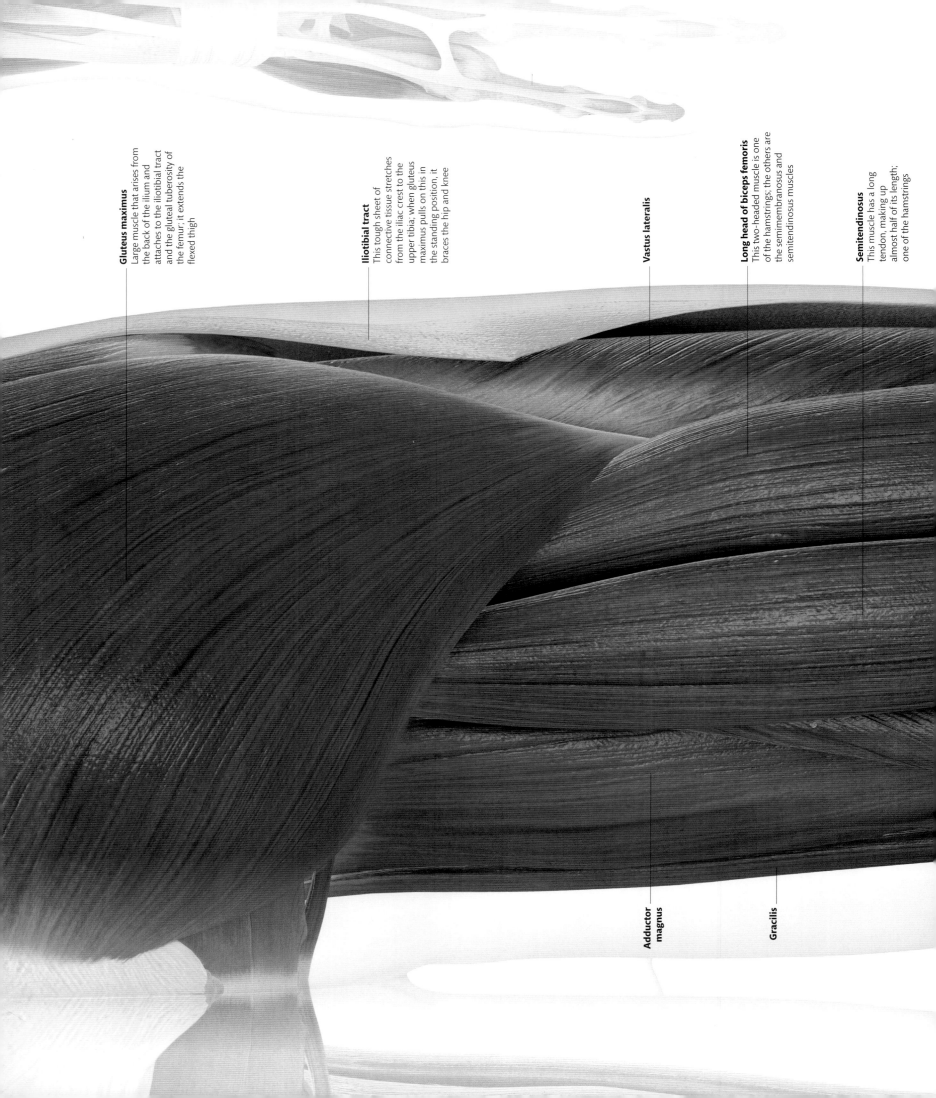

Gluteus maximus
Large muscle that arises from the back of the ilium and attaches to the iliotibial tract and the gluteal tuberosity of the femur; it extends the flexed thigh

Iliotibial tract
This tough sheet of connective tissue stretches from the iliac crest to the upper tibia; when gluteus maximus pulls on this in the standing position, it braces the hip and knee

Vastus lateralis

Long head of biceps femoris
This two-headed muscle is one of the hamstrings; the others are the semimembranosus and semitendinosus muscles

Semitendinosus
This muscle has a long tendon, making up almost half of its length; one of the hamstrings

Adductor magnus

Gracilis

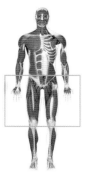

HIP AND THIGH

On the back of the hip and thigh, a superficial dissection reveals the large gluteus maximus, an extensor of the hip joint, and the three hamstrings. The gluteus maximus acts to extend the hip joint, swinging the leg backwards. While it doesn't really contribute to gentle walking, it is very important in running, and also when the hip is being extended from a flexed position, such as when getting up from sitting on the floor or when climbing the stairs. The hamstrings – the semimembranosus, semitendinosus, and biceps femoris muscles – attach from the ischial tuberosity of the pelvis and sweep down the back of the thigh to the tibia and fibula. They are the main flexors of the knee.

POSTERIOR (BACK)
SUPERFICIAL

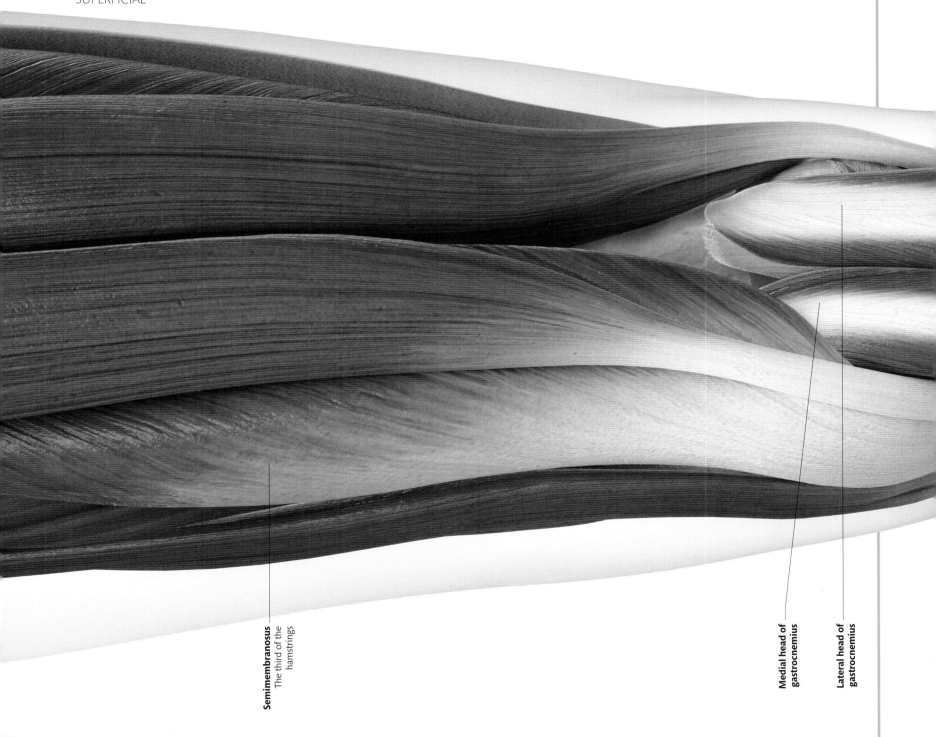

Semimembranosus
The third of the hamstrings

Medial head of gastrocnemius

Lateral head of gastrocnemius

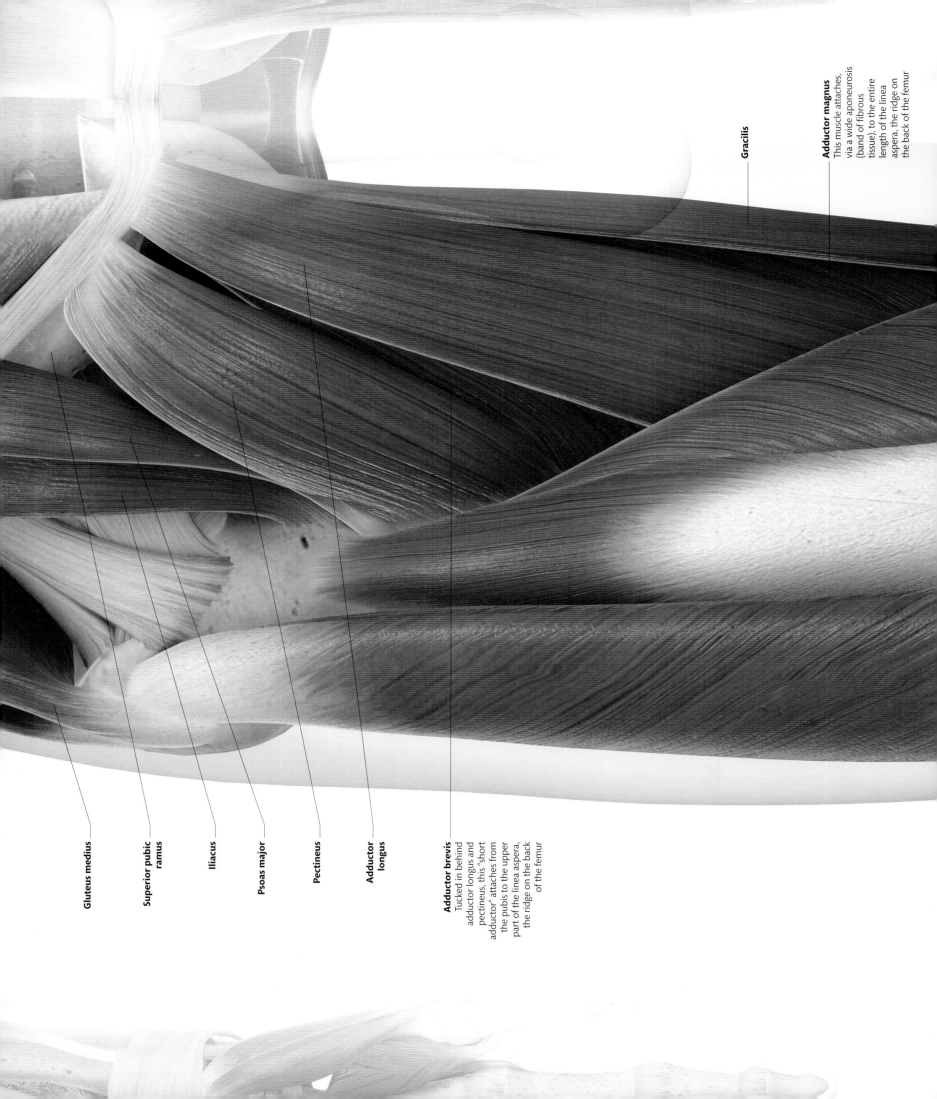

Gracilis

Adductor magnus
This muscle attaches, via a wide aponeurosis (band of fibrous tissue), to the entire length of the linea aspera, the ridge on the back of the femur

Gluteus medius

Superior pubic ramus

Iliacus

Psoas major

Pectineus

Adductor longus

Adductor brevis
Tucked in behind adductor longus and pectineus, this "short adductor" attaches from the pubis to the upper part of the linea aspera, the ridge on the back of the femur

Vastus intermedius
Sitting behind the rectus femoris, this muscle arises from the upper femur and attaches to the patella via the quadriceps tendon

Vastus medialis
With the rectus femoris removed, a separation between this muscle and the vastus intermedius can be seen

Quadriceps tendon

Bursa

Patella

Pre-patellar bursa

Bursa

Vastus lateralis
This muscle is the largest part of the quadriceps

ANTERIOR (FRONT)
DEEP

HIP AND THIGH

With the rectus femoris and sartorius muscles stripped away, we can see the deep, fourth head of the quadriceps, known as vastus intermedius. The adductor muscles that bring the thighs together can also be seen clearly, including the gracilis, which is long and slender, as its name suggests. The largest adductor muscle – the adductor magnus – has a hole in its tendon, through which the main artery of the leg (the femoral artery) passes. The adductor tendons attach from the pubis and ischium of the pelvis, and the sporting injuries referred to as "groin pulls" are often tears in these particular tendons.

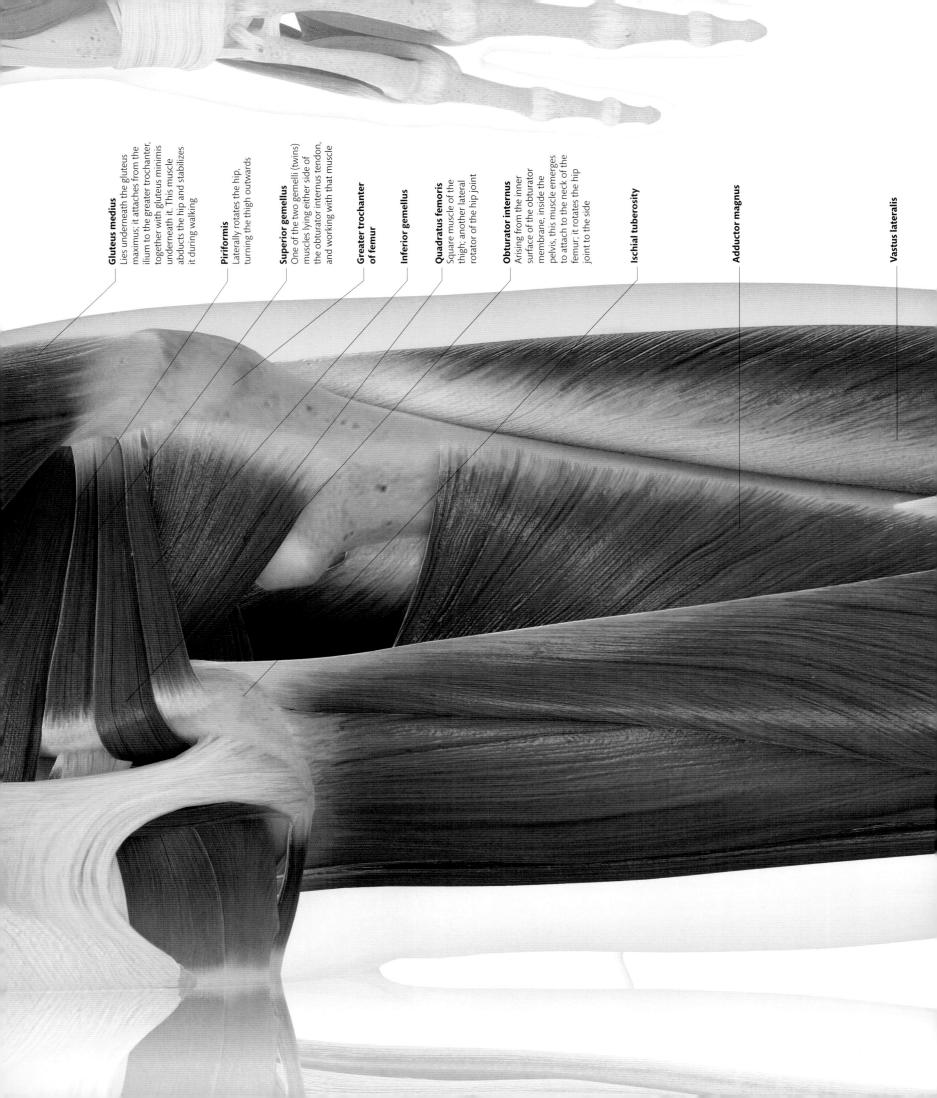

Gluteus medius
Lies underneath the gluteus maximus; it attaches from the ilium to the greater trochanter, together with gluteus minimis underneath it. This muscle abducts the hip and stabilizes it during walking

Piriformis
Laterally rotates the hip, turning the thigh outwards

Superior gemellus
One of the two gemelli (twins) muscles lying either side of the obturator internus tendon, and working with that muscle

Greater trochanter of femur

Inferior gemellus

Quadratus femoris
Square muscle of the thigh; another lateral rotator of the hip joint

Obturator internus
Arising from the inner surface of the obturator membrane, inside the pelvis, this muscle emerges to attach to the neck of the femur; it rotates the hip joint to the side

Ischial tuberosity

Adductor magnus

Vastus lateralis

HIP AND THIGH

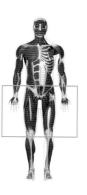

On the back of the hip, with the gluteus maximus removed, the short muscles that rotate the hip out to the side are clearly revealed. These include the piriformis, obturator internus, and quadratus femoris muscles. With the long head of the biceps femoris removed, we can now see the deeper, short head attaching to the linea aspera on the back of the femur. The semitendinosus muscle has also been cut away to reveal the semimembranosus underneath it, with its flat, membrane-like tendon at the top. Popliteus muscle is also visible at the back of the knee joint, as is one of the many fluid filled bursae around the knee.

POSTERIOR (BACK)
DEEP

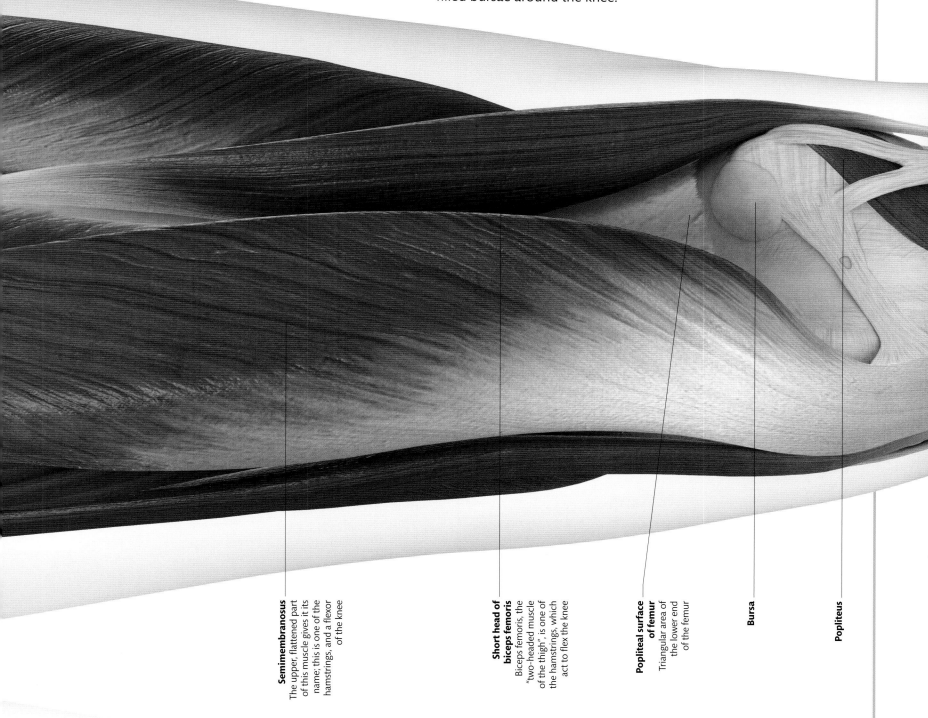

Semimembranosus
The upper, flattened part of this muscle gives it its name; this is one of the hamstrings, and a flexor of the knee

Short head of biceps femoris
Biceps femoris, the "two-headed muscle of the thigh", is one of the hamstrings, which act to flex knee

Popliteal surface of femur
Triangular area of the lower end of the femur

Bursa

Popliteus

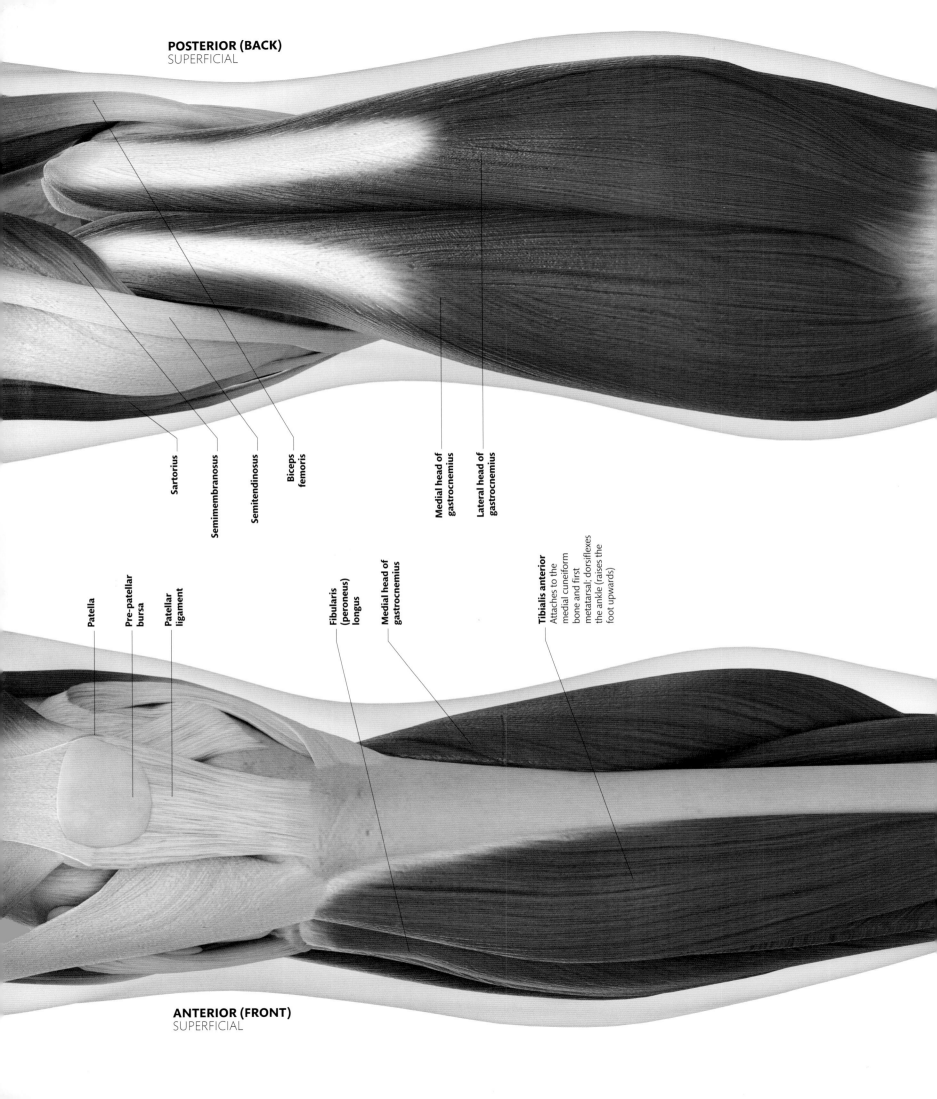

Sartorius

Semimembranosus

Semitendinosus

Biceps
femoris

Medial head of
gastrocnemius

Lateral head of
gastrocnemius

Patella

Pre-patellar
bursa

Patellar
ligament

Fibularis
(peroneus)
longus

Medial head of
gastrocnemius

Tibialis anterior
Attaches to the
medial cuneiform
bone and first
metatarsal; dorsiflexes
the ankle (raises the
foot upwards)

ANTERIOR (FRONT)
SUPERFICIAL

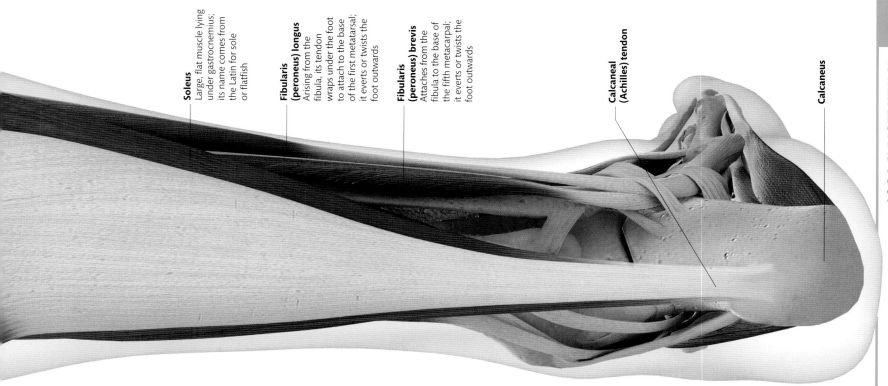

Soleus Large, flat muscle lying under gastrocnemius; its name comes from the Latin for sole or flatfish

Fibularis (peroneus) longus Arising from the fibula, its tendon wraps under the foot to attach to the base of the first metatarsal; it everts or twists the foot outwards

Fibularis (peroneus) brevis Attaches from the fibula to the base of the fifth metacarpal; it everts or twists the foot outwards

Calcaneal (Achilles) tendon

Calcaneus

LOWER LEG AND FOOT

You can feel the medial surface of the tibia easily, just under the skin on the front of your lower leg, on the inner side. Move your fingers outwards, and you feel the sharp border of the bone, and then a soft wedge of muscles alongside it. These muscles have tendons that run down to the foot. They can pull the foot upwards at the ankle, in a movement called dorsiflexion. Some extensor tendons continue all the way to the toes. There are much bulkier muscles on the back of the leg, and these form the calf. The gastrocnemius, and soleus underneath it, are large muscles that join together to form the Achilles tendon. They pull up on the lever of the calcaneus, pushing the ball of the foot down. They are involved as the foot pushes off from the ground during walking and running.

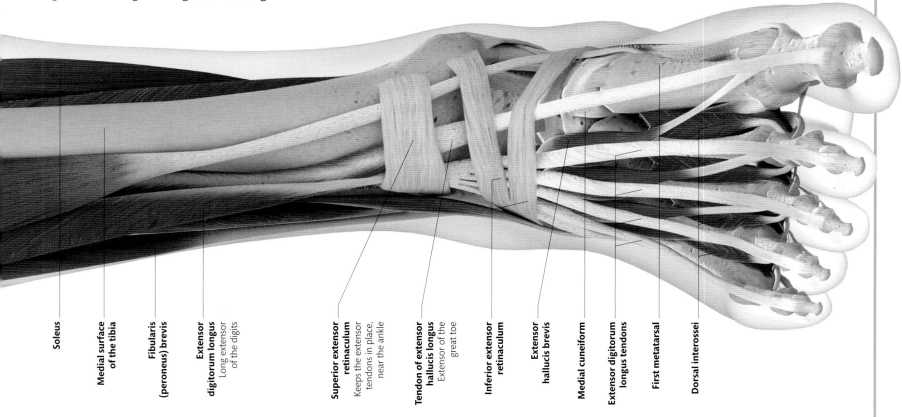

Soleus

Medial surface of the tibia

Fibularis (peroneus) brevis

Extensor digitorum longus Long extensor of the digits

Superior extensor retinaculum Keeps the extensor tendons in place, near the ankle

Tendon of extensor hallucis longus Extensor of the great toe

Inferior extensor retinaculum

Extensor hallucis brevis

Medial cuneiform

Extensor digitorum longus tendons

First metatarsal

Dorsal interossei

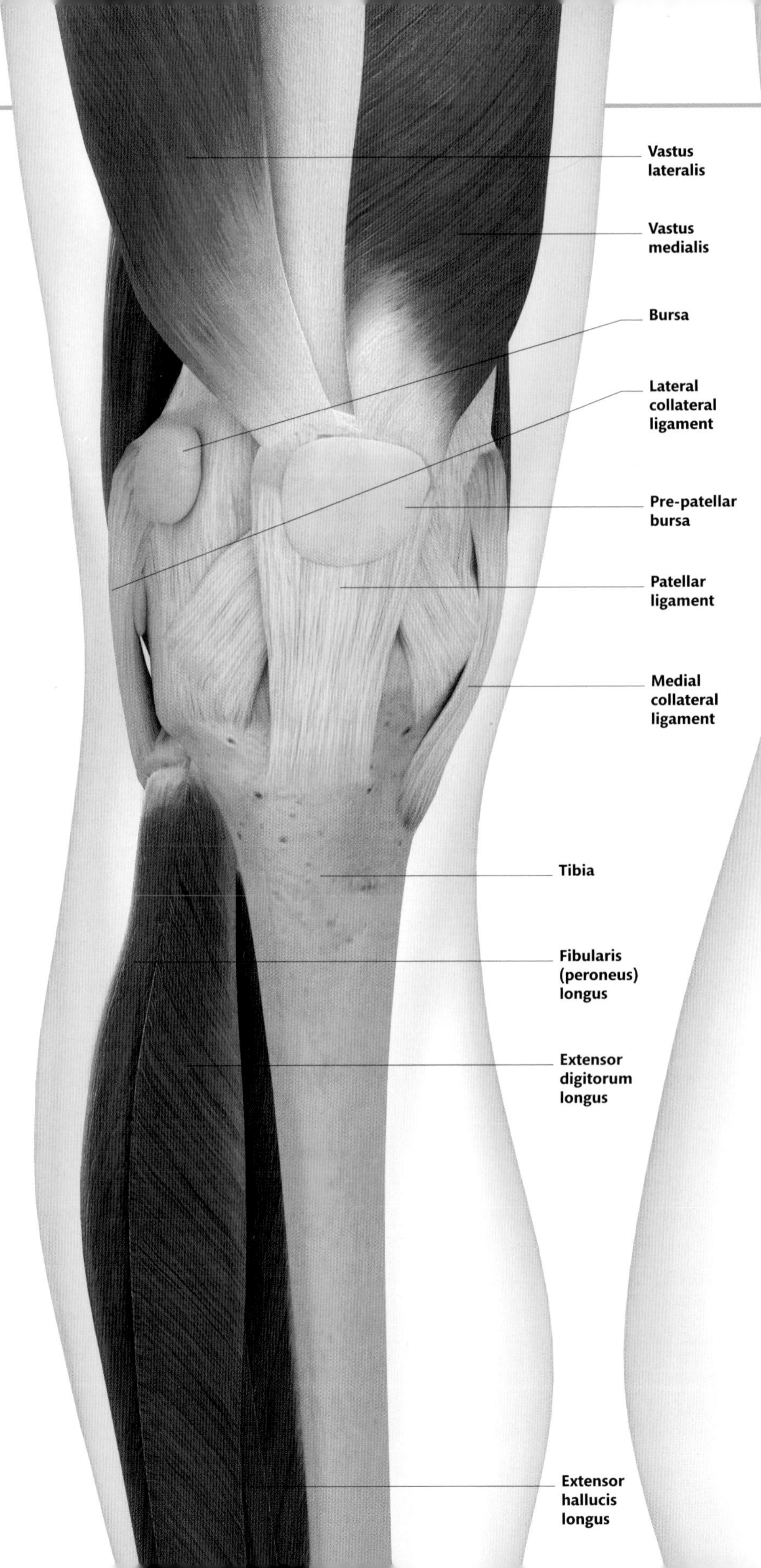

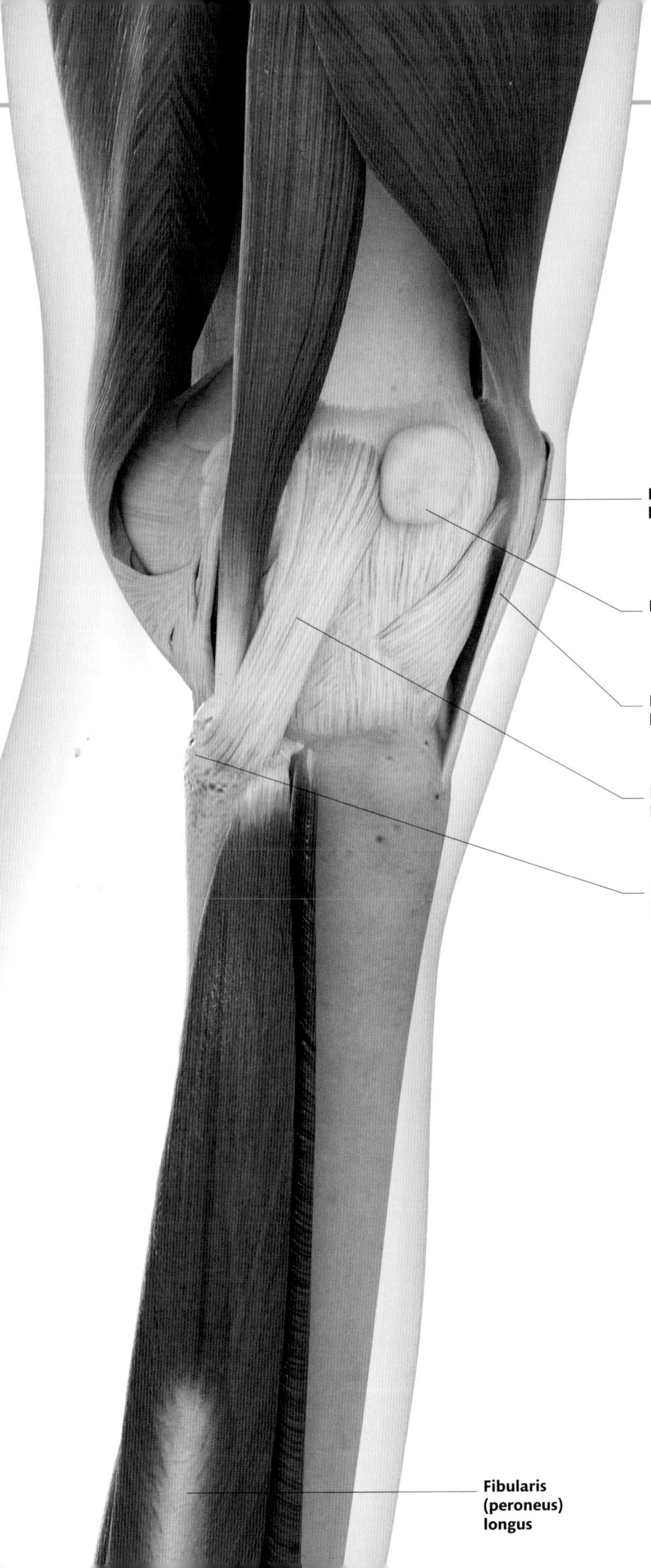

Pre-patellar bursa

Bursa

Patellar ligament

Fibular collateral ligament

Head of fibula

Fibularis (peroneus) longus

Vastus lateralis

Vastus medialis

Bursa

Lateral collateral ligament

Pre-patellar bursa

Patellar ligament

Medial collateral ligament

Tibia

Fibularis (peroneus) longus

Extensor digitorum longus

Extensor hallucis longus

LOWER LEG AND FOOT

Two muscles run along the outer, or lateral, side of the leg, down into the foot: the fibularis longus and fibularis brevis (see pp.104–05). These muscles pull the outer side of the foot upwards, in a movement called eversion. The tendon of fibularis longus runs right underneath the foot, to attach on the inner side, and helps to maintain the transverse arch of the foot. The flexor hallucis longus arises from the fibula and interosseous membrane, and sends its tendon down, behind the medial malleolus and into the sole of the foot, to attach to the distal phalanx of the great toe.

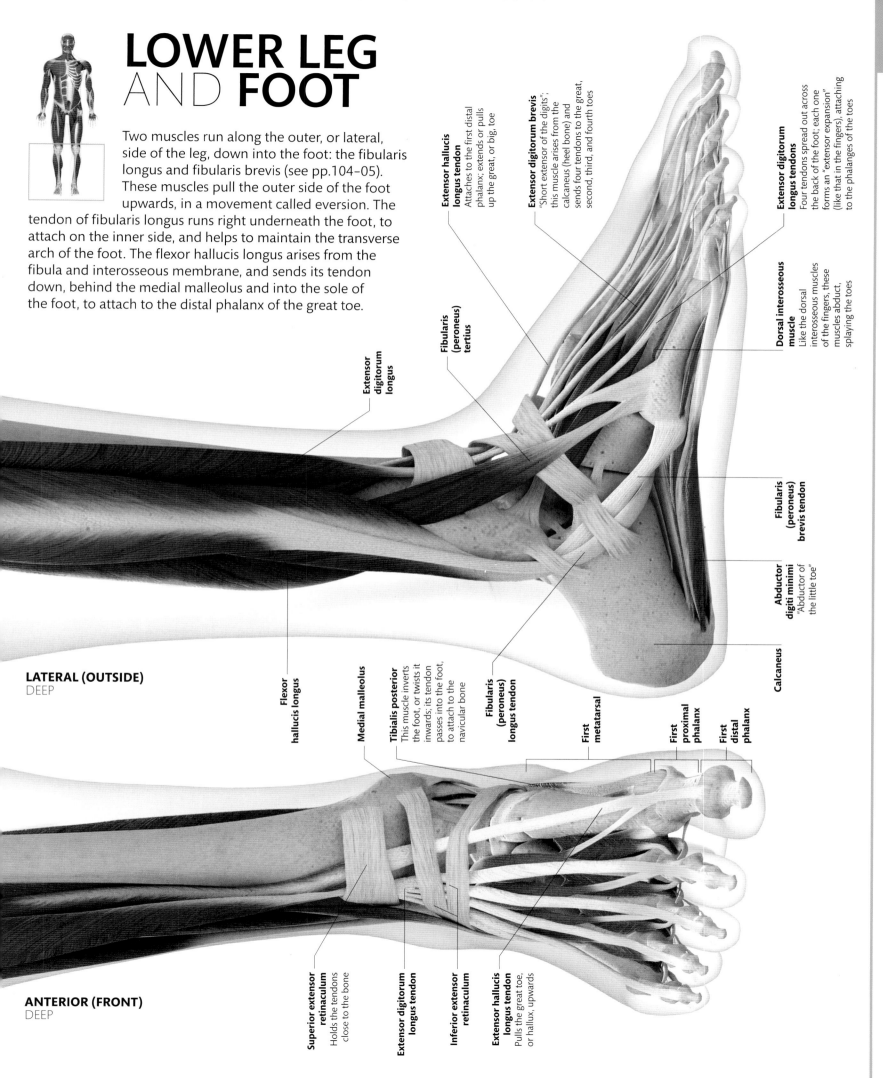

Extensor hallucis longus tendon
Attaches to the first distal phalanx; extends or pulls up the great, or big, toe

Extensor digitorum brevis
"Short extensor of the digits"; this muscle arises from the calcaneus (heel bone) and sends four tendons to the great, second, third, and fourth toes

Extensor digitorum longus tendons
Four tendons spread out across the back of the foot; each one forms an "extensor expansion" (like that in the fingers), attaching to the phalanges of the toes

Dorsal interosseous muscle
Like the dorsal interosseous muscles of the fingers, these muscles abduct, splaying the toes

Fibularis (peroneus) tertius

Extensor digitorum longus

Fibularis (peroneus) brevis tendon

Abductor digiti minimi
"Abductor of the little toe"

Calcaneus

LATERAL (OUTSIDE)
DEEP

Flexor hallucis longus

Medial malleolus

Tibialis posterior
This muscle inverts the foot, or twists it inwards; its tendon passes into the foot, to attach to the navicular bone

Fibularis (peroneus) longus tendon

First metatarsal

First proximal phalanx

First distal phalanx

Superior extensor retinaculum
Holds the tendons close to the bone

Extensor digitorum longus tendon

Inferior extensor retinaculum

Extensor hallucis longus tendon
Pulls the great toe, or hallux, upwards

ANTERIOR (FRONT)
DEEP

NERVOUS SYSTEM OVERVIEW

The nervous system contains billions of intercommunicating nerve cells, or neurons. It can be broadly divided into the central nervous system (brain and spinal cord) and the peripheral nervous system (cranial and spinal nerves and their branches). The brain and spinal cord are protected by the skull and vertebral column respectively. Twelve cranial nerves emerge from the brain and exit through holes in the skull to supply the head and neck; thirty one pairs of spinal nerves leave via gaps between vertebrae to supply the rest of the body. You can also divide the nervous system by function. The part that deals more with the way we sense and interact with our surroundings is called the somatic nervous system. The part involved with sensing and controlling our internal environments – affecting glands or heart rate, for example – is the autonomic nervous system.

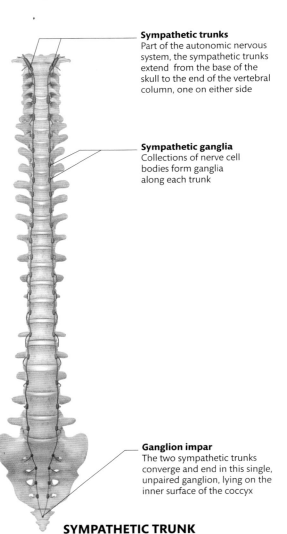

Sympathetic trunks
Part of the autonomic nervous system, the sympathetic trunks extend from the base of the skull to the end of the vertebral column, one on either side

Sympathetic ganglia
Collections of nerve cell bodies form ganglia along each trunk

Ganglion impar
The two sympathetic trunks converge and end in this single, unpaired ganglion, lying on the inner surface of the coccyx

SYMPATHETIC TRUNK

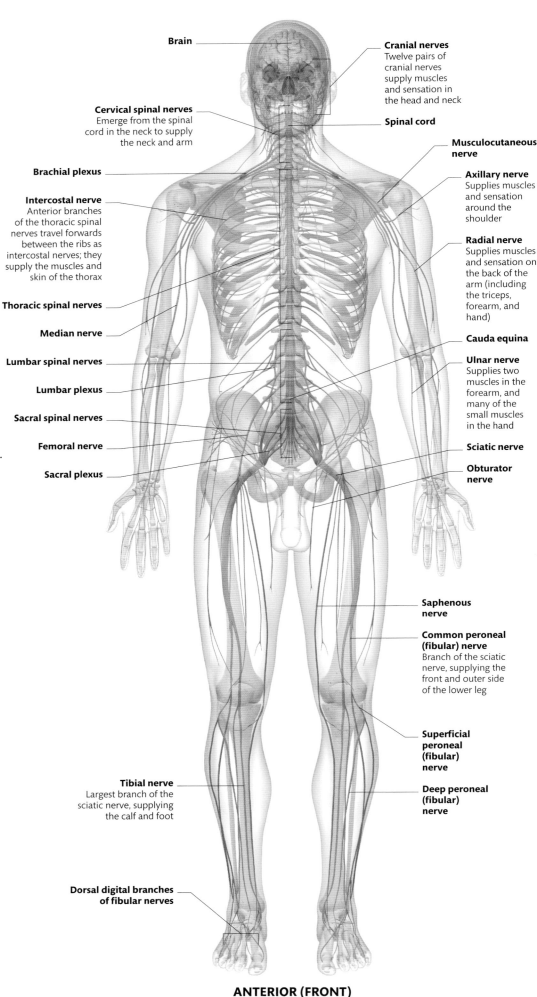

Brain

Cranial nerves
Twelve pairs of cranial nerves supply muscles and sensation in the head and neck

Cervical spinal nerves
Emerge from the spinal cord in the neck to supply the neck and arm

Spinal cord

Brachial plexus

Musculocutaneous nerve

Axillary nerve
Supplies muscles and sensation around the shoulder

Intercostal nerve
Anterior branches of the thoracic spinal nerves travel forwards between the ribs as intercostal nerves; they supply the muscles and skin of the thorax

Radial nerve
Supplies muscles and sensation on the back of the arm (including the triceps, forearm, and hand)

Thoracic spinal nerves

Median nerve

Cauda equina

Ulnar nerve
Supplies two muscles in the forearm, and many of the small muscles in the hand

Lumbar spinal nerves

Lumbar plexus

Sacral spinal nerves

Sciatic nerve

Femoral nerve

Obturator nerve

Sacral plexus

Saphenous nerve

Common peroneal (fibular) nerve
Branch of the sciatic nerve, supplying the front and outer side of the lower leg

Superficial peroneal (fibular) nerve

Tibial nerve
Largest branch of the sciatic nerve, supplying the calf and foot

Deep peroneal (fibular) nerve

Dorsal digital branches of fibular nerves

ANTERIOR (FRONT)

Cerebrum

Cerebellum
Literally little brain in Latin, this part of the brain is involved with balance and coordination of movement

Cranial nerves

Spinal cord
The continuation of the brainstem, lying protected within the vertebral canal of the spine

Musculocutaneous nerve
Supplies the muscles in the front of the upper arm (including the biceps) as well as sensation to the skin of the outer side of the forearm

Axillary nerve

Brachial plexus
Anterior branches of the lower cervical spinal nerves, together with the first thoracic spinal nerve, form a network, or plexus, from which branches emerge to supply the arm, forearm, and hand

Intercostal nerve

Median nerve
Supplies most of the muscles in the front of the forearm, and also some in the hand

Radial nerve

Ulnar nerve
This nerve lies on the ulnar, or inner, side of the arm and forearm

Femoral nerve
Supplies sensation over the thigh and inner leg, and muscles in the front of the thigh, including the quadriceps

Obturator nerve
Supplies the muscles and skin of the inner thigh

Common fibular (peroneal) nerve
Lies on the outer side of the leg and is named after the bone around which it wraps; perona is an alternative Latin name for fibula

Tibial nerve
Named after the other bone of the lower leg – the tibia, or shinbone

Brainstem
Emerges from the foramen magnum in the base of the skull

Cervical spinal nerves
Cervical means of the neck; cervix is Latin for neck

Thoracic spinal nerves
Thorax is Latin for chest so the term thoracic means of the chest

Lumbar plexus
Anterior branches of the lumbar spinal nerves form a network here, from which nerves emerge to supply the leg

Sacral plexus
Anterior branches of sacral spinal nerves come together here as a network; the network provides nerves to the buttock and leg

Cauda equina
Below the end of the spinal cord, the lumbar and sacral nerve roots continue for some way inside the vertebral canal, before emerging from the spine

Sciatic nerve
Largest nerve in the body, which supplies the hamstrings in the back of the thigh; its branches supply muscles and sensation in the lower leg and foot

SIDE

NEURON STRUCTURE

A single neuron such as the cell shown below from the central nervous system can make contact with hundreds of other neurons, creating an incredibly complex network of connections. Each neuron's cell body has projections, or dendrites. One is usually longer and thinner than the rest, and this is the axon. Some axons within the brain are less than 1mm (1/32in) in length; others, stretching from the spinal cord to limb muscles, can measure over 1m (39in) long.

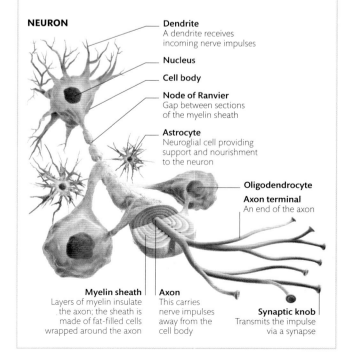

NEURON

Dendrite
A dendrite receives incoming nerve impulses

Nucleus

Cell body

Node of Ranvier
Gap between sections of the myelin sheath

Astrocyte
Neuroglial cell providing support and nourishment to the neuron

Oligodendrocyte

Axon terminal
An end of the axon

Myelin sheath
Layers of myelin insulate the axon; the sheath is made of fat-filled cells wrapped around the axon

Axon
This carries nerve impulses away from the cell body

Synaptic knob
Transmits the impulse via a synapse

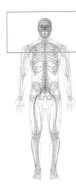

BRAIN

Compared to other animals, we humans have massive brains for the size of our bodies. The human brain has grown larger and larger over the course of evolution, and it is now so overblown that the frontal lobes of the brain lie right over the top of the orbits that contain the eyes. Think about any other mammal, perhaps a dog or a cat for easy reference, and you will quickly realize what an odd shape the human head is – and most of that is down to our huge brains. Looking at a side view of the brain, you can see all the lobes that make up each cerebral hemisphere: the frontal, parietal, temporal, and occipital lobes (individually coloured, below). Tucked under the cerebral hemispheres at the back of the brain is the cerebellum (Latin for little brain). The brainstem leads down, through the foramen magnum of the skull, to the spinal cord.

Superior frontal gyrus
The word gyrus comes from the Latin for ring or convolution, and is a term used for the scroll-like folds of the cerebral cortex

Middle frontal gyrus

Inferior frontal gyrus
Includes Broca's area, part of the cerebral cortex that is involved with generating speech

Olfactory bulb

Optic nerve
The second cranial nerve. It carries nerve fibres from the retina to the optic chiasma

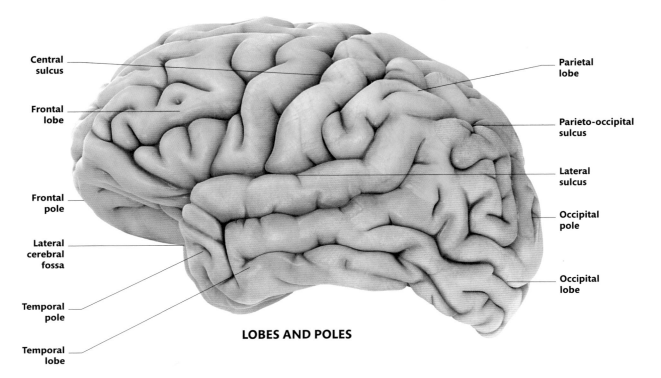

Central sulcus

Frontal lobe

Frontal pole

Lateral cerebral fossa

Temporal pole

Temporal lobe

Parietal lobe

Parieto-occipital sulcus

Lateral sulcus

Occipital pole

Occipital lobe

LOBES AND POLES

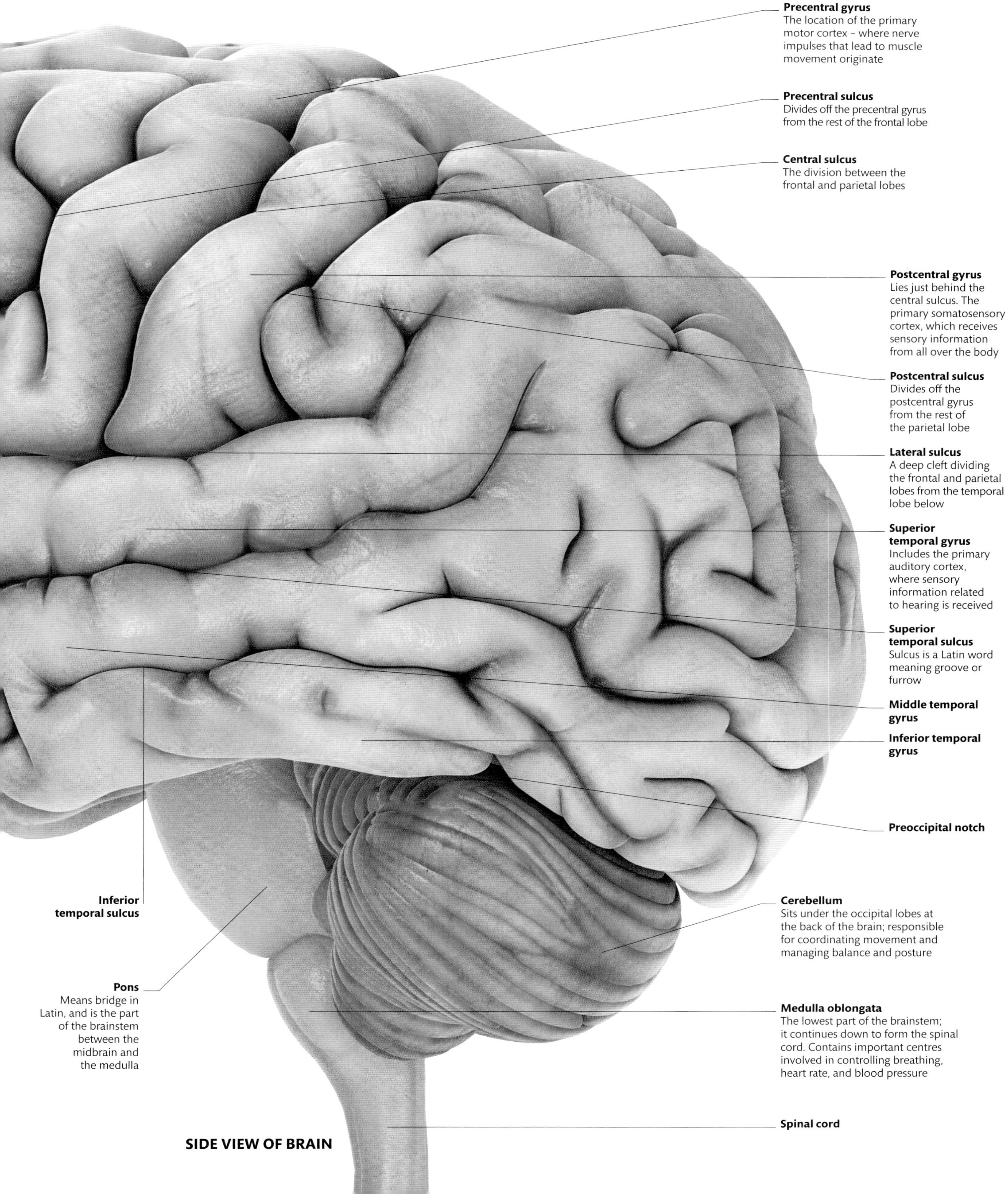

Precentral gyrus
The location of the primary motor cortex – where nerve impulses that lead to muscle movement originate

Precentral sulcus
Divides off the precentral gyrus from the rest of the frontal lobe

Central sulcus
The division between the frontal and parietal lobes

Postcentral gyrus
Lies just behind the central sulcus. The primary somatosensory cortex, which receives sensory information from all over the body

Postcentral sulcus
Divides off the postcentral gyrus from the rest of the parietal lobe

Lateral sulcus
A deep cleft dividing the frontal and parietal lobes from the temporal lobe below

Superior temporal gyrus
Includes the primary auditory cortex, where sensory information related to hearing is received

Superior temporal sulcus
Sulcus is a Latin word meaning groove or furrow

Middle temporal gyrus

Inferior temporal gyrus

Preoccipital notch

Inferior temporal sulcus

Pons
Means bridge in Latin, and is the part of the brainstem between the midbrain and the medulla

Cerebellum
Sits under the occipital lobes at the back of the brain; responsible for coordinating movement and managing balance and posture

Medulla oblongata
The lowest part of the brainstem; it continues down to form the spinal cord. Contains important centres involved in controlling breathing, heart rate, and blood pressure

Spinal cord

SIDE VIEW OF BRAIN

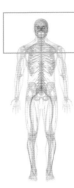

BRAIN

From an anatomist's point of view, the brain is quite an ugly and unprepossessing organ. It looks rather like a large, pinkish grey, wrinkled walnut – especially when viewed from above. The outer layer of grey matter, called the cortex, is highly folded. Underneath the brain we see some more detail, including some of the cranial nerves that emerge from the brain itself. To the naked eye, there is little to suggest that the brain is the most complicated organ in the human body. Its true complexity is only visible through a microscope, revealing billions of neurons that connect with each other to form the pathways that carry our senses, govern our actions, and create our minds.

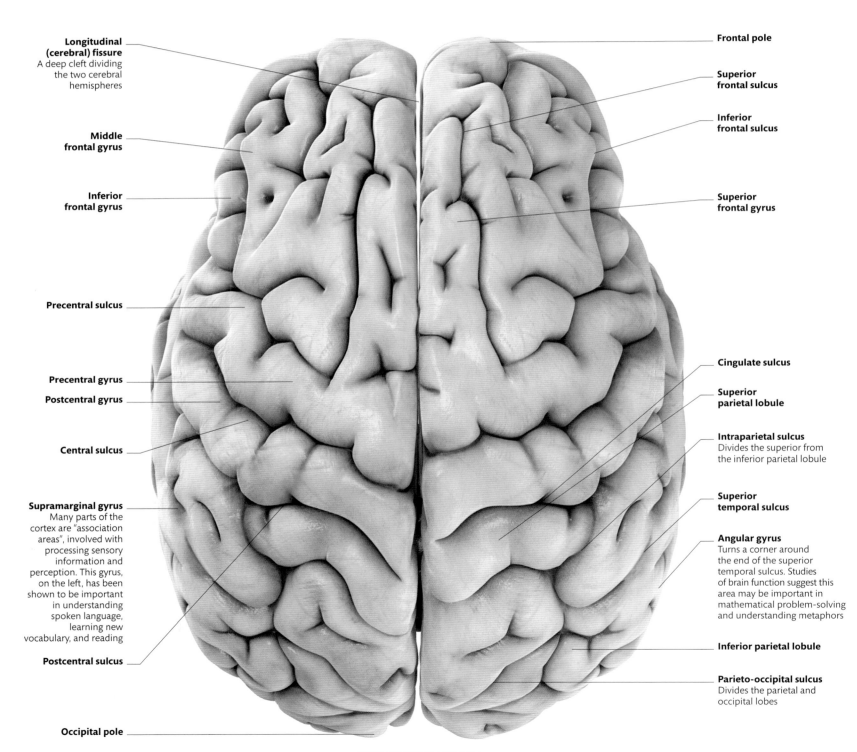

Longitudinal (cerebral) fissure
A deep cleft dividing the two cerebral hemispheres

Middle frontal gyrus

Inferior frontal gyrus

Precentral sulcus

Precentral gyrus

Postcentral gyrus

Central sulcus

Supramarginal gyrus
Many parts of the cortex are "association areas", involved with processing sensory information and perception. This gyrus, on the left, has been shown to be important in understanding spoken language, learning new vocabulary, and reading

Postcentral sulcus

Occipital pole

Frontal pole

Superior frontal sulcus

Inferior frontal sulcus

Superior frontal gyrus

Cingulate sulcus

Superior parietal lobule

Intraparietal sulcus
Divides the superior from the inferior parietal lobule

Superior temporal sulcus

Angular gyrus
Turns a corner around the end of the superior temporal sulcus. Studies of brain function suggest this area may be important in mathematical problem-solving and understanding metaphors

Inferior parietal lobule

Parieto-occipital sulcus
Divides the parietal and occipital lobes

TOP VIEW OF BRAIN

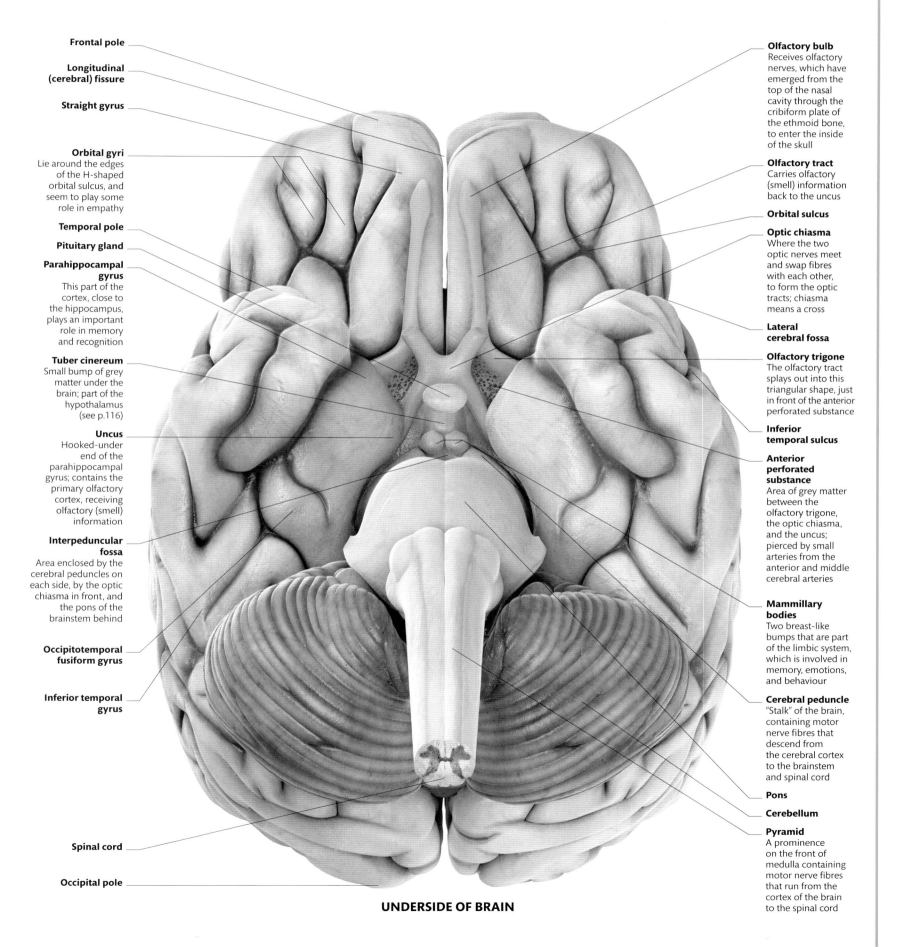

Frontal pole

Longitudinal (cerebral) fissure

Straight gyrus

Orbital gyri
Lie around the edges of the H-shaped orbital sulcus, and seem to play some role in empathy

Temporal pole

Pituitary gland

Parahippocampal gyrus
This part of the cortex, close to the hippocampus, plays an important role in memory and recognition

Tuber cinereum
Small bump of grey matter under the brain; part of the hypothalamus (see p.116)

Uncus
Hooked-under end of the parahippocampal gyrus; contains the primary olfactory cortex, receiving olfactory (smell) information

Interpeduncular fossa
Area enclosed by the cerebral peduncles on each side, by the optic chiasma in front, and the pons of the brainstem behind

Occipitotemporal fusiform gyrus

Inferior temporal gyrus

Spinal cord

Occipital pole

Olfactory bulb
Receives olfactory nerves, which have emerged from the top of the nasal cavity through the cribiform plate of the ethmoid bone, to enter the inside of the skull

Olfactory tract
Carries olfactory (smell) information back to the uncus

Orbital sulcus

Optic chiasma
Where the two optic nerves meet and swap fibres with each other, to form the optic tracts; chiasma means a cross

Lateral cerebral fossa

Olfactory trigone
The olfactory tract splays out into this triangular shape, just in front of the anterior perforated substance

Inferior temporal sulcus

Anterior perforated substance
Area of grey matter between the olfactory trigone, the optic chiasma, and the uncus; pierced by small arteries from the anterior and middle cerebral arteries

Mammillary bodies
Two breast-like bumps that are part of the limbic system, which is involved in memory, emotions, and behaviour

Cerebral peduncle
"Stalk" of the brain, containing motor nerve fibres that descend from the cerebral cortex to the brainstem and spinal cord

Pons

Cerebellum

Pyramid
A prominence on the front of medulla containing motor nerve fibres that run from the cortex of the brain to the spinal cord

UNDERSIDE OF BRAIN

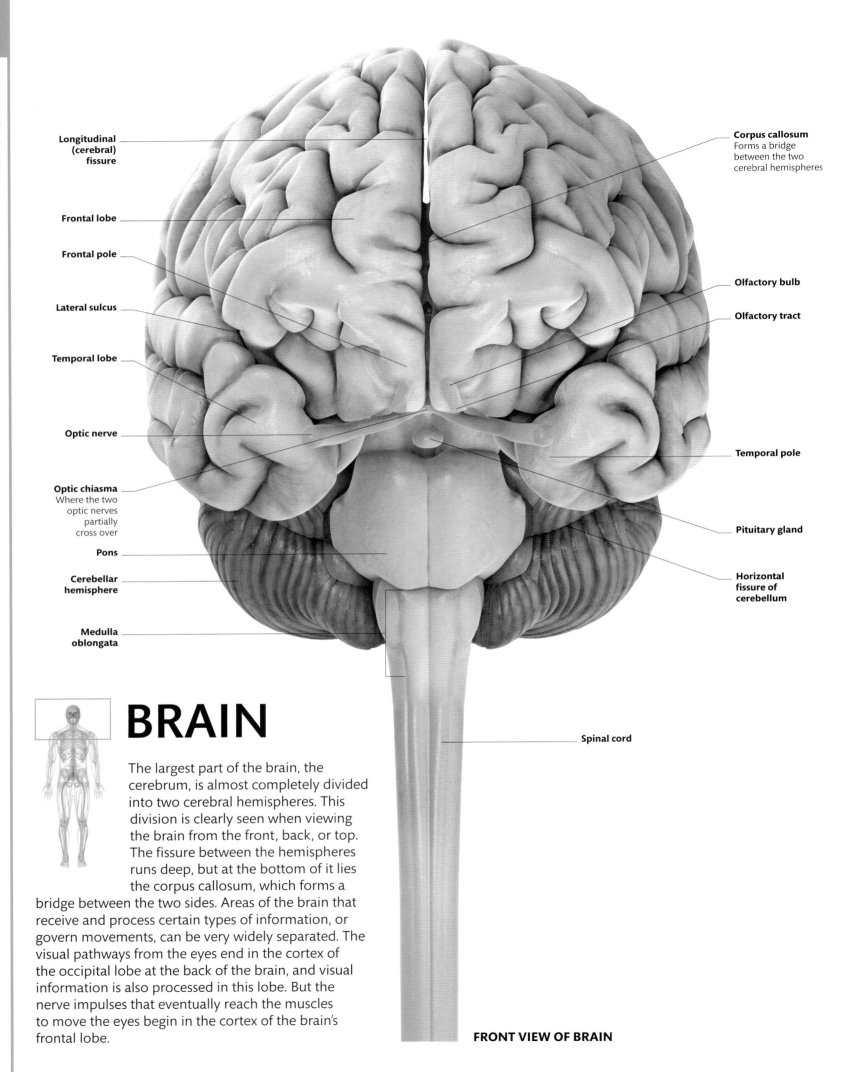

Longitudinal (cerebral) fissure

Frontal lobe

Frontal pole

Lateral sulcus

Temporal lobe

Optic nerve

Optic chiasma
Where the two optic nerves partially cross over

Pons

Cerebellar hemisphere

Medulla oblongata

Corpus callosum
Forms a bridge between the two cerebral hemispheres

Olfactory bulb

Olfactory tract

Temporal pole

Pituitary gland

Horizontal fissure of cerebellum

Spinal cord

BRAIN

The largest part of the brain, the cerebrum, is almost completely divided into two cerebral hemispheres. This division is clearly seen when viewing the brain from the front, back, or top. The fissure between the hemispheres runs deep, but at the bottom of it lies the corpus callosum, which forms a bridge between the two sides. Areas of the brain that receive and process certain types of information, or govern movements, can be very widely separated. The visual pathways from the eyes end in the cortex of the occipital lobe at the back of the brain, and visual information is also processed in this lobe. But the nerve impulses that eventually reach the muscles to move the eyes begin in the cortex of the brain's frontal lobe.

FRONT VIEW OF BRAIN

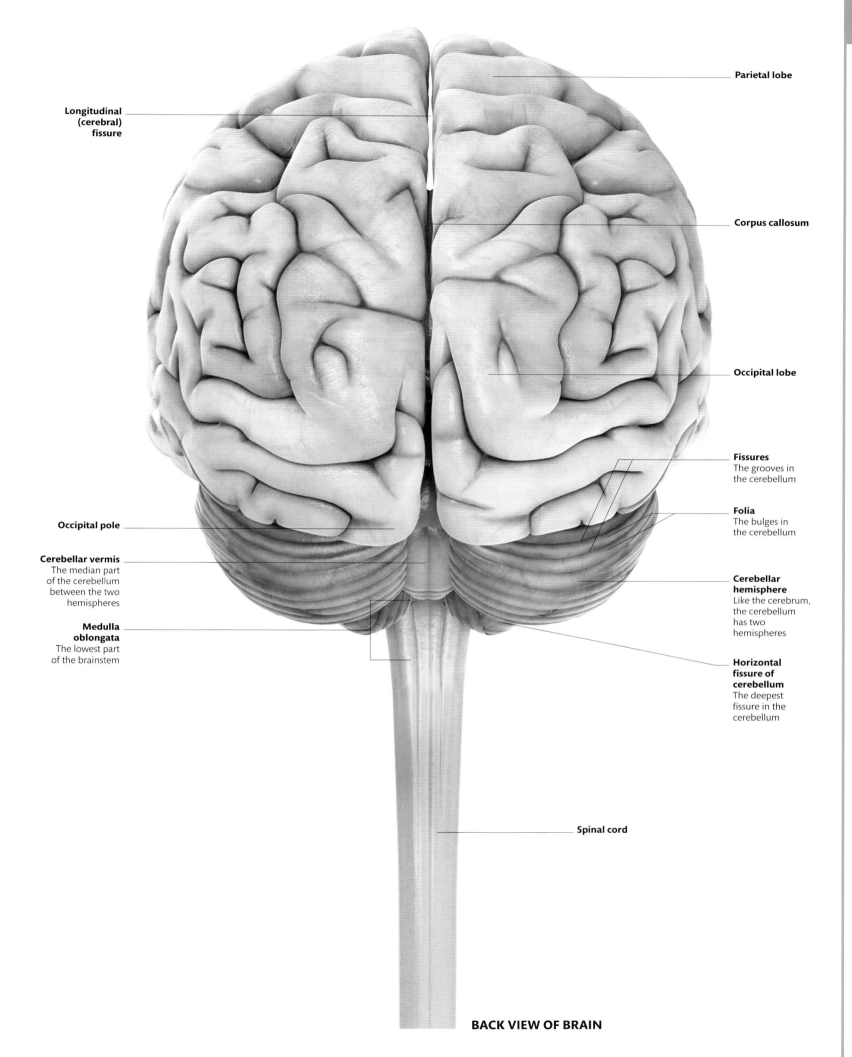

Longitudinal (cerebral) fissure

Parietal lobe

Corpus callosum

Occipital lobe

Fissures
The grooves in the cerebellum

Folia
The bulges in the cerebellum

Occipital pole

Cerebellar vermis
The median part of the cerebellum between the two hemispheres

Cerebellar hemisphere
Like the cerebrum, the cerebellum has two hemispheres

Medulla oblongata
The lowest part of the brainstem

Horizontal fissure of cerebellum
The deepest fissure in the cerebellum

Spinal cord

BACK VIEW OF BRAIN

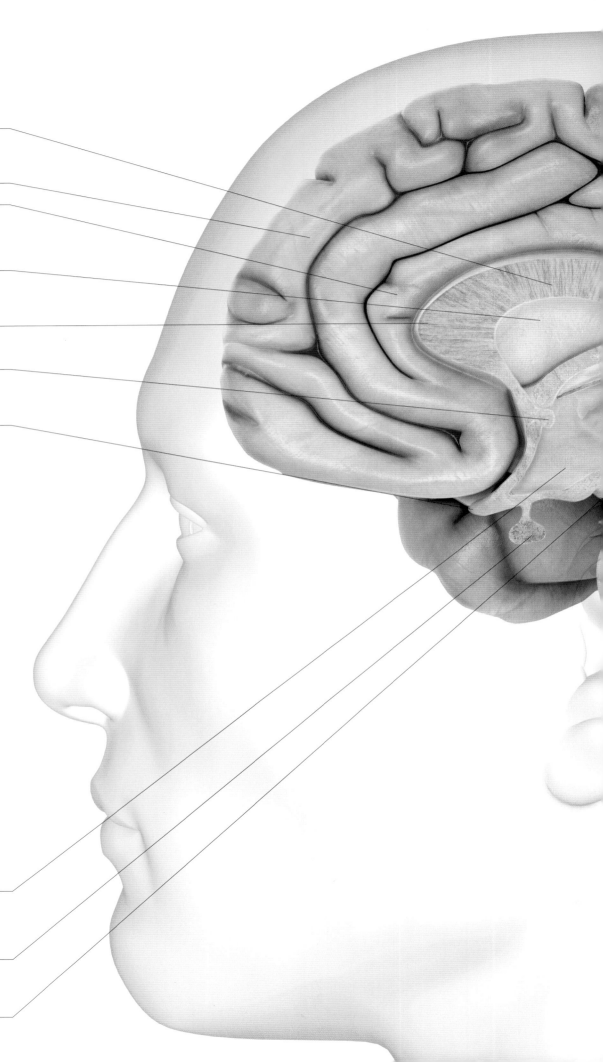

Body of corpus callosum
The largest commissure (or bundle of connecting nerve fibres) between the two hemispheres, this forms the roofs of the lateral ventricles

Superior frontal gyrus

Cingulate gyrus
Cingulum is the Latin for girdle and this gyrus wraps closely around the corpus callosum; it is part of the limbic system, which is involved with emotional responses and behaviours

Septum pellucidum
This translucent partition is a thin dividing wall between the two lateral ventricles

Genu of corpus callosum
The anterior (front) end of the corpus callosum is bent over – genu means knee in Latin

Anterior commissure
A bundle of nerve fibres connecting parts of the two cerebral hemispheres

Optic chiasma
The crossover point where the two optic nerves meet and swap fibres, then part company as the optic tracts, which continue on each side of the brain towards the thalamus

**SAGITTAL SECTION
THROUGH BRAIN**

Hypothalamus
Plays an important role in regulating the internal environment of the body, by keeping a check on body temperature, blood pressure, and blood sugar level, for instance

Pituitary gland
Produces many hormones and forms a link between the brain and endocrine system

Mammillary body
Part of the limbic system of the brain

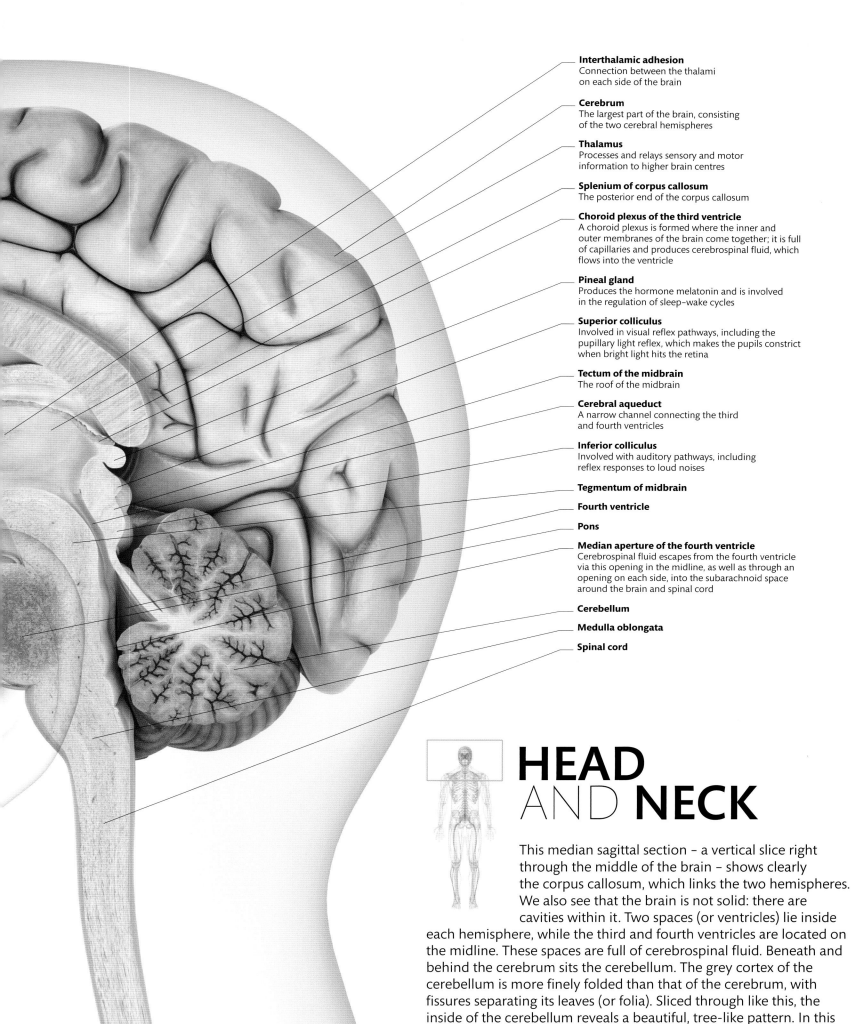

Interthalamic adhesion
Connection between the thalami
on each side of the brain

Cerebrum
The largest part of the brain, consisting
of the two cerebral hemispheres

Thalamus
Processes and relays sensory and motor
information to higher brain centres

Splenium of corpus callosum
The posterior end of the corpus callosum

Choroid plexus of the third ventricle
A choroid plexus is formed where the inner and
outer membranes of the brain come together; it is full
of capillaries and produces cerebrospinal fluid, which
flows into the ventricle

Pineal gland
Produces the hormone melatonin and is involved
in the regulation of sleep–wake cycles

Superior colliculus
Involved in visual reflex pathways, including the
pupillary light reflex, which makes the pupils constrict
when bright light hits the retina

Tectum of the midbrain
The roof of the midbrain

Cerebral aqueduct
A narrow channel connecting the third
and fourth ventricles

Inferior colliculus
Involved with auditory pathways, including
reflex responses to loud noises

Tegmentum of midbrain

Fourth ventricle

Pons

Median aperture of the fourth ventricle
Cerebrospinal fluid escapes from the fourth ventricle
via this opening in the midline, as well as through an
opening on each side, into the subarachnoid space
around the brain and spinal cord

Cerebellum

Medulla oblongata

Spinal cord

HEAD
AND NECK

This median sagittal section – a vertical slice right
through the middle of the brain – shows clearly
the corpus callosum, which links the two hemispheres.
We also see that the brain is not solid: there are
cavities within it. Two spaces (or ventricles) lie inside
each hemisphere, while the third and fourth ventricles are located on
the midline. These spaces are full of cerebrospinal fluid. Beneath and
behind the cerebrum sits the cerebellum. The grey cortex of the
cerebellum is more finely folded than that of the cerebrum, with
fissures separating its leaves (or folia). Sliced through like this, the
inside of the cerebellum reveals a beautiful, tree-like pattern. In this
section, we can also see clearly all the parts of the brainstem – the
midbrain, pons, and medulla.

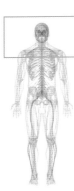

BRAIN

The brain is protected by three membranes called the meninges (which become inflamed in meningitis). The tough dura mater layer is the outermost covering, which surrounds the brain and the spinal cord. Under the dura mater is the cobweb-like arachnoid mater layer. The delicate pia mater is a thin membrane on the surface of the brain. Between the pia mater and the arachnoid mater there is a slim gap – the subarachnoid space – which contains cerebrospinal fluid (CSF). Mainly produced by the choroid plexus in the brain's lateral ventricles, CSF flows through the third ventricle into the fourth, where it can escape via small apertures into the subarachnoid space.

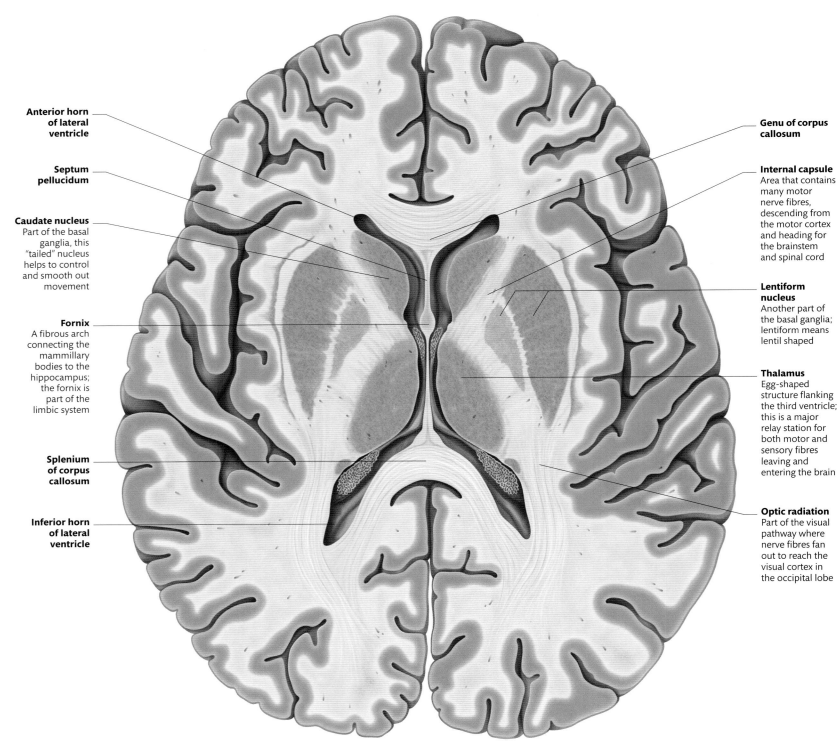

Anterior horn of lateral ventricle

Septum pellucidum

Caudate nucleus
Part of the basal ganglia, this "tailed" nucleus helps to control and smooth out movement

Fornix
A fibrous arch connecting the mammillary bodies to the hippocampus; the fornix is part of the limbic system

Splenium of corpus callosum

Inferior horn of lateral ventricle

Genu of corpus callosum

Internal capsule
Area that contains many motor nerve fibres, descending from the motor cortex and heading for the brainstem and spinal cord

Lentiform nucleus
Another part of the basal ganglia; lentiform means lentil shaped

Thalamus
Egg-shaped structure flanking the third ventricle; this is a major relay station for both motor and sensory fibres leaving and entering the brain

Optic radiation
Part of the visual pathway where nerve fibres fan out to reach the visual cortex in the occipital lobe

TRANSVERSE SECTION OF BRAIN

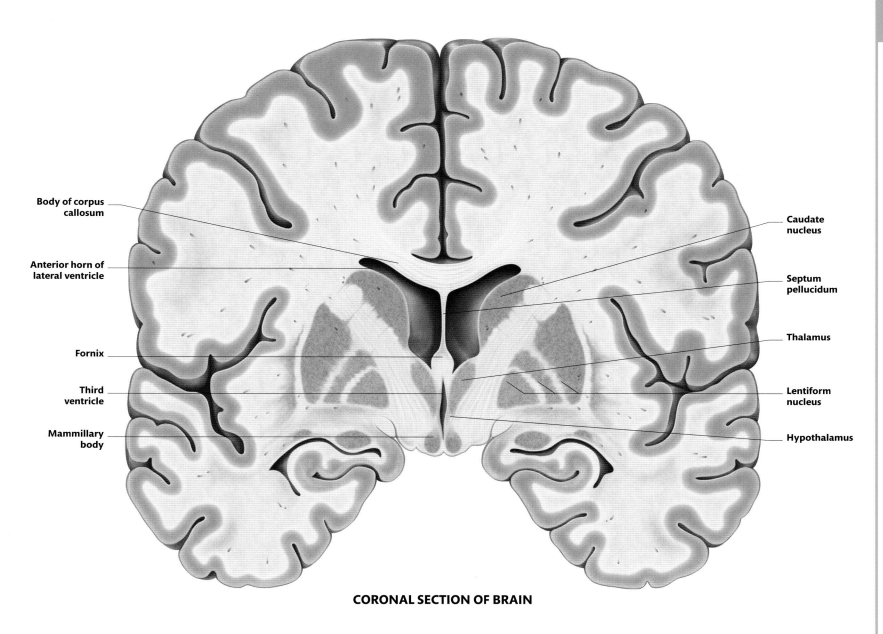

Body of corpus callosum

Anterior horn of lateral ventricle

Fornix

Third ventricle

Mammillary body

Caudate nucleus

Septum pellucidum

Thalamus

Lentiform nucleus

Hypothalamus

CORONAL SECTION OF BRAIN

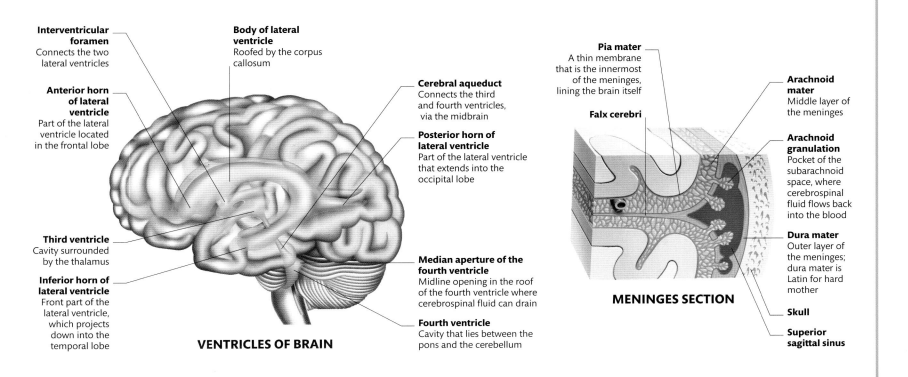

Interventricular foramen
Connects the two lateral ventricles

Anterior horn of lateral ventricle
Part of the lateral ventricle located in the frontal lobe

Third ventricle
Cavity surrounded by the thalamus

Inferior horn of lateral ventricle
Front part of the lateral ventricle, which projects down into the temporal lobe

Body of lateral ventricle
Roofed by the corpus callosum

Cerebral aqueduct
Connects the third and fourth ventricles, via the midbrain

Posterior horn of lateral ventricle
Part of the lateral ventricle that extends into the occipital lobe

Median aperture of the fourth ventricle
Midline opening in the roof of the fourth ventricle where cerebrospinal fluid can drain

Fourth ventricle
Cavity that lies between the pons and the cerebellum

VENTRICLES OF BRAIN

Pia mater
A thin membrane that is the innermost of the meninges, lining the brain itself

Falx cerebri

Arachnoid mater
Middle layer of the meninges

Arachnoid granulation
Pocket of the subarachnoid space, where cerebrospinal fluid flows back into the blood

Dura mater
Outer layer of the meninges; dura mater is Latin for hard mother

Skull

Superior sagittal sinus

MENINGES SECTION

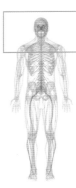

HEAD AND NECK

The 12 pairs of cranial nerves (the standard abbreviation for which is CN) emerge from the brain and brainstem, leaving through holes, or "foramina", in the base of the skull. Some nerves are purely sensory, some just have motor functions, but most contain a mixture of motor and sensory fibres. A few also contain autonomic nerve fibres. The olfactory nerve and the optic nerve attach to the brain itself. The other 10 pairs of cranial nerves emerge from the brainstem. All the cranial nerves supply parts of the head and neck, with the exception of the vagus nerve. This has branches in the neck, but then carries on to supply organs in the thorax and right down in the abdomen. Careful testing of cranial nerves, including tests of sight, eye and head movement, taste, and so on, can help doctors to pinpoint neurological problems in the head and neck.

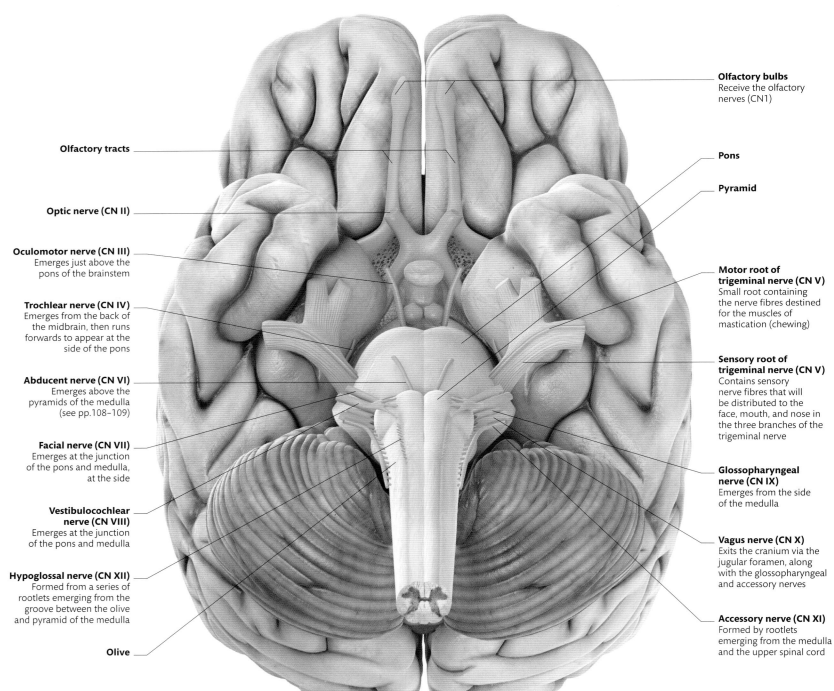

Olfactory bulbs
Receive the olfactory nerves (CN1)

Pons

Pyramid

Motor root of trigeminal nerve (CN V)
Small root containing the nerve fibres destined for the muscles of mastication (chewing)

Sensory root of trigeminal nerve (CN V)
Contains sensory nerve fibres that will be distributed to the face, mouth, and nose in the three branches of the trigeminal nerve

Glossopharyngeal nerve (CN IX)
Emerges from the side of the medulla

Vagus nerve (CN X)
Exits the cranium via the jugular foramen, along with the glossopharyngeal and accessory nerves

Accessory nerve (CN XI)
Formed by rootlets emerging from the medulla and the upper spinal cord

Olfactory tracts

Optic nerve (CN II)

Oculomotor nerve (CN III)
Emerges just above the pons of the brainstem

Trochlear nerve (CN IV)
Emerges from the back of the midbrain, then runs forwards to appear at the side of the pons

Abducent nerve (CN VI)
Emerges above the pyramids of the medulla (see pp.108–109)

Facial nerve (CN VII)
Emerges at the junction of the pons and medulla, at the side

Vestibulocochlear nerve (CN VIII)
Emerges at the junction of the pons and medulla

Hypoglossal nerve (CN XII)
Formed from a series of rootlets emerging from the groove between the olive and pyramid of the medulla

Olive

ORIGIN OF CRANIAL NERVES (UNDERSIDE OF BRAIN)

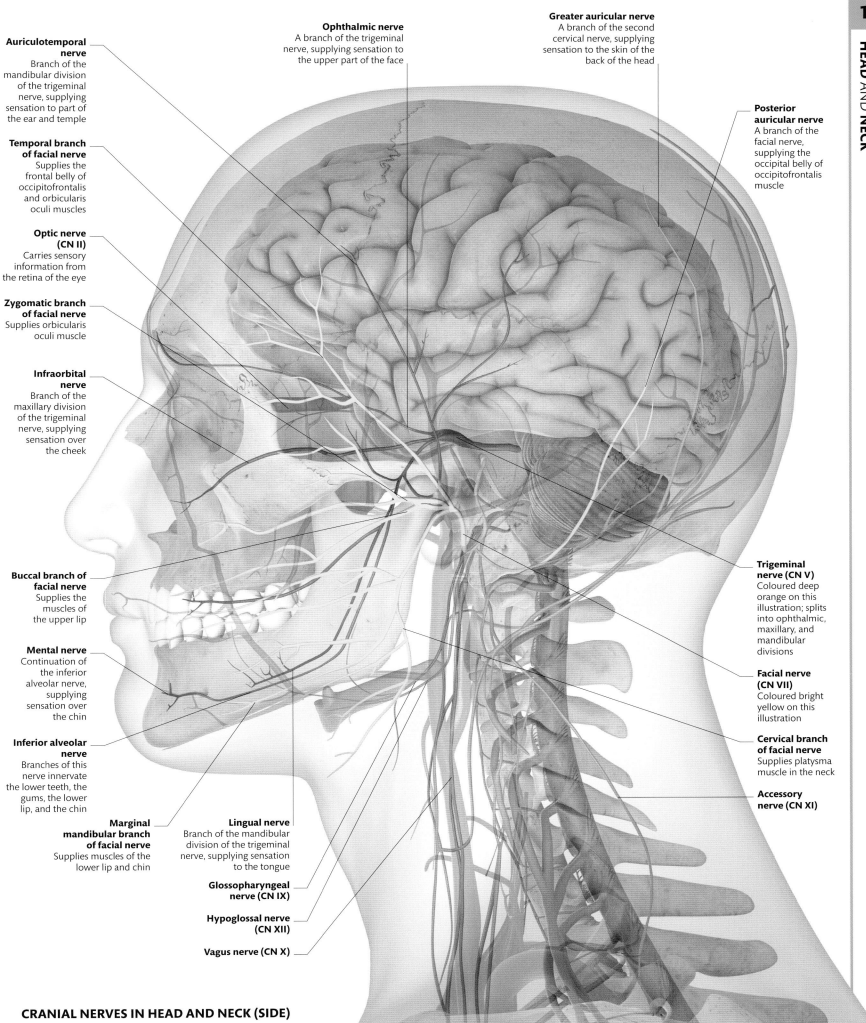

Auriculotemporal nerve
Branch of the mandibular division of the trigeminal nerve, supplying sensation to part of the ear and temple

Temporal branch of facial nerve
Supplies the frontal belly of occipitofrontalis and orbicularis oculi muscles

Optic nerve (CN II)
Carries sensory information from the retina of the eye

Zygomatic branch of facial nerve
Supplies orbicularis oculi muscle

Infraorbital nerve
Branch of the maxillary division of the trigeminal nerve, supplying sensation over the cheek

Buccal branch of facial nerve
Supplies the muscles of the upper lip

Mental nerve
Continuation of the inferior alveolar nerve, supplying sensation over the chin

Inferior alveolar nerve
Branches of this nerve innervate the lower teeth, the gums, the lower lip, and the chin

Marginal mandibular branch of facial nerve
Supplies muscles of the lower lip and chin

Ophthalmic nerve
A branch of the trigeminal nerve, supplying sensation to the upper part of the face

Greater auricular nerve
A branch of the second cervical nerve, supplying sensation to the skin of the back of the head

Posterior auricular nerve
A branch of the facial nerve, supplying the occipital belly of occipitofrontalis muscle

Trigeminal nerve (CN V)
Coloured deep orange on this illustration; splits into ophthalmic, maxillary, and mandibular divisions

Facial nerve (CN VII)
Coloured bright yellow on this illustration

Cervical branch of facial nerve
Supplies platysma muscle in the neck

Accessory nerve (CN XI)

Lingual nerve
Branch of the mandibular division of the trigeminal nerve, supplying sensation to the tongue

Glossopharyngeal nerve (CN IX)

Hypoglossal nerve (CN XII)

Vagus nerve (CN X)

CRANIAL NERVES IN HEAD AND NECK (SIDE)

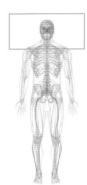

EYE

The eyes are precious organs. They are well protected inside the eye sockets, or bony orbits, of the skull. They are also protected by the eyelids, and bathed in tears produced by the lacrimal glands. Each eyeball is only 2.5cm (1in) in diameter. The orbit provides an anchor for the muscles that move the eye, and the rest of the space inside the orbit is largely filled up with fat. Holes and fissures at the back of this bony cavern transmit nerves and blood vessels, including the optic nerve, which carries sensory information from the retina to the brain. Other nerves supply the eye muscles and the lacrimal glands, and even continue on to the face to supply sensation to the skin of the eyelids and forehead.

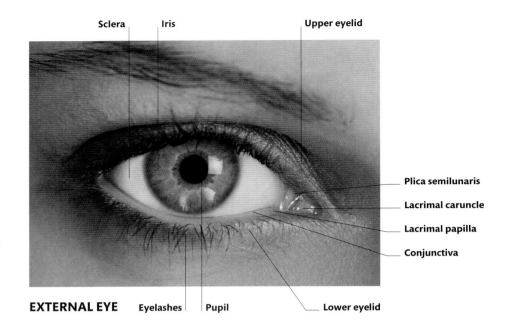

Sclera | Iris | Upper eyelid

Plica semilunaris

Lacrimal caruncle

Lacrimal papilla

Conjunctiva

EXTERNAL EYE | Eyelashes | Pupil | Lower eyelid

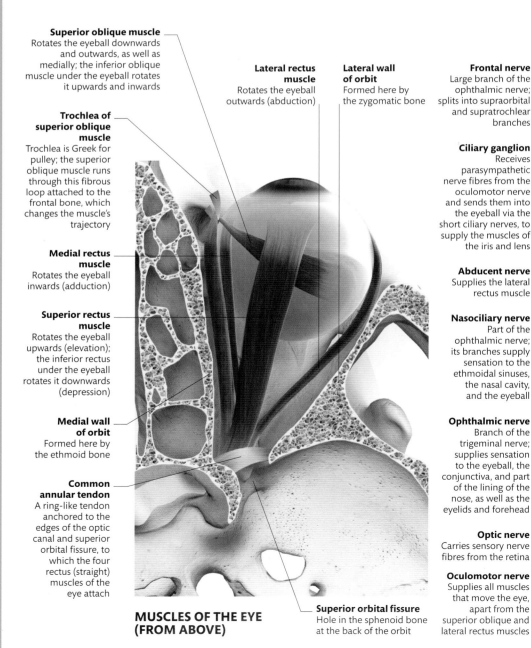

Superior oblique muscle
Rotates the eyeball downwards and outwards, as well as medially; the inferior oblique muscle under the eyeball rotates it upwards and inwards

Lateral rectus muscle
Rotates the eyeball outwards (abduction)

Lateral wall of orbit
Formed here by the zygomatic bone

Trochlea of superior oblique muscle
Trochlea is Greek for pulley; the superior oblique muscle runs through this fibrous loop attached to the frontal bone, which changes the muscle's trajectory

Medial rectus muscle
Rotates the eyeball inwards (adduction)

Superior rectus muscle
Rotates the eyeball upwards (elevation); the inferior rectus under the eyeball rotates it downwards (depression)

Medial wall of orbit
Formed here by the ethmoid bone

Common annular tendon
A ring-like tendon anchored to the edges of the optic canal and superior orbital fissure, to which the four rectus (straight) muscles of the eye attach

MUSCLES OF THE EYE (FROM ABOVE)

Superior orbital fissure
Hole in the sphenoid bone at the back of the orbit

Supratrochlear nerve
Runs over the eyeball and up, out of the orbit, to supply sensation to the middle of the forehead

Frontal nerve
Large branch of the ophthalmic nerve; splits into supraorbital and supratrochlear branches

Ciliary ganglion
Receives parasympathetic nerve fibres from the oculomotor nerve and sends them into the eyeball via the short ciliary nerves, to supply the muscles of the iris and lens

Abducent nerve
Supplies the lateral rectus muscle

Nasociliary nerve
Part of the ophthalmic nerve; its branches supply sensation to the ethmoidal sinuses, the nasal cavity, and the eyeball

Ophthalmic nerve
Branch of the trigeminal nerve; supplies sensation to the eyeball, the conjunctiva, and part of the lining of the nose, as well as the eyelids and forehead

Optic nerve
Carries sensory nerve fibres from the retina

Oculomotor nerve
Supplies all muscles that move the eye, apart from the superior oblique and lateral rectus muscles

Supraorbital nerve
Runs forwards, out of the orbit, and turns upwards on the frontal bone to supply the upper eyelid

Lacrimal nerve
Supplies skin over the upper eyelid and lateral forehead

Lacrimal gland

Trochlear nerve
Supplies the superior oblique muscle

NERVES OF THE ORBIT (FROM ABOVE)

Sclera
From the Greek for hard; the tough, outer coat of the eyeball

Conjunctiva
Thin mucous membrane covering the front of the eyeball, as well as the inner surfaces of the eyelids, but not the cornea

Iris
From the Greek for rainbow; contains smooth muscle: circular fibres constrict the pupil, while radial muscle fibres dilate it

Cornea
Transparent outer layer of the front of the eye; continuous with the sclera

Pupil

Aqueous humour
Watery fluid occupies the anterior and posterior chambers of the eye, either side of the iris

Lens
Made up of long, transparent cells called lens fibres; tends to become less clear in old age

Suspensory ligament
Attaches the lens to the ciliary body

Ciliary body
Contains smooth muscle fibres that pull to alter the shape of the lens in order to focus

Medial rectus muscle

Lateral rectus muscle

Vitreous humour
Means glassy fluid in Latin. The main filling of the eyeball, it is liquid in the centre but moregel-like at the edges

Choroid
This layer is packed with blood vessels

Optic disc
Retinal nerve fibres create a doughnut-like bulge where they gather to form the optic nerve

Optic nerve
Carries visual information from the retina back to the brain

Blind spot
Where retinal nerve fibres leave the back of the retina the eye has no sensory cells; the brain fills in the missing information, so that we are not aware of the tiny blind spot in each eye

Retina
Inner, sensory lining of the eyeball; forms as an outgrowth of the brain itself during embryological development

HORIZONTAL SECTION THROUGH THE EYEBALL

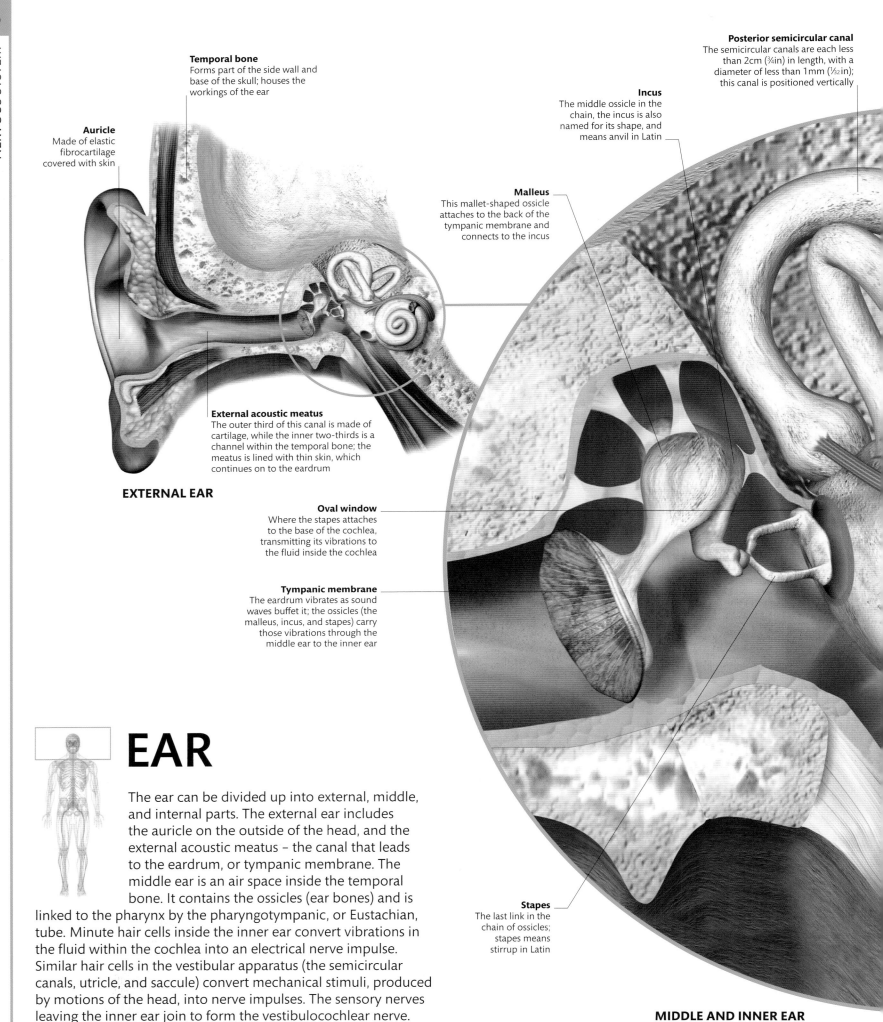

Temporal bone
Forms part of the side wall and base of the skull; houses the workings of the ear

Auricle
Made of elastic fibrocartilage covered with skin

Posterior semicircular canal
The semicircular canals are each less than 2cm (¾in) in length, with a diameter of less than 1mm (½in); this canal is positioned vertically

Incus
The middle ossicle in the chain, the incus is also named for its shape, and means anvil in Latin

Malleus
This mallet-shaped ossicle attaches to the back of the tympanic membrane and connects to the incus

External acoustic meatus
The outer third of this canal is made of cartilage, while the inner two-thirds is a channel within the temporal bone; the meatus is lined with thin skin, which continues on to the eardrum

EXTERNAL EAR

Oval window
Where the stapes attaches to the base of the cochlea, transmitting its vibrations to the fluid inside the cochlea

Tympanic membrane
The eardrum vibrates as sound waves buffet it; the ossicles (the malleus, incus, and stapes) carry those vibrations through the middle ear to the inner ear

Stapes
The last link in the chain of ossicles; stapes means stirrup in Latin

EAR

The ear can be divided up into external, middle, and internal parts. The external ear includes the auricle on the outside of the head, and the external acoustic meatus – the canal that leads to the eardrum, or tympanic membrane. The middle ear is an air space inside the temporal bone. It contains the ossicles (ear bones) and is linked to the pharynx by the pharyngotympanic, or Eustachian, tube. Minute hair cells inside the inner ear convert vibrations in the fluid within the cochlea into an electrical nerve impulse. Similar hair cells in the vestibular apparatus (the semicircular canals, utricle, and saccule) convert mechanical stimuli, produced by motions of the head, into nerve impulses. The sensory nerves leaving the inner ear join to form the vestibulocochlear nerve.

MIDDLE AND INNER EAR

Lateral semicircular canal
This is positioned horizontally

Anterior semicircular canal
Positioned vertically, but at right angles to the plane of the posterior semicircular canal

Vestibular nerve
Carries sensory information from the vestibular apparatus – including the semicircular canals

Helix
The outer rim of the auricle

External acoustic meatus

Antihelix
A curved prominence, parallel to the helix

Concha
This hollow is named after the Greek for shell

Tragus
This little flap overlaps the external acoustic meatus

Intertragic notch

Lobule

Antitragus
A small tubercle opposite the tragus

AURICLE

Section cut from cochlea
From top to bottom shows vestibular canal, cochlear duct, and tympanic canal

Vestibulocochlear nerve
The cochlear nerve conveys sensory information about sound from the cochlea. It joins the vestibular nerve to the vestibulocochlear nerve

Cochlea
Not surprisingly, cochlea means snail in Latin

Tympanic membrane
As seen with an otoscope, a healthy eardrum has a pearly, almost translucent appearance

Lateral process of malleus

Handle of malleus

Vestibule
Contains the utricle and sacule, organs of balance

Round window
Vibrations can travel in the fluid inside the cochlea, all the way up to its apex and back down to the round window

Pharyngotympanic tube
Passage connecting the middle ear to the back of the throat, and allowing air pressure either side of the eardrum to be equalized

Cone of light
Light is reflected in the front, lower quadrant of the eardrum

EARDRUM

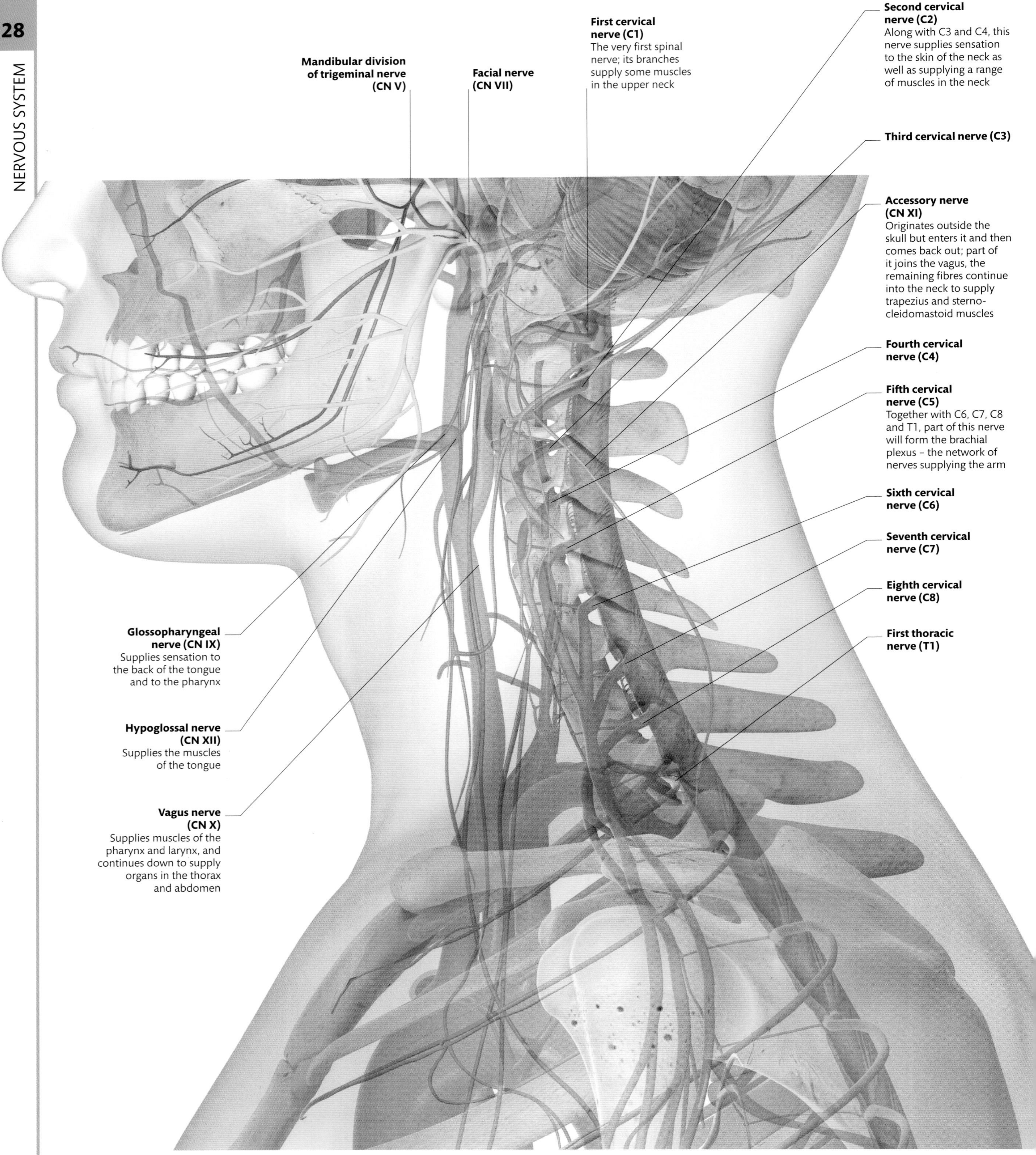

Mandibular division of trigeminal nerve (CN V)

Facial nerve (CN VII)

First cervical nerve (C1)
The very first spinal nerve; its branches supply some muscles in the upper neck

Second cervical nerve (C2)
Along with C3 and C4, this nerve supplies sensation to the skin of the neck as well as supplying a range of muscles in the neck

Third cervical nerve (C3)

Accessory nerve (CN XI)
Originates outside the skull but enters it and then comes back out; part of it joins the vagus, the remaining fibres continue into the neck to supply trapezius and sterno-cleidomastoid muscles

Fourth cervical nerve (C4)

Fifth cervical nerve (C5)
Together with C6, C7, C8 and T1, part of this nerve will form the brachial plexus – the network of nerves supplying the arm

Sixth cervical nerve (C6)

Seventh cervical nerve (C7)

Eighth cervical nerve (C8)

First thoracic nerve (T1)

Glossopharyngeal nerve (CN IX)
Supplies sensation to the back of the tongue and to the pharynx

Hypoglossal nerve (CN XII)
Supplies the muscles of the tongue

Vagus nerve (CN X)
Supplies muscles of the pharynx and larynx, and continues down to supply organs in the thorax and abdomen

NERVES OF THE NECK (SIDE)

NECK

The last four cranial nerves all appear in the neck. The glossopharyngeal nerve supplies the parotid gland and the back of the tongue, then runs down to the pharynx. The vagus nerve is sandwiched between the common carotid artery and the internal jugular vein, and it gives branches to the pharynx and larynx before continuing down into the thorax. The accessory nerve supplies the sternocleidomastoid and trapezius muscles in the neck, while the last cranial nerve, the hypoglossal, dips down below the mandible, then curves back up to supply the muscles of the tongue. We can also see spinal nerves in the neck. The upper four cervical nerves supply neck muscles and skin, while the lower four contribute to the brachial plexus and are destined for the arm.

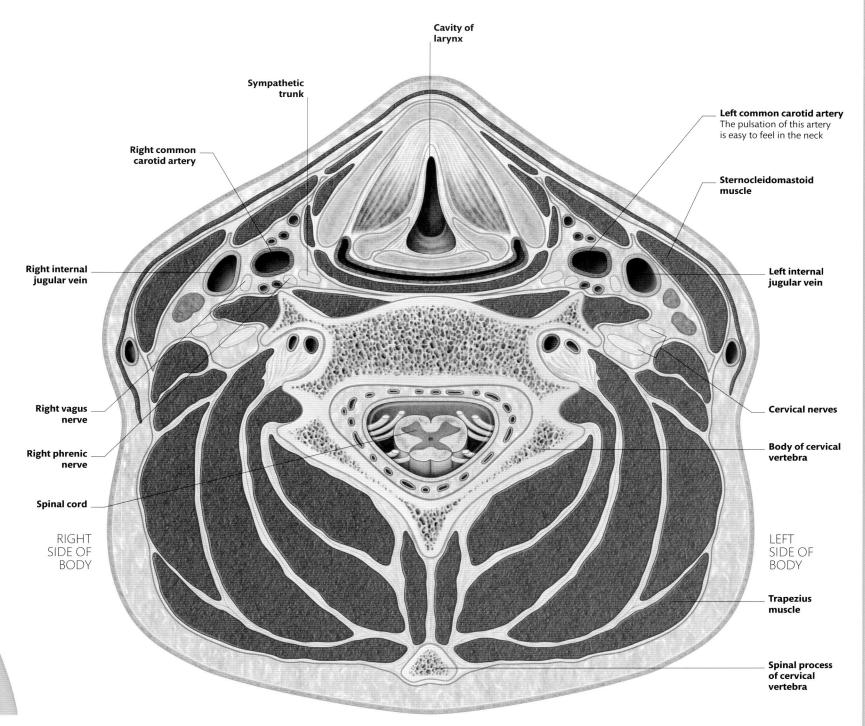

Cavity of larynx

Sympathetic trunk

Right common carotid artery

Right internal jugular vein

Right vagus nerve

Right phrenic nerve

Spinal cord

RIGHT SIDE OF BODY

Left common carotid artery
The pulsation of this artery is easy to feel in the neck

Sternocleidomastoid muscle

Left internal jugular vein

Cervical nerves

Body of cervical vertebra

LEFT SIDE OF BODY

Trapezius muscle

Spinal process of cervical vertebra

TRANSVERSE SECTION OF THE NECK

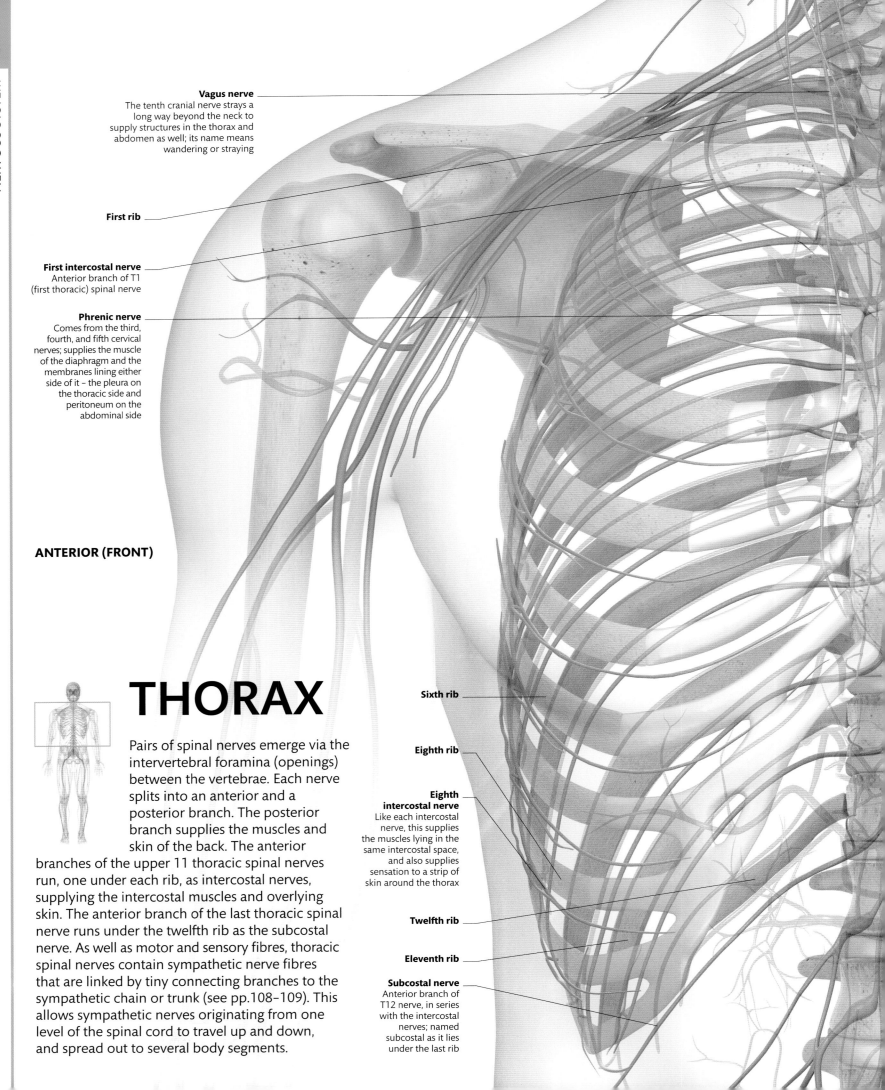

Vagus nerve
The tenth cranial nerve strays a long way beyond the neck to supply structures in the thorax and abdomen as well; its name means wandering or straying

First rib

First intercostal nerve
Anterior branch of T1 (first thoracic) spinal nerve

Phrenic nerve
Comes from the third, fourth, and fifth cervical nerves; supplies the muscle of the diaphragm and the membranes lining either side of it – the pleura on the thoracic side and peritoneum on the abdominal side

ANTERIOR (FRONT)

THORAX

Pairs of spinal nerves emerge via the intervertebral foramina (openings) between the vertebrae. Each nerve splits into an anterior and a posterior branch. The posterior branch supplies the muscles and skin of the back. The anterior branches of the upper 11 thoracic spinal nerves run, one under each rib, as intercostal nerves, supplying the intercostal muscles and overlying skin. The anterior branch of the last thoracic spinal nerve runs under the twelfth rib as the subcostal nerve. As well as motor and sensory fibres, thoracic spinal nerves contain sympathetic nerve fibres that are linked by tiny connecting branches to the sympathetic chain or trunk (see pp.108–109). This allows sympathetic nerves originating from one level of the spinal cord to travel up and down, and spread out to several body segments.

Sixth rib

Eighth rib

Eighth intercostal nerve
Like each intercostal nerve, this supplies the muscles lying in the same intercostal space, and also supplies sensation to a strip of skin around the thorax

Twelfth rib

Eleventh rib

Subcostal nerve
Anterior branch of T12 nerve, in series with the intercostal nerves; named subcostal as it lies under the last rib

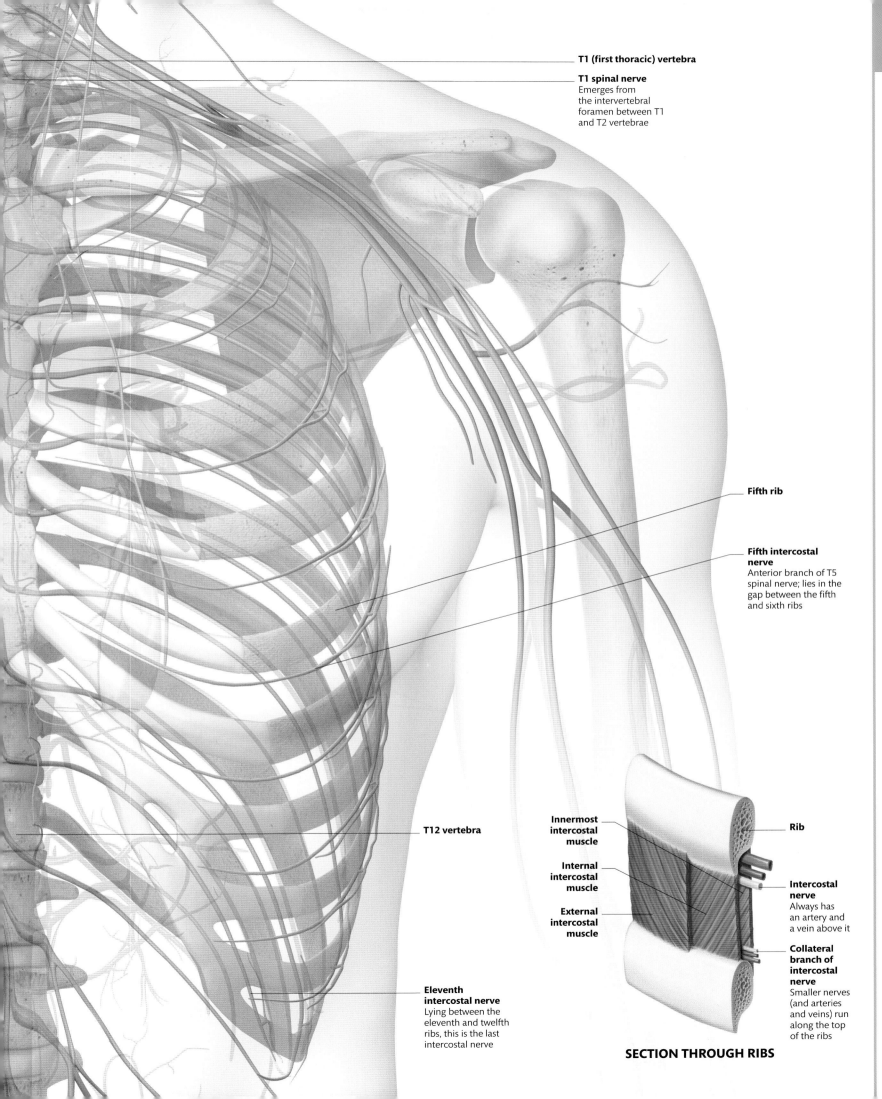

T1 (first thoracic) vertebra

T1 spinal nerve
Emerges from
the intervertebral
foramen between T1
and T2 vertebrae

Fifth rib

**Fifth intercostal
nerve**
Anterior branch of T5
spinal nerve; lies in the
gap between the fifth
and sixth ribs

T12 vertebra

**Innermost
intercostal
muscle**

**Internal
intercostal
muscle**

**External
intercostal
muscle**

Rib

**Intercostal
nerve**
Always has
an artery and
a vein above it

**Collateral
branch of
intercostal
nerve**
Smaller nerves
(and arteries
and veins) run
along the top
of the ribs

**Eleventh
intercostal nerve**
Lying between the
eleventh and twelfth
ribs, this is the last
intercostal nerve

SECTION THROUGH RIBS

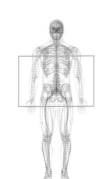

ABDOMEN AND PELVIS

The lower intercostal nerves continue past the lower edges of the ribcage at the front to supply the muscles and skin of the abdominal wall. The lower parts of the abdomen are supplied by the subcostal and iliohypogastric nerves. The abdominal portion of the sympathetic trunk receives nerves from the thoracic and first two lumbar spinal nerves, and sends nerves back to all the spinal nerves. The lumbar spinal nerves emerge from the spine and run into the psoas major muscle at the back of the abdomen. Inside the muscle, the nerves join up and swap fibres to form a network, or plexus. Branches of this lumbar plexus emerge around and through the psoas muscle and make their way into the thigh. Lower down, branches of the sacral plexus supply pelvic organs and enter the buttock. One of these branches, the sciatic nerve, is the largest nerve in the entire body. It supplies the back of the thigh, as well as the rest of the leg and foot.

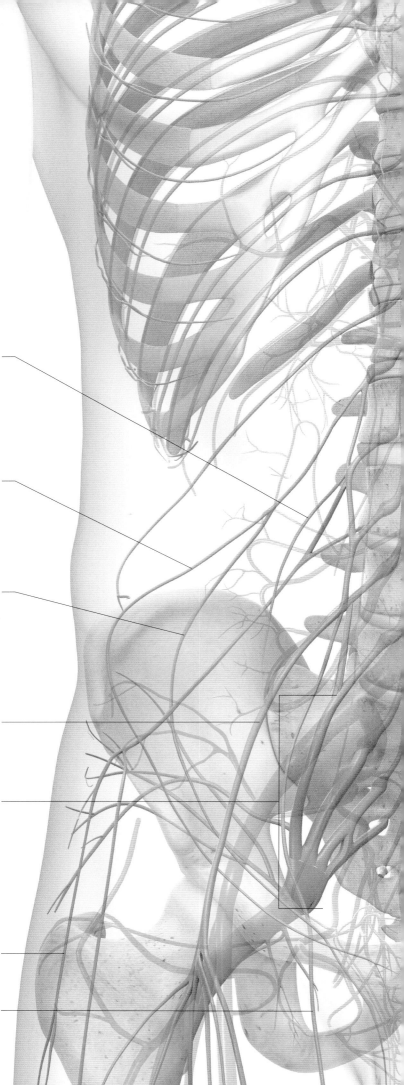

Genitofemoral nerve
Splits into two branches: the genital branch supplies some of the scrotum or labium majus, while the femoral branch supplies a small patch of skin at the top of the thigh

Iliohypogastric nerve
Runs around the side of the lower abdomen to supply the lowest parts of the muscles and skin of the abdominal wall

Ilioinguinal nerve
Travels through the layers of the abdominal wall, then down to supply sensation in the front of the scrotum in the male, or the labia majora in the female

Femoral nerve
Supplies the front of the thigh

Sacral plexus
Nerve roots from the fourth and fifth lumbar nerves join the upper four sacral nerves to form this network. Pelvic splanchnic nerves come from the second to fourth sacral nerve roots, and convey parasympathetic nerve fibres to the pelvic organs, via the pelvic plexus on each side

Lateral cutaneous nerve of the thigh
Supplies the skin of the side of the thigh

Obturator nerve
Travels along the inside of the pelvis, then emerges through the obturator foramen to supply the inner thigh

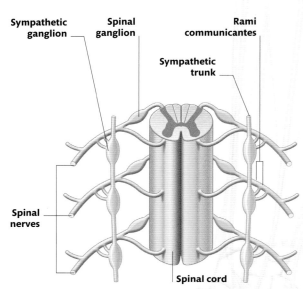

Sympathetic ganglion

Spinal ganglion

Rami communicantes

Sympathetic trunk

Spinal nerves

Spinal cord

Section of sympathetic trunk and spinal cord
Branches from the sympathetic trunk innervate the organs of the abdomen and pelvis

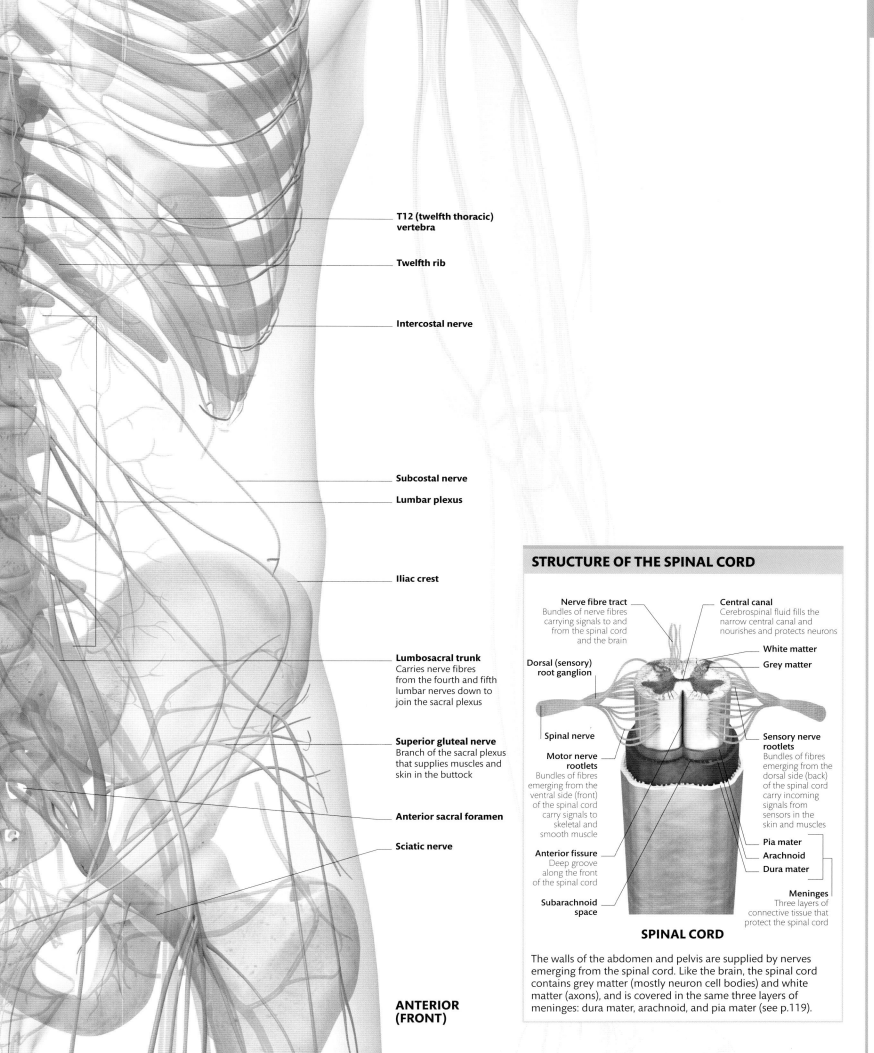

T12 (twelfth thoracic) vertebra

Twelfth rib

Intercostal nerve

Subcostal nerve

Lumbar plexus

Iliac crest

Lumbosacral trunk
Carries nerve fibres from the fourth and fifth lumbar nerves down to join the sacral plexus

Superior gluteal nerve
Branch of the sacral plexus that supplies muscles and skin in the buttock

Anterior sacral foramen

Sciatic nerve

ANTERIOR (FRONT)

STRUCTURE OF THE SPINAL CORD

Nerve fibre tract
Bundles of nerve fibres carrying signals to and from the spinal cord and the brain

Central canal
Cerebrospinal fluid fills the narrow central canal and nourishes and protects neurons

White matter

Grey matter

Dorsal (sensory) root ganglion

Spinal nerve

Motor nerve rootlets
Bundles of fibres emerging from the ventral side (front) of the spinal cord carry signals to skeletal and smooth muscle

Sensory nerve rootlets
Bundles of fibres emerging from the dorsal side (back) of the spinal cord carry incoming signals from sensors in the skin and muscles

Pia mater

Arachnoid

Dura mater

Anterior fissure
Deep groove along the front of the spinal cord

Subarachnoid space

Meninges
Three layers of connective tissue that protect the spinal cord

SPINAL CORD

The walls of the abdomen and pelvis are supplied by nerves emerging from the spinal cord. Like the brain, the spinal cord contains grey matter (mostly neuron cell bodies) and white matter (axons), and is covered in the same three layers of meninges: dura mater, arachnoid, and pia mater (see p.119).

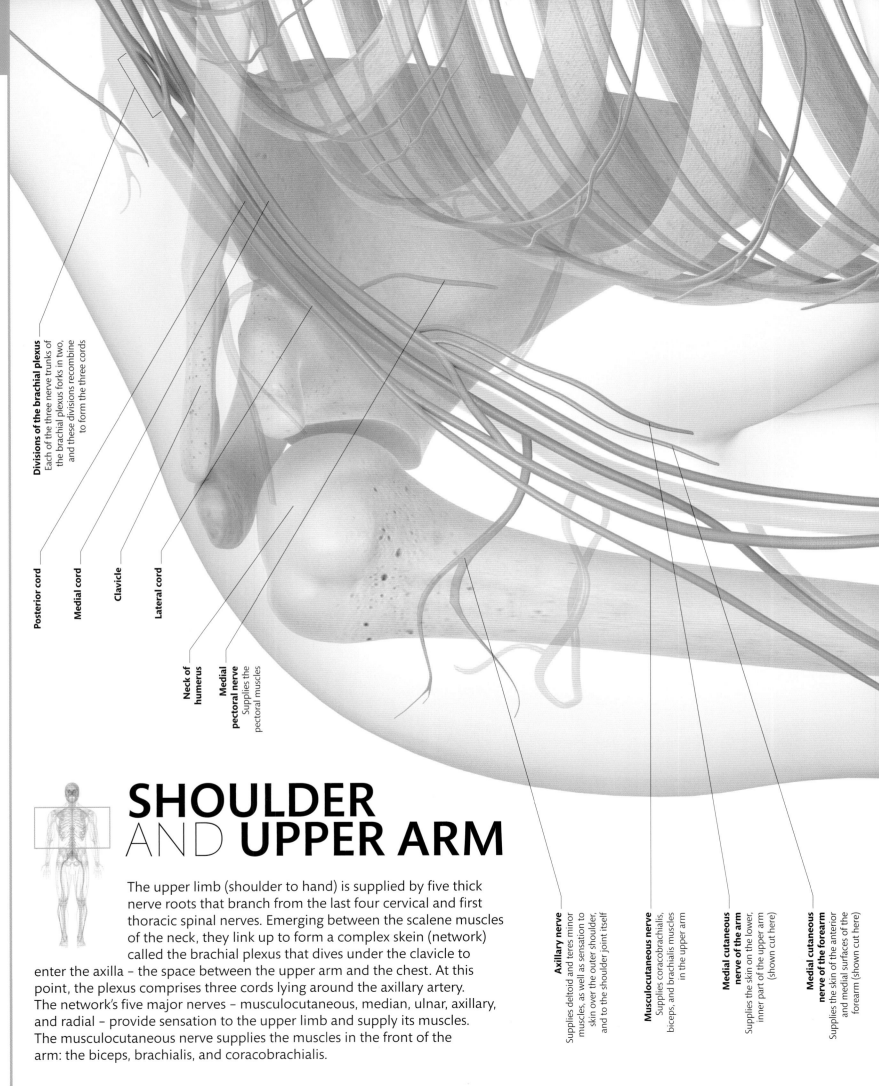

Divisions of the brachial plexus
Each of the three nerve trunks of the brachial plexus forks in two, and these divisions recombine to form the three cords

Posterior cord

Medial cord

Clavicle

Lateral cord

Neck of humerus

Medial pectoral nerve
Supplies the pectoral muscles

SHOULDER
AND **UPPER ARM**

The upper limb (shoulder to hand) is supplied by five thick nerve roots that branch from the last four cervical and first thoracic spinal nerves. Emerging between the scalene muscles of the neck, they link up to form a complex skein (network) called the brachial plexus that dives under the clavicle to enter the axilla – the space between the upper arm and the chest. At this point, the plexus comprises three cords lying around the axillary artery. The network's five major nerves – musculocutaneous, median, ulnar, axillary, and radial – provide sensation to the upper limb and supply its muscles. The musculocutaneous nerve supplies the muscles in the front of the arm: the biceps, brachialis, and coracobrachialis.

Axillary nerve
Supplies deltoid and teres minor muscles, as well as sensation to skin over the outer shoulder, and to the shoulder joint itself

Musculocutaneous nerve
Supplies coracobrachialis, biceps, and brachialis muscles in the upper arm

Medial cutaneous nerve of the arm
Supplies the skin on the lower, inner part of the upper arm (shown cut here)

Medial cutaneous nerve of the forearm
Supplies the skin of the anterior and medial surfaces of the forearm (shown cut here)

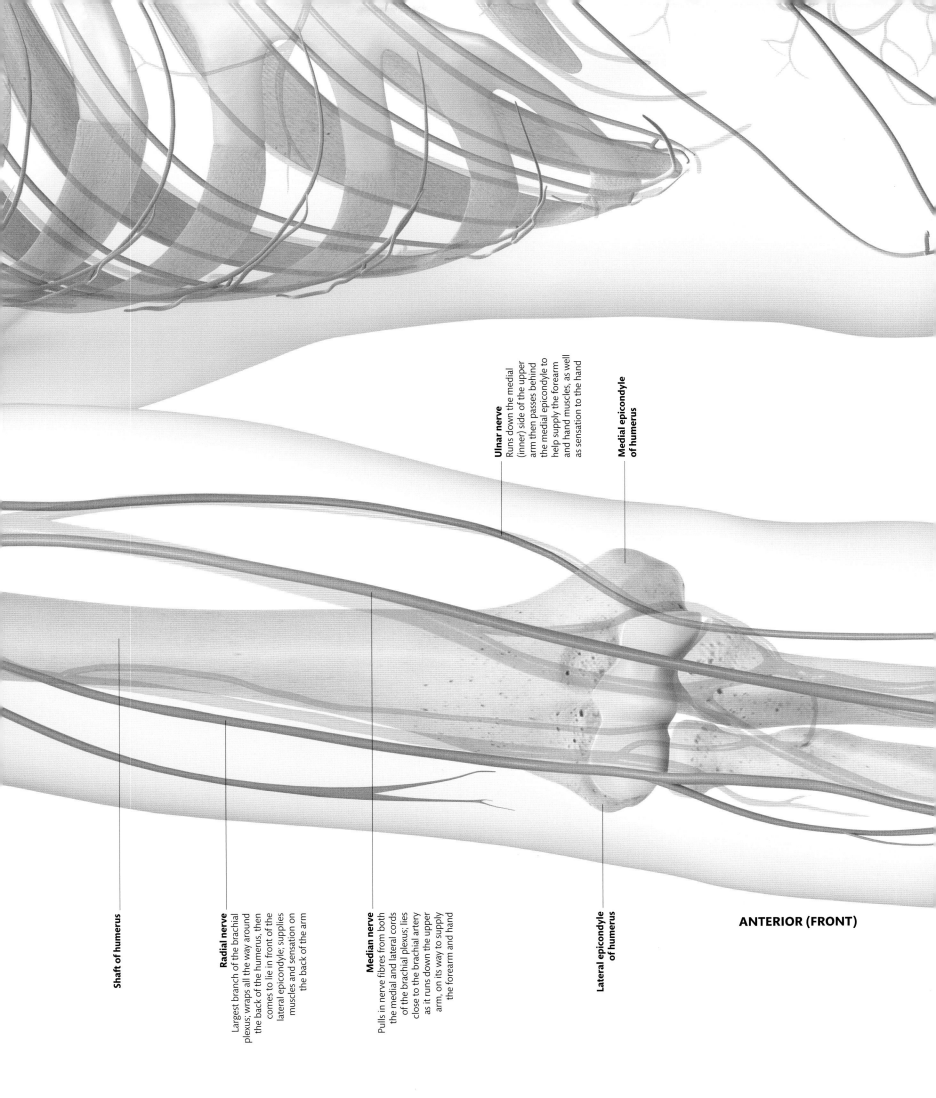

Shaft of humerus

Radial nerve
Largest branch of the brachial plexus; wraps all the way around the back of the humerus, then comes to lie in front of the lateral epicondyle; supplies muscles and sensation on the back of the arm

Median nerve
Pulls in nerve fibres from both the medial and lateral cords of the brachial plexus; lies close to the brachial artery as it runs down the upper arm, on its way to supply the forearm and hand

Lateral epicondyle of humerus

Ulnar nerve
Runs down the medial (inner) side of the upper arm then passes behind the medial epicondyle to help supply the forearm and hand muscles, as well as sensation to the hand

Medial epicondyle of humerus

ANTERIOR (FRONT)

SHOULDER
AND UPPER ARM

The axillary and radial nerves emerge from the back of the brachial plexus and run behind the humerus. The axillary nerve wraps around the neck of the humerus, just underneath the shoulder joint, and supplies the deltoid muscle. The radial nerve – the largest branch of the brachial plexus – supplies all the extensor muscles in the upper arm and in the forearm. It spirals around the back of the humerus, lying right against the bone, and sends branches to supply the heads of the triceps. The radial nerve then continues in its spiral, running forwards to lie just in front of the medial epicondyle of the humerus at the elbow.

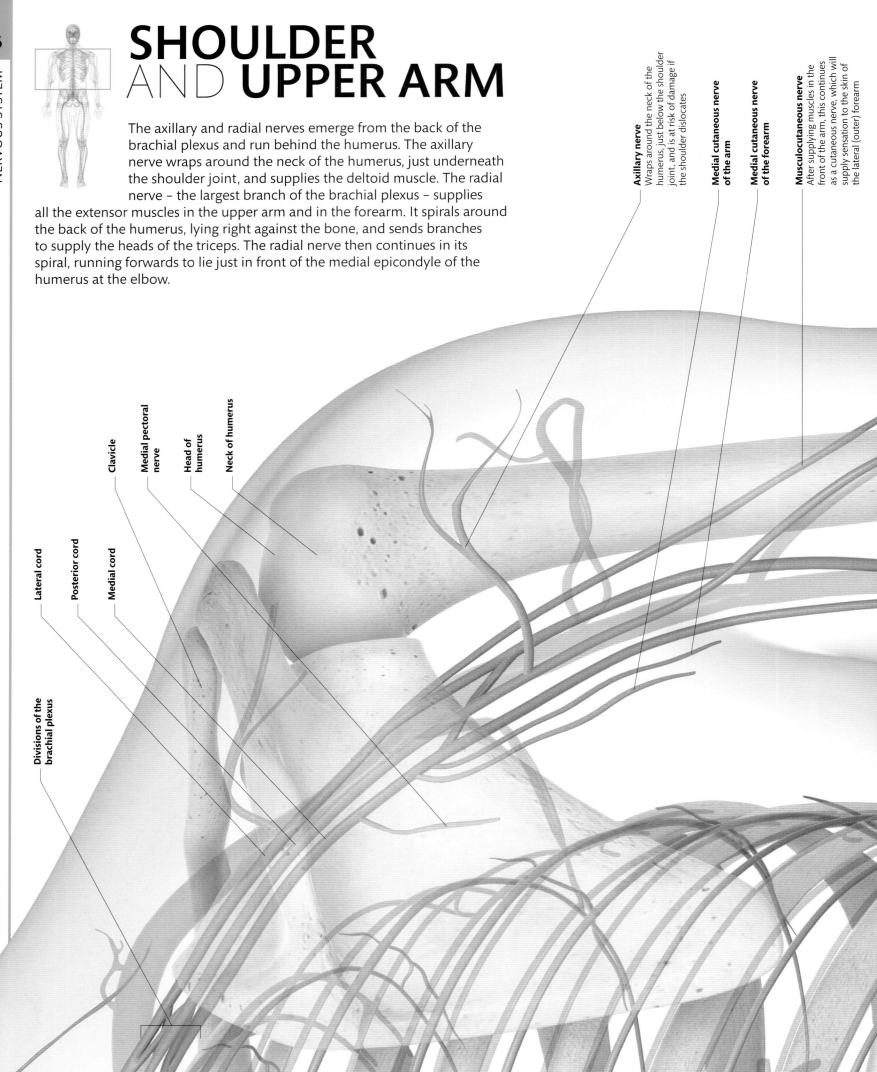

Axillary nerve
Wraps around the neck of the humerus, just below the shoulder joint, and is at risk of damage if the shoulder dislocates

Medial cutaneous nerve of the arm

Medial cutaneous nerve of the forearm

Musculocutaneous nerve
After supplying muscles in the front of the arm, this continues as a cutaneous nerve, which will supply sensation to the skin of the lateral (outer) forearm

Neck of humerus

Head of humerus

Medial pectoral nerve

Clavicle

Medial cord

Posterior cord

Lateral cord

Divisions of the brachial plexus

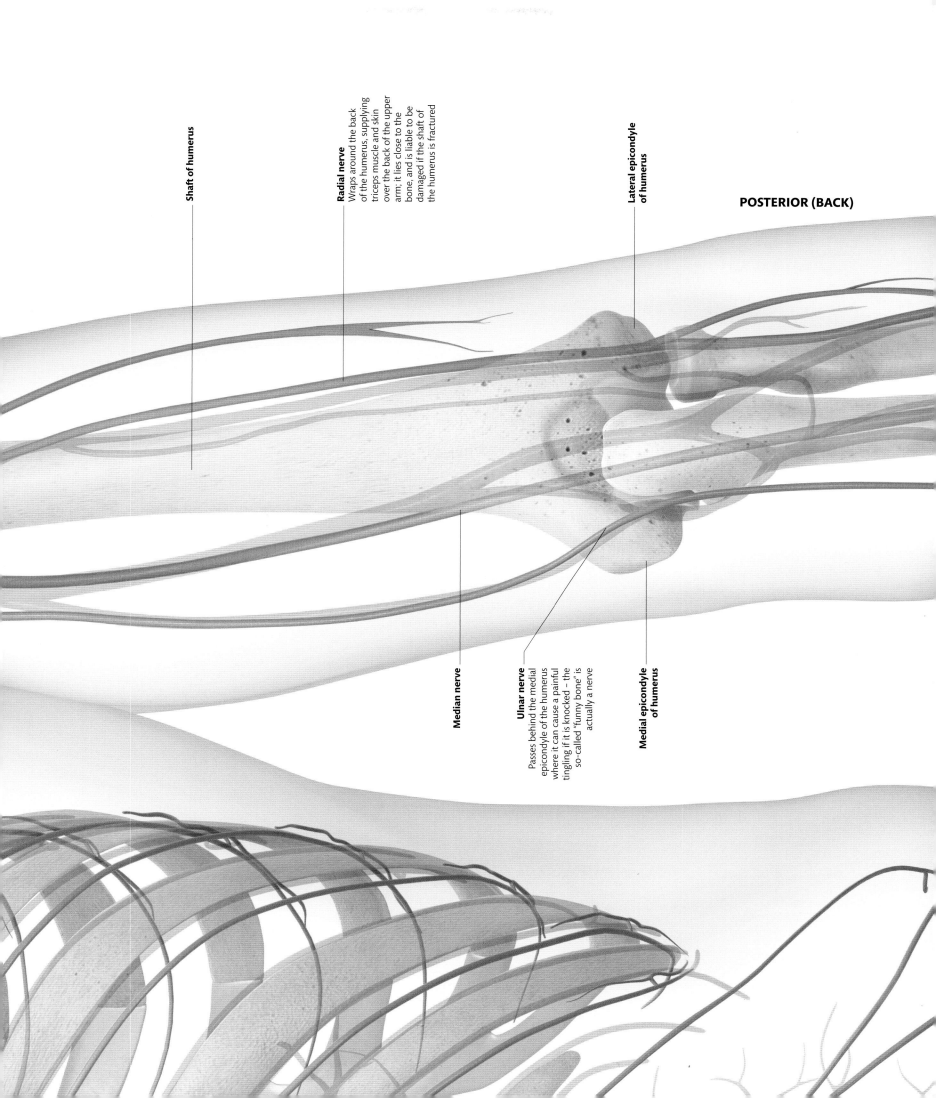

Shaft of humerus

Radial nerve
Wraps around the back of the humerus, supplying triceps muscle and skin over the back of the upper arm; it lies close to the bone, and is liable to be damaged if the shaft of the humerus is fractured

Lateral epicondyle of humerus

POSTERIOR (BACK)

Median nerve

Ulnar nerve
Passes behind the medial epicondyle of the humerus where it can cause a painful tingling if it is knocked – the so-called "funny bone" is actually a nerve

Medial epicondyle of humerus

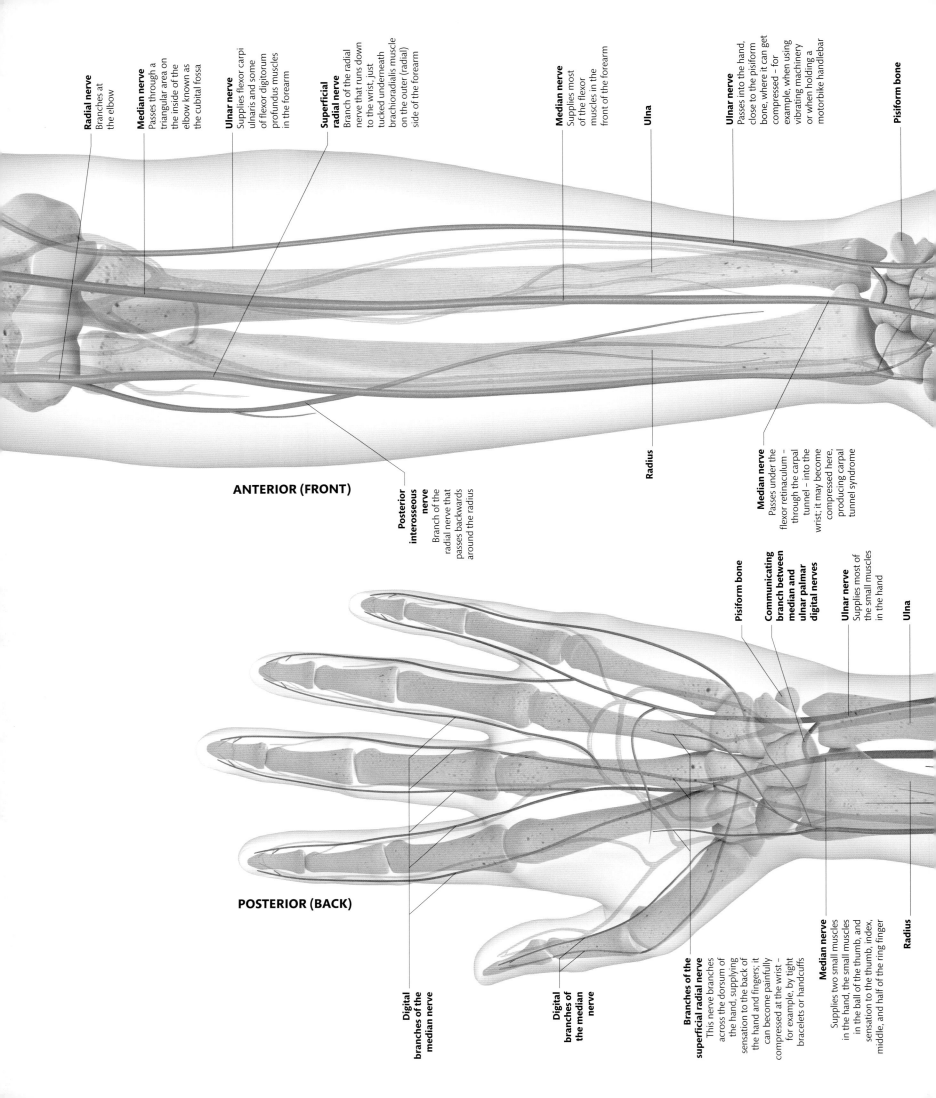

Radial nerve
Branches at the elbow

Median nerve
Passes through a triangular area on the inside of the elbow known as the cubital fossa

Ulnar nerve
Supplies flexor carpi ulnaris and some of flexor digitorum profundus muscles in the forearm

Superficial radial nerve
Branch of the radial nerve that runs down to the wrist, just tucked underneath brachioradialis muscle on the outer (radial) side of the forearm

Median nerve
Supplies most of the flexor muscles in the front of the forearm

Ulna

Ulnar nerve
Passes into the hand, close to the pisiform bone, where it can get compressed – for example, when using vibrating machinery or when holding a motorbike handlebar

Pisiform bone

ANTERIOR (FRONT)

Posterior interosseous nerve
Branch of the radial nerve that passes backwards around the radius

Radius

Median nerve
Passes under the flexor retinaculum – through the carpal tunnel – into the wrist; it may become compressed here, producing carpal tunnel syndrome

POSTERIOR (BACK)

Pisiform bone

Communicating branch between median and ulnar palmar digital nerves

Ulnar nerve
Supplies most of the small muscles in the hand

Ulna

Digital branches of the median nerve

Digital branches of the median nerve

Branches of the superficial radial nerve
This nerve branches across the dorsum of the hand, supplying sensation to the back of the hand and fingers; it can become painfully compressed at the wrist – for example, by tight bracelets or handcuffs

Median nerve
Supplies two small muscles in the hand, the small muscles in the ball of the thumb, and sensation to the thumb, index, middle, and half of the ring finger

Radius

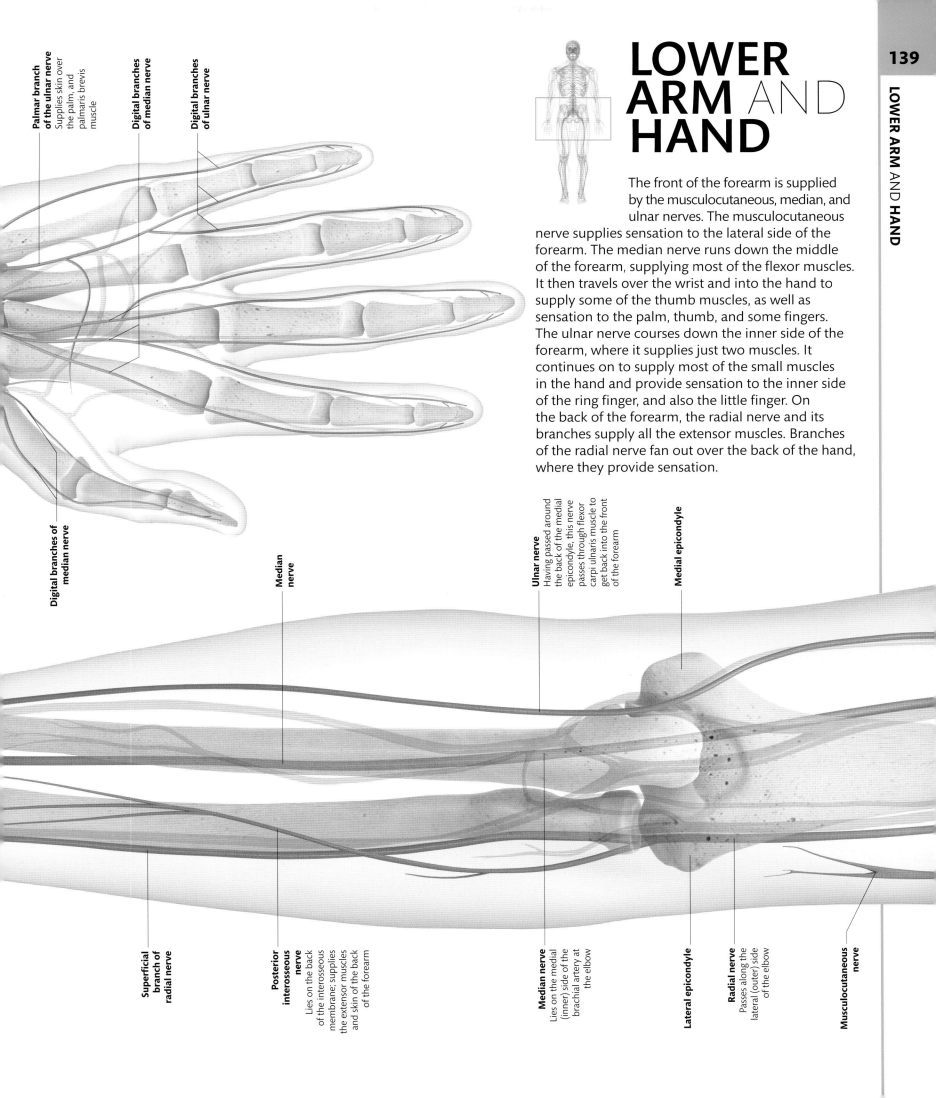

Palmar branch of the ulnar nerve
Supplies skin over the palm, and palmaris brevis muscle

Digital branches of median nerve

Digital branches of ulnar nerve

Digital branches of median nerve

LOWER ARM AND HAND

The front of the forearm is supplied by the musculocutaneous, median, and ulnar nerves. The musculocutaneous nerve supplies sensation to the lateral side of the forearm. The median nerve runs down the middle of the forearm, supplying most of the flexor muscles. It then travels over the wrist and into the hand to supply some of the thumb muscles, as well as sensation to the palm, thumb, and some fingers. The ulnar nerve courses down the inner side of the forearm, where it supplies just two muscles. It continues on to supply most of the small muscles in the hand and provide sensation to the inner side of the ring finger, and also the little finger. On the back of the forearm, the radial nerve and its branches supply all the extensor muscles. Branches of the radial nerve fan out over the back of the hand, where they provide sensation.

Median nerve

Ulnar nerve
Having passed around the back of the medial epicondyle, this nerve passes through flexor carpi ulnaris muscle to get back into the front of the forearm

Medial epicondyle

Superficial branch of radial nerve

Posterior interosseous nerve
Lies on the back of the interosseous membrane; supplies the extensor muscles and skin of the back of the forearm

Median nerve
Lies on the medial (inner) side of the brachial artery at the elbow

Lateral epicondyle

Radial nerve
Passes along the lateral (outer) side of the elbow

Musculocutaneous nerve

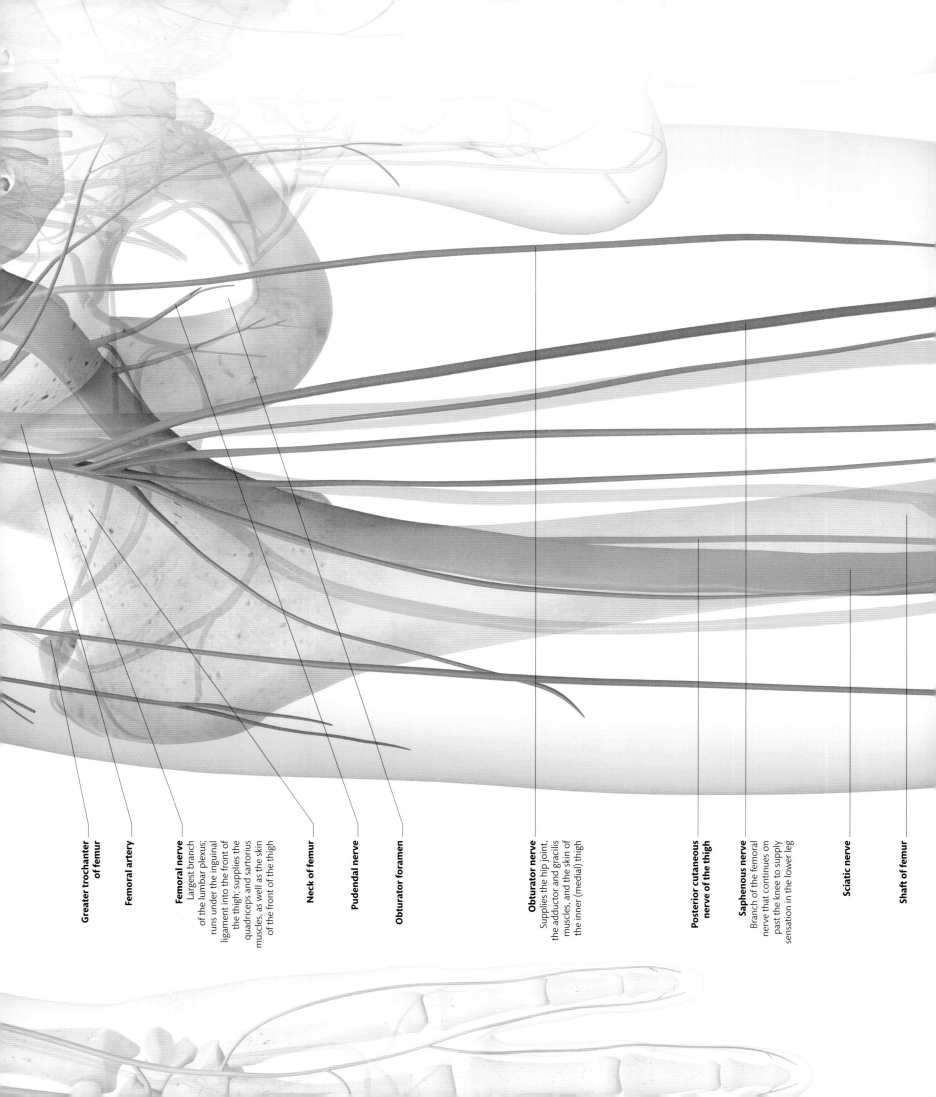

Greater trochanter of femur

Femoral artery

Femoral nerve
Largest branch of the lumbar plexus; runs under the inguinal ligament into the front of the thigh; supplies the quadriceps and sartorius muscles, as well as the skin of the front of the thigh

Neck of femur

Pudendal nerve

Obturator foramen

Obturator nerve
Supplies the hip joint, the adductor and gracilis muscles, and the skin of the inner (medial) thigh

Posterior cutaneous nerve of the thigh

Saphenous nerve
Branch of the femoral nerve that continues on past the knee to supply sensation in the lower leg

Sciatic nerve

Shaft of femur

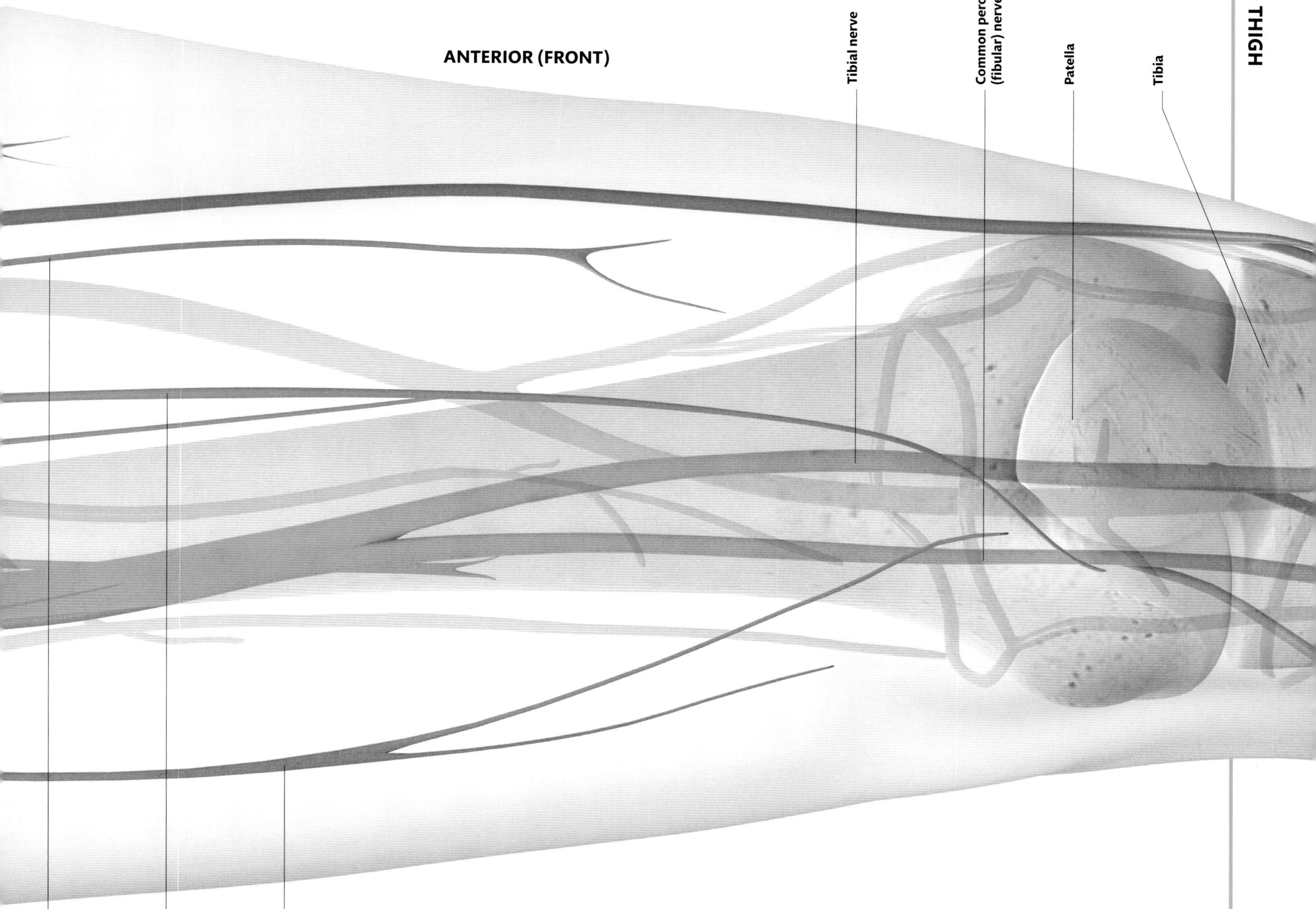

ANTERIOR (FRONT)

Tibial nerve

Common peroneal (fibular) nerve

Patella

Tibia

Medial femoral cutaneous nerve
A branch of the femoral nerve

Intermediate femoral cutaneous nerve
Also a branch of the femoral nerve

Lateral femoral cutaneous nerve
Emerges under or through the inguinal ligament, to supply the skin of the upper, outer thigh

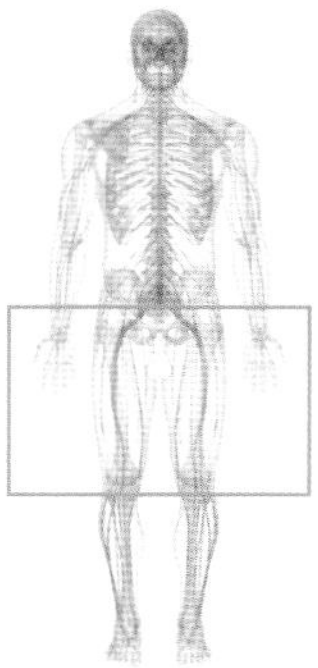

HIP AND THIGH

The lower limb (hip, thigh, leg, and foot) receives nerves from the lumbar and sacral plexuses. Three main nerves supply the thigh muscles: the femoral, obturator, and sciatic nerves (the last in the back). The femoral nerve runs over the pubic bone to supply the quadriceps and sartorius muscles in the front. The saphenous nerve, a slender branch of the femoral, continues past the knee and supplies skin on the inside of the lower leg and the inner side of the foot. The obturator nerve passes through the obturator foramen in the pelvic bone to supply the adductor muscles of the inner thigh and provide sensation to the skin there. Some smaller nerves just supply skin, such as the femoral cutaneous nerves.

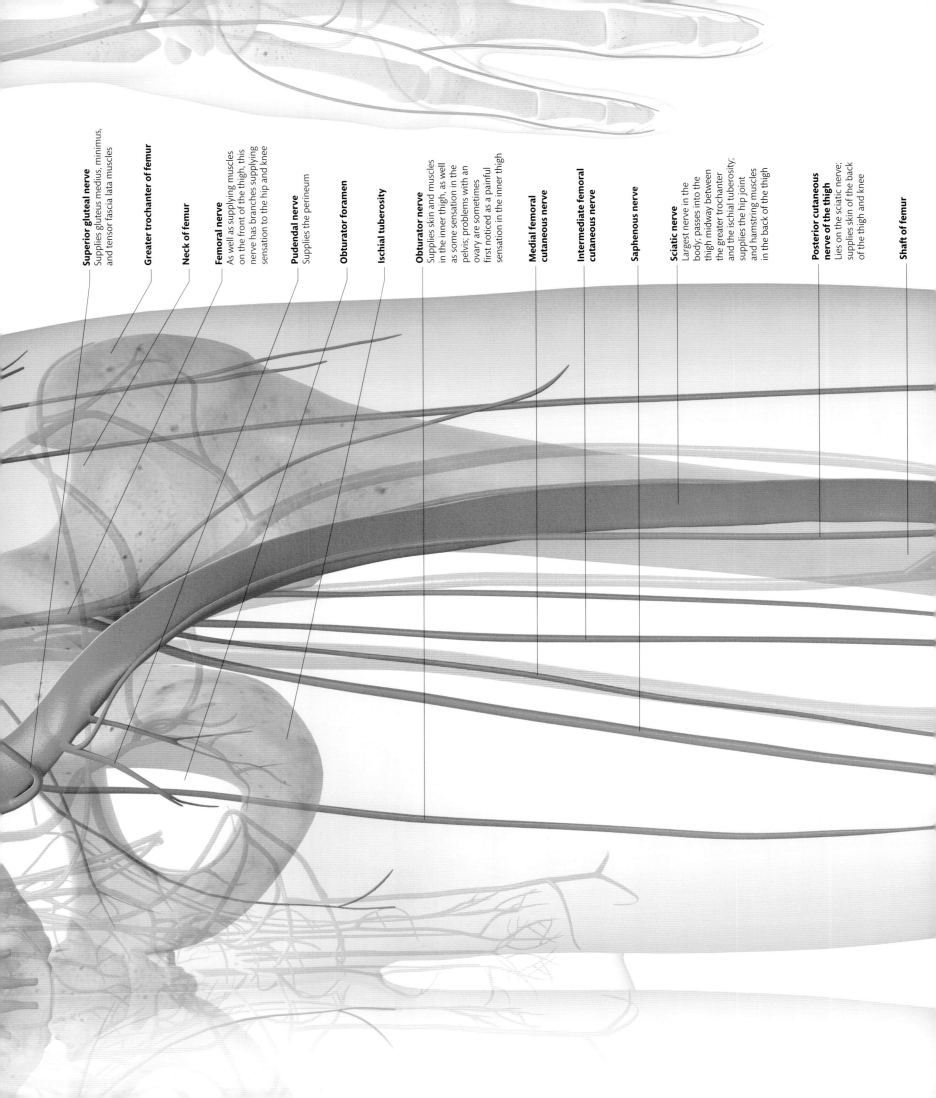

Superior gluteal nerve
Supplies gluteus medius, minimus, and tensor fascia lata muscles

Greater trochanter of femur

Neck of femur

Femoral nerve
As well as supplying muscles on the front of the thigh, this nerve has branches supplying sensation to the hip and knee

Pudendal nerve
Supplies the perineum

Obturator foramen

Ischial tuberosity

Obturator nerve
Supplies skin and muscles in the inner thigh, as well as some sensation in the pelvis; problems with an ovary are sometimes first noticed as a painful sensation in the inner thigh

Medial femoral cutaneous nerve

Intermediate femoral cutaneous nerve

Saphenous nerve

Sciatic nerve
Largest nerve in the body; passes into the thigh midway between the greater trochanter and the ischial tuberosity; supplies the hip joint and hamstring muscles in the back of the thigh

Posterior cutaneous nerve of the thigh
Lies on the sciatic nerve; supplies skin of the back of the thigh and knee

Shaft of femur

HIP AND THIGH

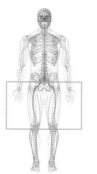

Gluteal nerves from the sacral plexus emerge via the greater sciatic foramen, at the back of the pelvis, to supply the muscles and skin of the buttock. The sciatic nerve also emerges through the greater sciatic foramen into the buttock. The gluteus maximus is a good site for injections into a muscle, but these should always be given in the upper, outer part of the buttock to ensure the needle is well away from the sciatic nerve. The sciatic nerve runs down the back of the thigh, supplying the hamstrings. In most people, the sciatic nerve runs halfway down the thigh then splits into two branches, the tibial and common peroneal nerves. These continue into the popliteal fossa (back of the knee) and on into the lower leg.

Lateral femoral cutaneous nerve
May become compressed at the inguinal ligament, causing a painful tingling in the thigh, called meralgia paraesthetica

POSTERIOR (BACK)

Tibial nerve
One of the main branches of the sciatic nerve, the tibial nerve passes straight down through the popliteal fossa at the back of the knee

Common peroneal (fibular) nerve
The other main branch from the sciatic nerve, the common peroneal nerve diverges from the tibial nerve, lying on the lateral (outer) side of the popliteal fossa

Popliteal surface of the femur

Tibia

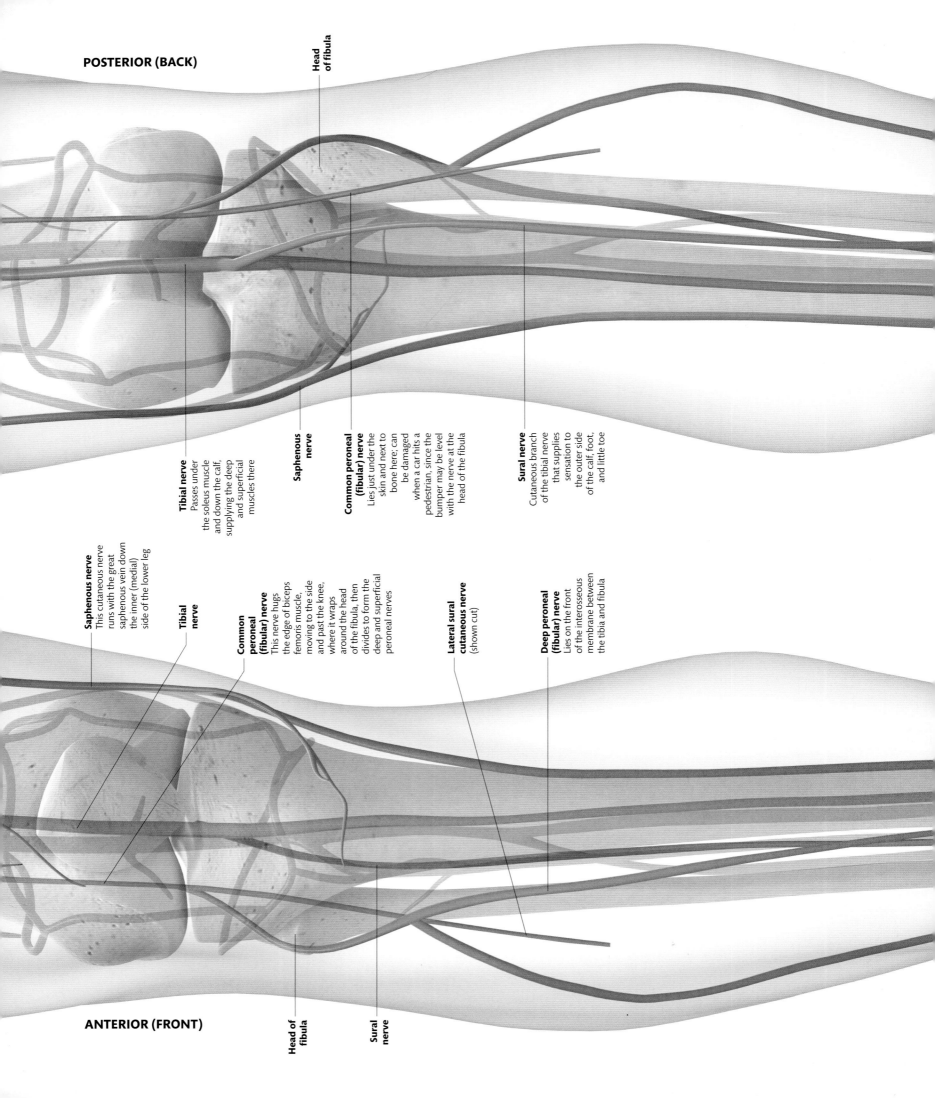

POSTERIOR (BACK)

Head of fibula

Tibial nerve
Passes under the soleus muscle and down the calf, supplying the deep and superficial muscles there

Saphenous nerve

Common peroneal (fibular) nerve
Lies just under the skin and next to bone here; can be damaged when a car hits a pedestrian, since the bumper may be level with the nerve at the head of the fibula

Sural nerve
Cutaneous branch of the tibial nerve that supplies sensation to the outer side of the calf, foot, and little toe

Saphenous nerve
This cutaneous nerve runs with the great saphenous vein down the inner (medial) side of the lower leg

Tibial nerve

Common peroneal (fibular) nerve
This nerve hugs the edge of biceps femoris muscle, moving to the side and past the knee, where it wraps around the head of the fibula, then divides to form the deep and superficial peroneal nerves

Lateral sural cutaneous nerve
(shown cut)

Deep peroneal (fibular) nerve
Lies on the front of the interosseous membrane between the tibia and fibula

ANTERIOR (FRONT)

Head of fibula

Sural nerve

LOWER LEG AND FOOT

The common peroneal nerve runs past the knee and wraps around the neck of the fibula. Then it splits into the deep and superficial peroneal nerves. The deep peroneal nerve supplies the extensor muscles of the shin, then fans out to provide sensation to the skin at the back of the foot. The superficial peroneal nerve stays on the side of the leg and supplies the peroneal muscles. The tibial nerve runs through the popliteal fossa (back of the knee), under the soleus muscle, and between the deep and superficial calf muscles, which it supplies. It continues behind the medial malleolus and under the foot, then splits into two plantar nerves that supply the small muscles of the foot and the skin of the sole.

Superficial peroneal (fibular) nerve

Deep peroneal (fibular) nerve
Supplies the extensor muscles in the front of the leg, as well as the ankle joint

Saphenous nerve

Tibial nerve
May become trapped under the retinaculum, which holds the flexor tendons close to the front of the ankle, causing the rare tarsal tunnel syndrome

Deep peroneal (fibular) nerve

Sural nerve

Dorsal digital nerves

Medial plantar nerve
One of the terminal branches of the tibial nerve, supplying the sole and toes

Calcaneal branch of tibial nerve
Supplies the heel and inner (medial) sole

Medial malleolus

Lateral plantar nerve
With the medial plantar nerve, supplies the muscles and skin of the sole and toes

Superficial peroneal (fibular) nerve
Supplies the peroneus longus and brevis muscles in the lower leg

Tibial nerve
Runs behind the medial malleolus

Saphenous nerve
Runs in front of the medial malleolus, to supply sensation to the inner (medial) side of the foot

Lateral branch of superficial peroneal nerve
With the medial branch, supplies skin over the top of the foot and toes

Medial branch of superficial peroneal nerve

Deep peroneal (fibular) nerve
Runs with the dorsal artery of the foot, and supplies the skin of the first web-space

Dorsal digital nerves
Branches of the superficial peroneal nerve

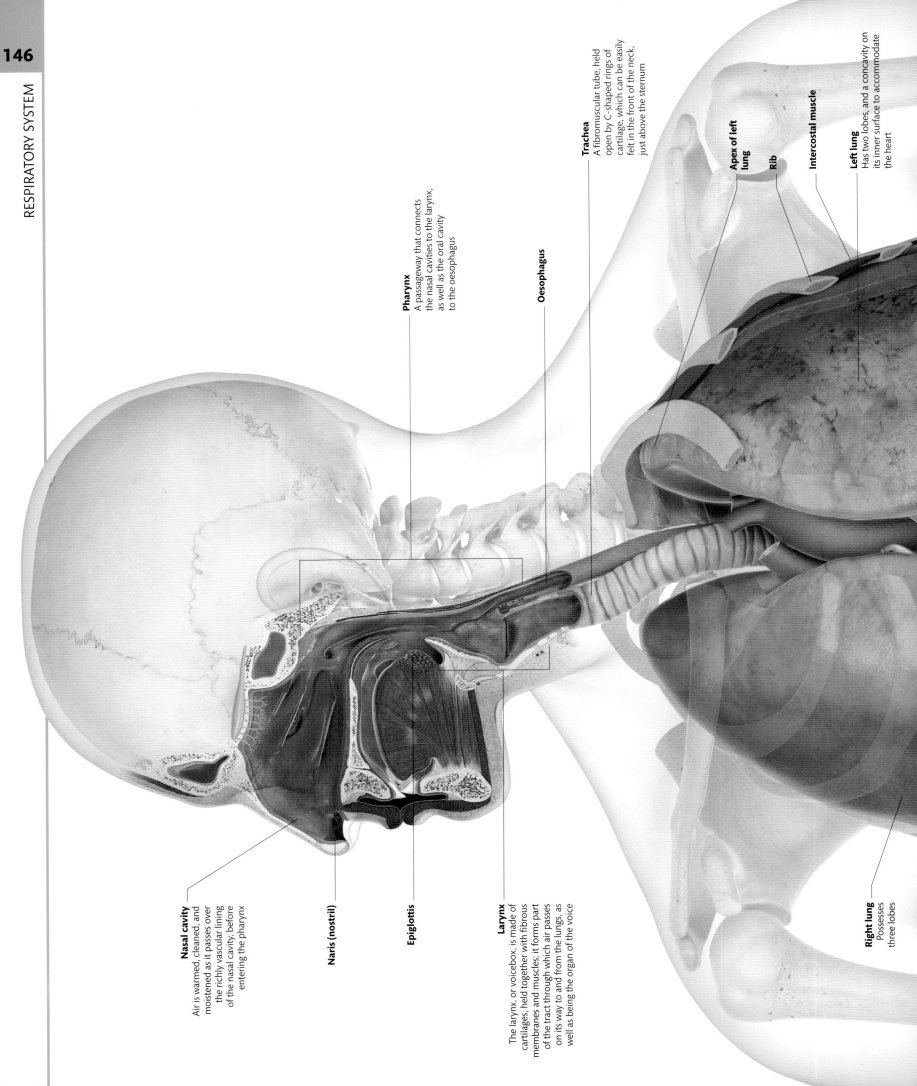

Nasal cavity
Air is warmed, cleaned, and moistened as it passes over the richly vascular lining of the nasal cavity, before entering the pharynx

Naris (nostril)

Epiglottis

Larynx
The larynx, or voicebox, is made of cartilages, held together with fibrous membranes and muscles; it forms part of the tract through which air passes on its way to and from the lungs, as well as being the organ of the voice

Right lung
Possesses three lobes

Pharynx
A passageway that connects the nasal cavities to the larynx, as well as the oral cavity to the oesophagus

Oesophagus

Trachea
A fibromuscular tube, held open by C-shaped rings of cartilage, which can be easily felt in the front of the neck, just above the sternum

Apex of left lung

Rib

Intercostal muscle

Left lung
Has two lobes, and a concavity on its inner surface to accommodate the heart

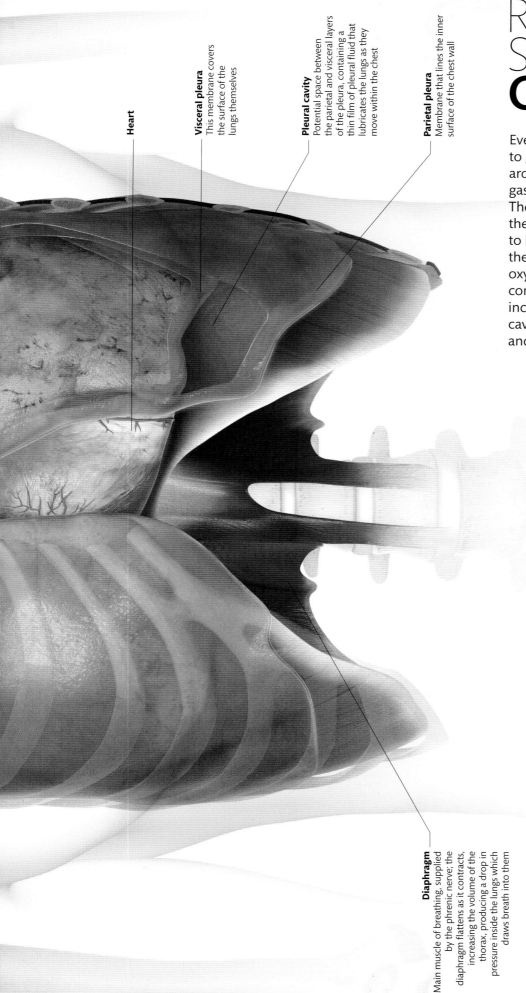

RESPIRATORY SYSTEM
OVERVIEW

Every cell in the human body needs to get oxygen, and to get rid of carbon dioxide. These gases are transported around the body in the blood, but the actual transfer of gases between the air and the blood occurs in the lungs. The lungs have extremely thin membranes that allow the gases to pass across easily. However, air also needs to be regularly drawn in and out of the lungs, to expel the building carbon dioxide and to bring in fresh oxygen, and this is brought about by respiration – commonly called breathing. The respiratory system includes the airways on the way to the lungs: the nasal cavities, parts of the pharynx, the larynx, the trachea, and the bronchi (see p.149).

Heart

Visceral pleura
This membrane covers the surface of the lungs themselves

Pleural cavity
Potential space between the parietal and visceral layers of the pleura, containing a thin film of pleural fluid that lubricates the lungs as they move within the chest

Parietal pleura
Membrane that lines the inner surface of the chest wall

Diaphragm
Main muscle of breathing, supplied by the phrenic nerve; the diaphragm flattens as it contracts, increasing the volume of the thorax, producing a drop in pressure inside the lungs which draws breath into them

ANTERIOR (FRONT)

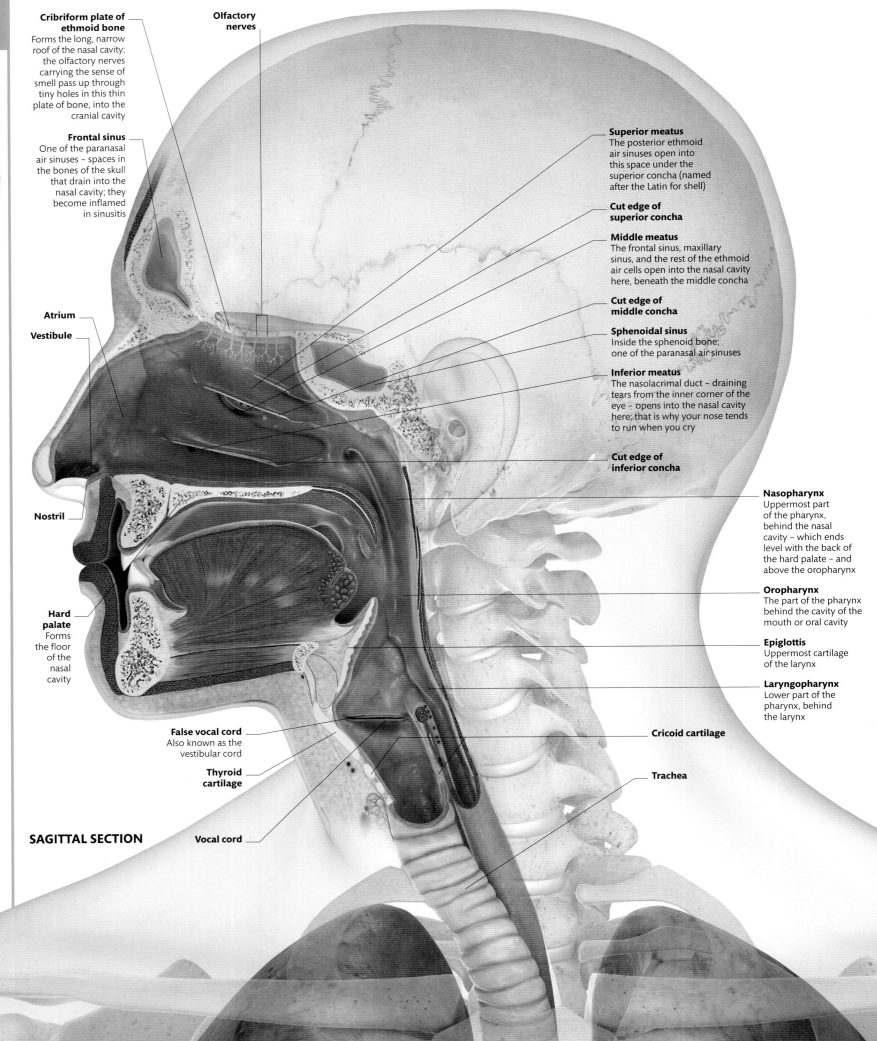

Cribriform plate of ethmoid bone
Forms the long, narrow roof of the nasal cavity; the olfactory nerves carrying the sense of smell pass up through tiny holes in this thin plate of bone, into the cranial cavity

Olfactory nerves

Frontal sinus
One of the paranasal air sinuses – spaces in the bones of the skull that drain into the nasal cavity; they become inflamed in sinusitis

Atrium

Vestibule

Nostril

Hard palate
Forms the floor of the nasal cavity

SAGITTAL SECTION

False vocal cord
Also known as the vestibular cord

Thyroid cartilage

Vocal cord

Superior meatus
The posterior ethmoid air sinuses open into this space under the superior concha (named after the Latin for shell)

Cut edge of superior concha

Middle meatus
The frontal sinus, maxillary sinus, and the rest of the ethmoid air cells open into the nasal cavity here, beneath the middle concha

Cut edge of middle concha

Sphenoidal sinus
Inside the sphenoid bone; one of the paranasal air sinuses

Inferior meatus
The nasolacrimal duct – draining tears from the inner corner of the eye – opens into the nasal cavity here; that is why your nose tends to run when you cry

Cut edge of inferior concha

Nasopharynx
Uppermost part of the pharynx, behind the nasal cavity – which ends level with the back of the hard palate – and above the oropharynx

Oropharynx
The part of the pharynx behind the cavity of the mouth or oral cavity

Epiglottis
Uppermost cartilage of the larynx

Laryngopharynx
Lower part of the pharynx, behind the larynx

Cricoid cartilage

Trachea

HEAD AND NECK

When we take a breath, air is pulled in through our nostrils, into the nasal cavities. Here the air is cleaned, warmed, and moistened before its onward journey. The nasal cavities are divided by the thin partition of the nasal septum, which is composed of plates of cartilage and bone. The lateral walls of the nasal cavity are more elaborate, with bony curls (conchae) that increase the surface area over which the air flows. The nasal cavity is lined with mucosa, which produces mucus. This often undervalued substance does an important job of trapping particles and moistening the air. The nasal sinuses, also lined with mucosa, open via tiny orifices into the nasal cavity. Below and in front of the pharynx is the larynx – the organ of speech. The way that air passes through this can be modulated to produce sound.

Ethmoid sinus   Frontal sinus

X-RAY OF HEAD SHOWING SINUSES

Nasal cavity   Nasal septum   Maxillary sinus

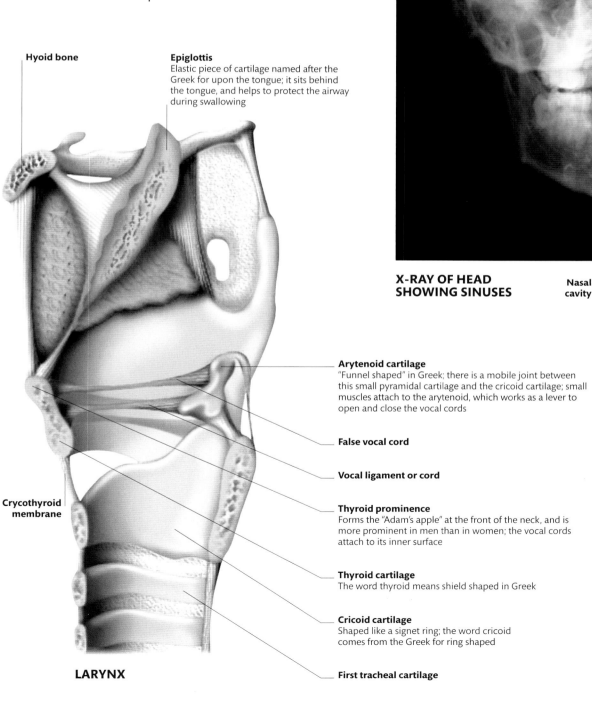

Hyoid bone

Epiglottis
Elastic piece of cartilage named after the Greek for upon the tongue; it sits behind the tongue, and helps to protect the airway during swallowing

Arytenoid cartilage
"Funnel shaped" in Greek; there is a mobile joint between this small pyramidal cartilage and the cricoid cartilage; small muscles attach to the arytenoid, which works as a lever to open and close the vocal cords

False vocal cord

Vocal ligament or cord

Thyroid prominence
Forms the "Adam's apple" at the front of the neck, and is more prominent in men than in women; the vocal cords attach to its inner surface

Crycothyroid membrane

Thyroid cartilage
The word thyroid means shield shaped in Greek

Cricoid cartilage
Shaped like a signet ring; the word cricoid comes from the Greek for ring shaped

LARYNX

First tracheal cartilage

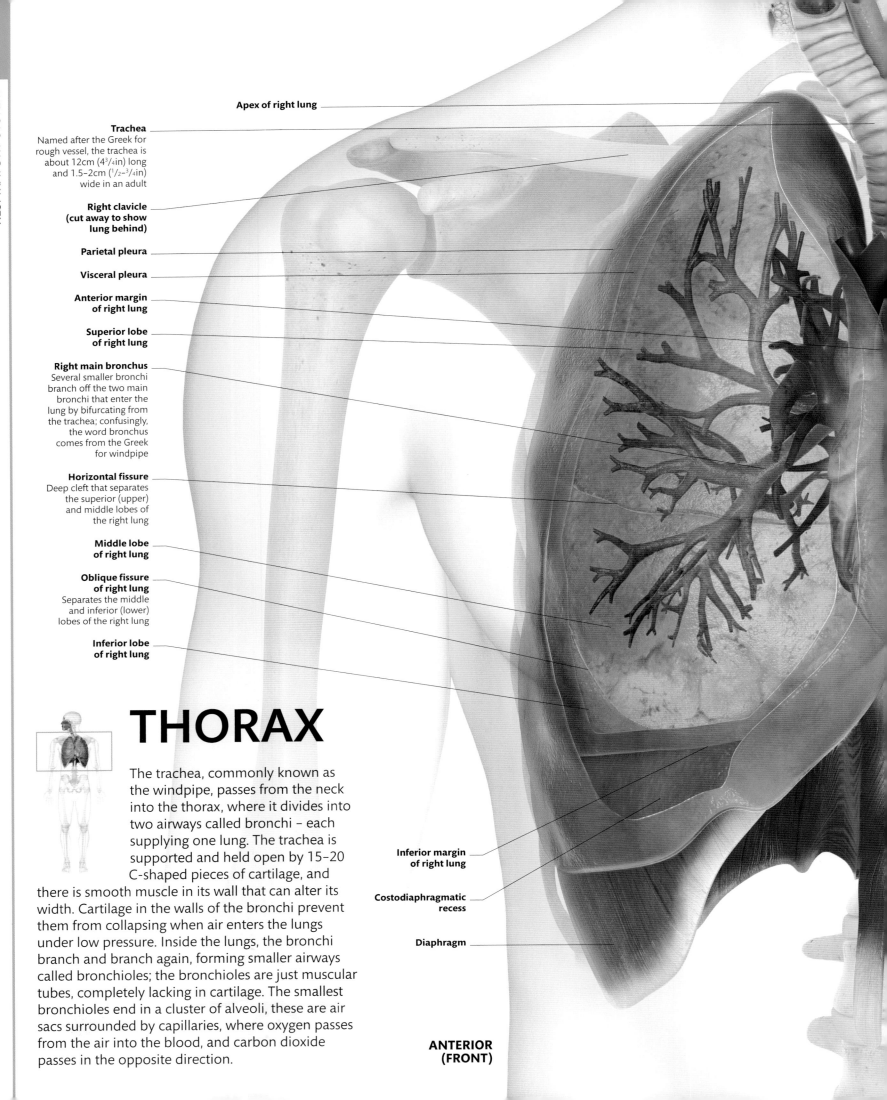

Apex of right lung

Trachea
Named after the Greek for rough vessel, the trachea is about 12cm (4³/₄in) long and 1.5–2cm (¹/₂–³/₄in) wide in an adult

Right clavicle (cut away to show lung behind)

Parietal pleura

Visceral pleura

Anterior margin of right lung

Superior lobe of right lung

Right main bronchus
Several smaller bronchi branch off the two main bronchi that enter the lung by bifurcating from the trachea; confusingly, the word bronchus comes from the Greek for windpipe

Horizontal fissure
Deep cleft that separates the superior (upper) and middle lobes of the right lung

Middle lobe of right lung

Oblique fissure of right lung
Separates the middle and inferior (lower) lobes of the right lung

Inferior lobe of right lung

Inferior margin of right lung

Costodiaphragmatic recess

Diaphragm

THORAX

The trachea, commonly known as the windpipe, passes from the neck into the thorax, where it divides into two airways called bronchi – each supplying one lung. The trachea is supported and held open by 15–20 C-shaped pieces of cartilage, and there is smooth muscle in its wall that can alter its width. Cartilage in the walls of the bronchi prevent them from collapsing when air enters the lungs under low pressure. Inside the lungs, the bronchi branch and branch again, forming smaller airways called bronchioles; the bronchioles are just muscular tubes, completely lacking in cartilage. The smallest bronchioles end in a cluster of alveoli, these are air sacs surrounded by capillaries, where oxygen passes from the air into the blood, and carbon dioxide passes in the opposite direction.

ANTERIOR (FRONT)

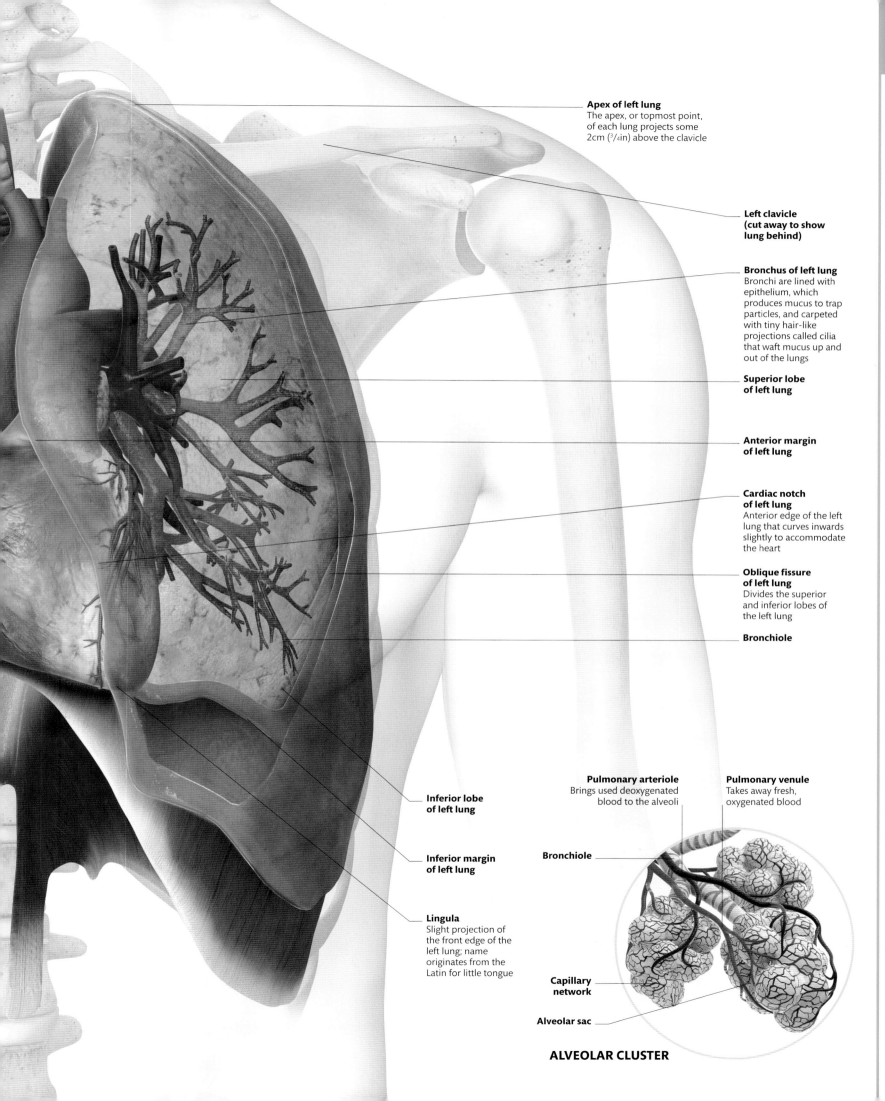

Apex of left lung
The apex, or topmost point, of each lung projects some 2cm (³/₄in) above the clavicle

Left clavicle (cut away to show lung behind)

Bronchus of left lung
Bronchi are lined with epithelium, which produces mucus to trap particles, and carpeted with tiny hair-like projections called cilia that waft mucus up and out of the lungs

Superior lobe of left lung

Anterior margin of left lung

Cardiac notch of left lung
Anterior edge of the left lung that curves inwards slightly to accommodate the heart

Oblique fissure of left lung
Divides the superior and inferior lobes of the left lung

Bronchiole

Inferior lobe of left lung

Inferior margin of left lung

Lingula
Slight projection of the front edge of the left lung; name originates from the Latin for little tongue

Pulmonary arteriole
Brings used deoxygenated blood to the alveoli

Pulmonary venule
Takes away fresh, oxygenated blood

Bronchiole

Capillary network

Alveolar sac

ALVEOLAR CLUSTER

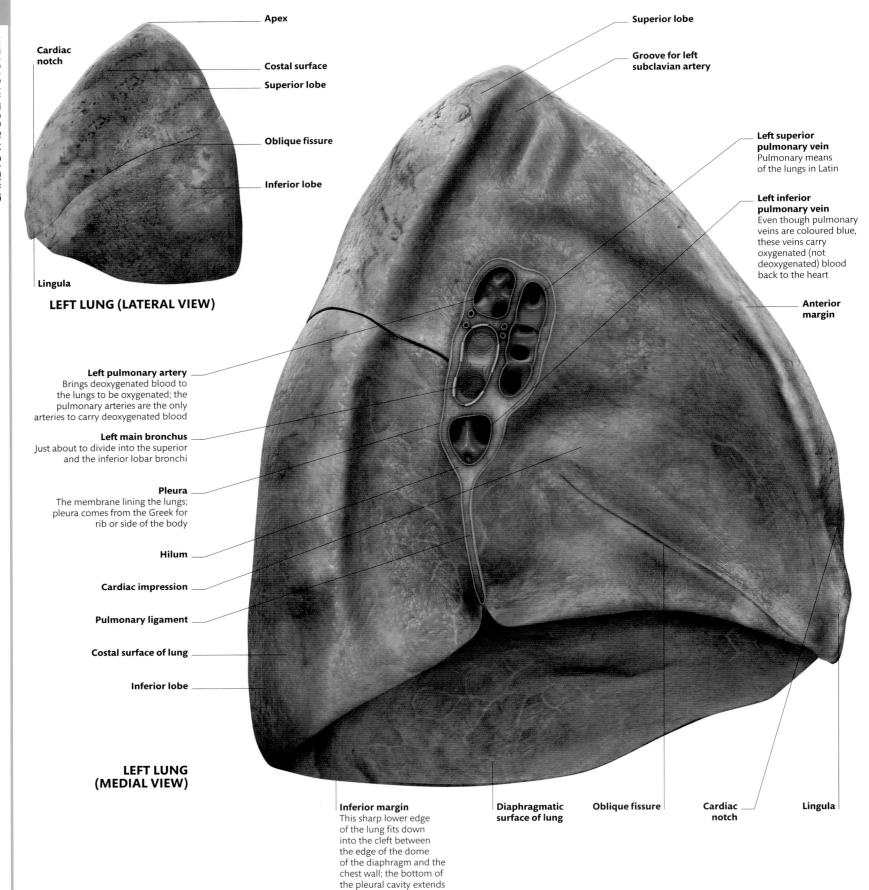

Apex

Cardiac notch

Costal surface

Superior lobe

Oblique fissure

Inferior lobe

Lingula

LEFT LUNG (LATERAL VIEW)

Superior lobe

Groove for left subclavian artery

Left superior pulmonary vein
Pulmonary means of the lungs in Latin

Left inferior pulmonary vein
Even though pulmonary veins are coloured blue, these veins carry oxygenated (not deoxygenated) blood back to the heart

Anterior margin

Left pulmonary artery
Brings deoxygenated blood to the lungs to be oxygenated; the pulmonary arteries are the only arteries to carry deoxygenated blood

Left main bronchus
Just about to divide into the superior and the inferior lobar bronchi

Pleura
The membrane lining the lungs; pleura comes from the Greek for rib or side of the body

Hilum

Cardiac impression

Pulmonary ligament

Costal surface of lung

Inferior lobe

LEFT LUNG (MEDIAL VIEW)

Inferior margin
This sharp lower edge of the lung fits down into the cleft between the edge of the dome of the diaphragm and the chest wall; the bottom of the pleural cavity extends a few more centimetres below the edge of the lung

Diaphragmatic surface of lung

Oblique fissure

Cardiac notch

Lingula

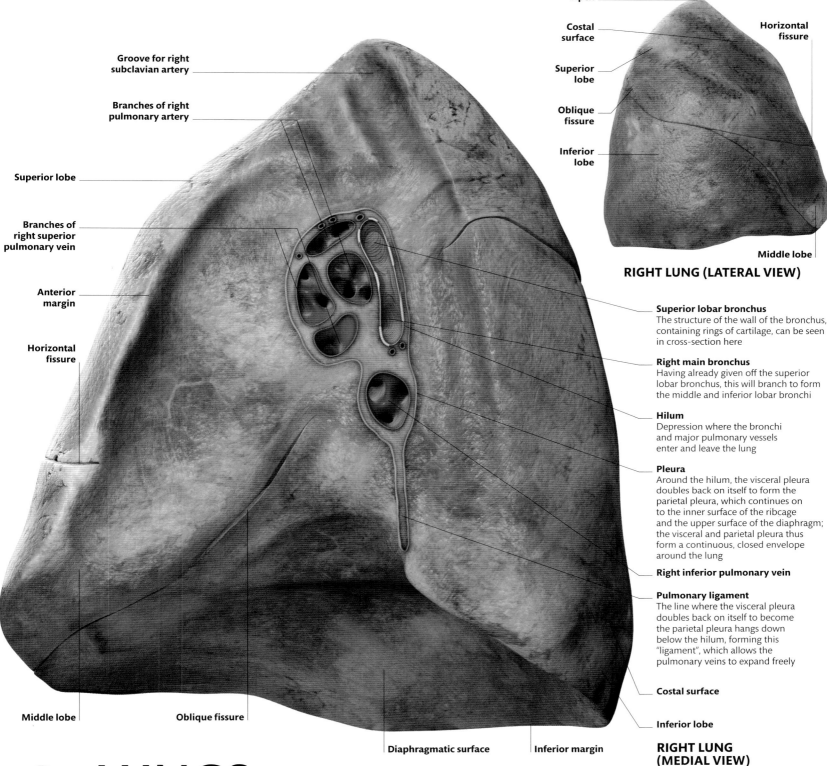

Apex

Costal surface

Horizontal fissure

Superior lobe

Oblique fissure

Inferior lobe

Middle lobe

RIGHT LUNG (LATERAL VIEW)

Groove for right subclavian artery

Branches of right pulmonary artery

Superior lobe

Branches of right superior pulmonary vein

Anterior margin

Horizontal fissure

Middle lobe

Oblique fissure

Diaphragmatic surface

Inferior margin

Superior lobar bronchus
The structure of the wall of the bronchus, containing rings of cartilage, can be seen in cross-section here

Right main bronchus
Having already given off the superior lobar bronchus, this will branch to form the middle and inferior lobar bronchi

Hilum
Depression where the bronchi and major pulmonary vessels enter and leave the lung

Pleura
Around the hilum, the visceral pleura doubles back on itself to form the parietal pleura, which continues on to the inner surface of the ribcage and the upper surface of the diaphragm; the visceral and parietal pleura thus form a continuous, closed envelope around the lung

Right inferior pulmonary vein

Pulmonary ligament
The line where the visceral pleura doubles back on itself to become the parietal pleura hangs down below the hilum, forming this "ligament", which allows the pulmonary veins to expand freely

Costal surface

Inferior lobe

RIGHT LUNG (MEDIAL VIEW)

LUNGS

Each lung fits snugly inside its half of the thoracic cavity. The surface of each lung is covered with a thin pleural membrane (visceral pleura), and the inside of the chest wall is also lined with pleura (parietal pleura). Between the two pleural layers lies a thin film of lubricating fluid that allows the lungs to slide against the chest wall during breathing movements, but it also creates a fluid seal, effectively sticking the lungs to the ribs and the diaphragm. When you breathe in, this seal means the lungs are pulled outwards in all directions, and air rushes into them. The bronchi and blood vessels enter each lung at the hilum on its inner or medial surface. Although the two lungs may appear to be similar at first glance, there is some asymmetry. The left lung is concave to fit around the heart and has only two lobes, whereas the right lung has three lobes, marked out by two deep fissures.

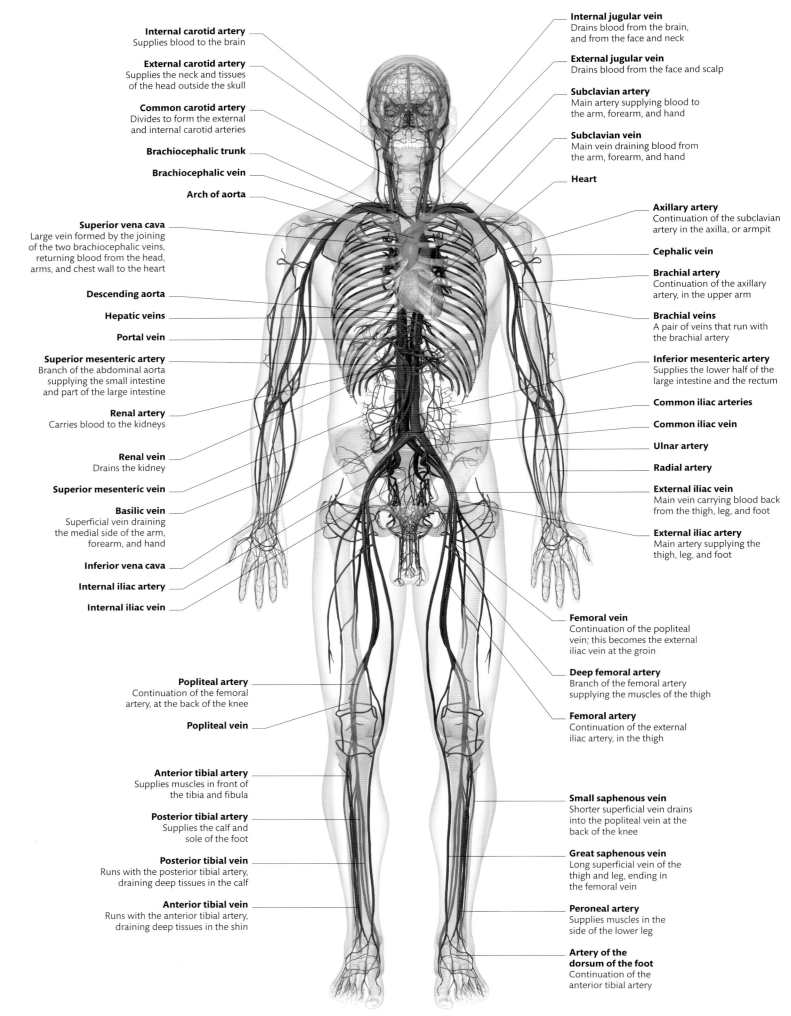

Internal carotid artery
Supplies blood to the brain

External carotid artery
Supplies the neck and tissues
of the head outside the skull

Common carotid artery
Divides to form the external
and internal carotid arteries

Brachiocephalic trunk

Brachiocephalic vein

Arch of aorta

Superior vena cava
Large vein formed by the joining
of the two brachiocephalic veins,
returning blood from the head,
arms, and chest wall to the heart

Descending aorta

Hepatic veins

Portal vein

Superior mesenteric artery
Branch of the abdominal aorta
supplying the small intestine
and part of the large intestine

Renal artery
Carries blood to the kidneys

Renal vein
Drains the kidney

Superior mesenteric vein

Basilic vein
Superficial vein draining
the medial side of the arm,
forearm, and hand

Inferior vena cava

Internal iliac artery

Internal iliac vein

Popliteal artery
Continuation of the femoral
artery, at the back of the knee

Popliteal vein

Anterior tibial artery
Supplies muscles in front of
the tibia and fibula

Posterior tibial artery
Supplies the calf and
sole of the foot

Posterior tibial vein
Runs with the posterior tibial artery,
draining deep tissues in the calf

Anterior tibial vein
Runs with the anterior tibial artery,
draining deep tissues in the shin

Internal jugular vein
Drains blood from the brain,
and from the face and neck

External jugular vein
Drains blood from the face and scalp

Subclavian artery
Main artery supplying blood to
the arm, forearm, and hand

Subclavian vein
Main vein draining blood from
the arm, forearm, and hand

Heart

Axillary artery
Continuation of the subclavian
artery in the axilla, or armpit

Cephalic vein

Brachial artery
Continuation of the axillary
artery, in the upper arm

Brachial veins
A pair of veins that run with
the brachial artery

Inferior mesenteric artery
Supplies the lower half of the
large intestine and the rectum

Common iliac arteries

Common iliac vein

Ulnar artery

Radial artery

External iliac vein
Main vein carrying blood back
from the thigh, leg, and foot

External iliac artery
Main artery supplying the
thigh, leg, and foot

Femoral vein
Continuation of the popliteal
vein; this becomes the external
iliac vein at the groin

Deep femoral artery
Branch of the femoral artery
supplying the muscles of the thigh

Femoral artery
Continuation of the external
iliac artery, in the thigh

Small saphenous vein
Shorter superficial vein drains
into the popliteal vein at the
back of the knee

Great saphenous vein
Long superficial vein of the
thigh and leg, ending in
the femoral vein

Peroneal artery
Supplies muscles in the
side of the lower leg

**Artery of the
dorsum of the foot**
Continuation of the
anterior tibial artery

ANTERIOR (FRONT)

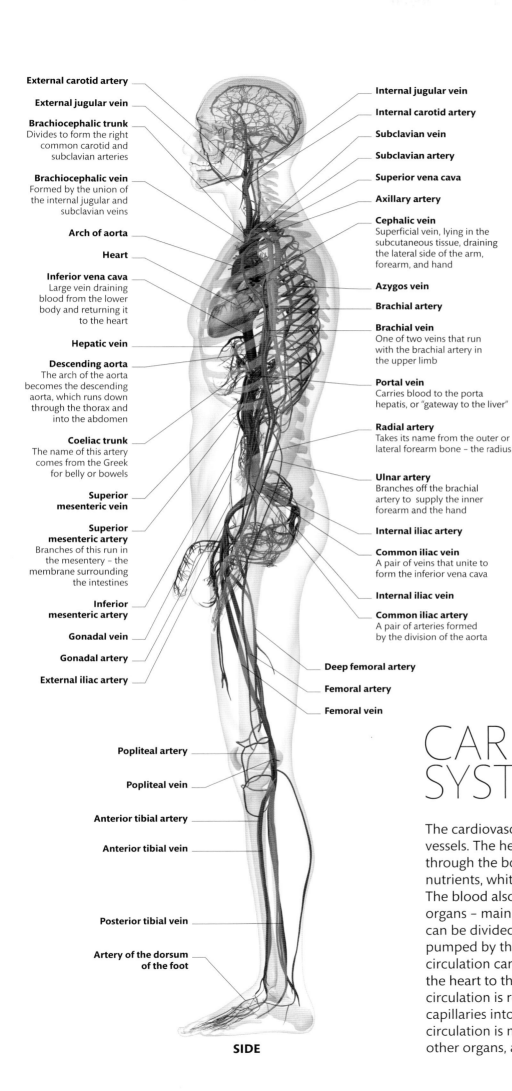

External carotid artery

External jugular vein

Brachiocephalic trunk
Divides to form the right common carotid and subclavian arteries

Brachiocephalic vein
Formed by the union of the internal jugular and subclavian veins

Arch of aorta

Heart

Inferior vena cava
Large vein draining blood from the lower body and returning it to the heart

Hepatic vein

Descending aorta
The arch of the aorta becomes the descending aorta, which runs down through the thorax and into the abdomen

Coeliac trunk
The name of this artery comes from the Greek for belly or bowels

Superior mesenteric vein

Superior mesenteric artery
Branches of this run in the mesentery – the membrane surrounding the intestines

Inferior mesenteric artery

Gonadal vein

Gonadal artery

External iliac artery

Popliteal artery

Popliteal vein

Anterior tibial artery

Anterior tibial vein

Posterior tibial vein

Artery of the dorsum of the foot

SIDE

Internal jugular vein

Internal carotid artery

Subclavian vein

Subclavian artery

Superior vena cava

Axillary artery

Cephalic vein
Superficial vein, lying in the subcutaneous tissue, draining the lateral side of the arm, forearm, and hand

Azygos vein

Brachial artery

Brachial vein
One of two veins that run with the brachial artery in the upper limb

Portal vein
Carries blood to the porta hepatis, or "gateway to the liver"

Radial artery
Takes its name from the outer or lateral forearm bone – the radius

Ulnar artery
Branches off the brachial artery to supply the inner forearm and the hand

Internal iliac artery

Common iliac vein
A pair of veins that unite to form the inferior vena cava

Internal iliac vein

Common iliac artery
A pair of arteries formed by the division of the aorta

Deep femoral artery

Femoral artery

Femoral vein

BLOOD VESSELS

The heart contracts to keep blood moving through a vast network of blood vessels – arteries, arterioles, capillaries, venules, and veins. A thick elastic wall in arteries helps them to carry high-pressure blood from the heart to organs and tissues; veins contain valves that prevent backflow of blood when carrying it back to the heart. Arteries and smaller vessels branch into capillaries – the smallest blood vessels. The endothelial wall of a capillary is one cell thick.

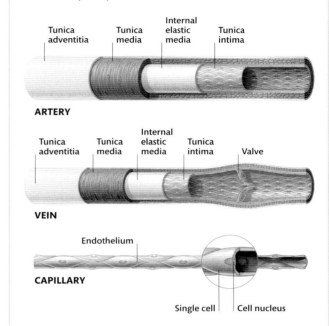

Tunica adventitia   Tunica media   Internal elastic media   Tunica intima

ARTERY

Tunica adventitia   Tunica media   Internal elastic media   Tunica intima   Valve

VEIN

Endothelium

CAPILLARY

Single cell   Cell nucleus

CARDIOVASCULAR SYSTEM **OVERVIEW**

The cardiovascular system consists of the heart, blood, and blood vessels. The heart – a muscular pump – contracts to push blood through the body's network of vessels in order to deliver oxygen, nutrients, white blood cells, and hormones to the tissues of the body. The blood also removes waste products and takes them to other organs – mainly the liver and kidneys – for excretion. The circulation can be divided in two: the pulmonary circulation carries blood pumped by the right side of the heart to the lungs, and the systemic circulation carries blood pumped by the more powerful left side of the heart to the rest of the body. The pressure in the pulmonary circulation is relatively low, to prevent fluid being forced out of capillaries into the alveoli of the lungs. The pressure in the systemic circulation is much higher, to push blood up to the brain, into all other organs, and out into the fingers and toes.

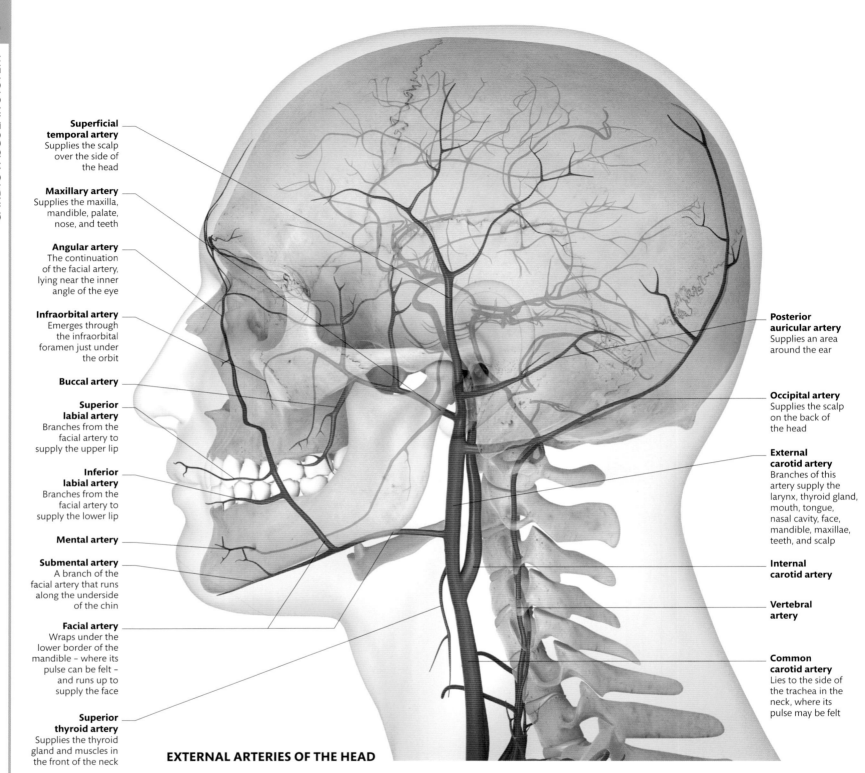

Superficial temporal artery
Supplies the scalp over the side of the head

Maxillary artery
Supplies the maxilla, mandible, palate, nose, and teeth

Angular artery
The continuation of the facial artery, lying near the inner angle of the eye

Infraorbital artery
Emerges through the infraorbital foramen just under the orbit

Buccal artery

Superior labial artery
Branches from the facial artery to supply the upper lip

Inferior labial artery
Branches from the facial artery to supply the lower lip

Mental artery

Submental artery
A branch of the facial artery that runs along the underside of the chin

Facial artery
Wraps under the lower border of the mandible – where its pulse can be felt – and runs up to supply the face

Superior thyroid artery
Supplies the thyroid gland and muscles in the front of the neck

Posterior auricular artery
Supplies an area around the ear

Occipital artery
Supplies the scalp on the back of the head

External carotid artery
Branches of this artery supply the larynx, thyroid gland, mouth, tongue, nasal cavity, face, mandible, maxillae, teeth, and scalp

Internal carotid artery

Vertebral artery

Common carotid artery
Lies to the side of the trachea in the neck, where its pulse may be felt

EXTERNAL ARTERIES OF THE HEAD

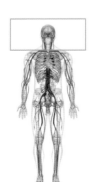

HEAD AND NECK

The main vessels supplying oxygenated blood to the head and neck are the common carotid and vertebral arteries. The vertebral artery runs up through holes in the cervical vertebrae and eventually enters the skull through the foramen magnum. The common carotid artery runs up the neck and divides into two – the internal carotid artery supplies the brain, and the external carotid artery gives rise to a profusion of branches, some of which supply the thyroid gland, the mouth, tongue, and nasal cavity. Veins of the head and neck come together like river tributaries, draining into the large internal jugular vein, behind the sternocleidomastoid muscle, and into the subclavian vein, low in the neck.

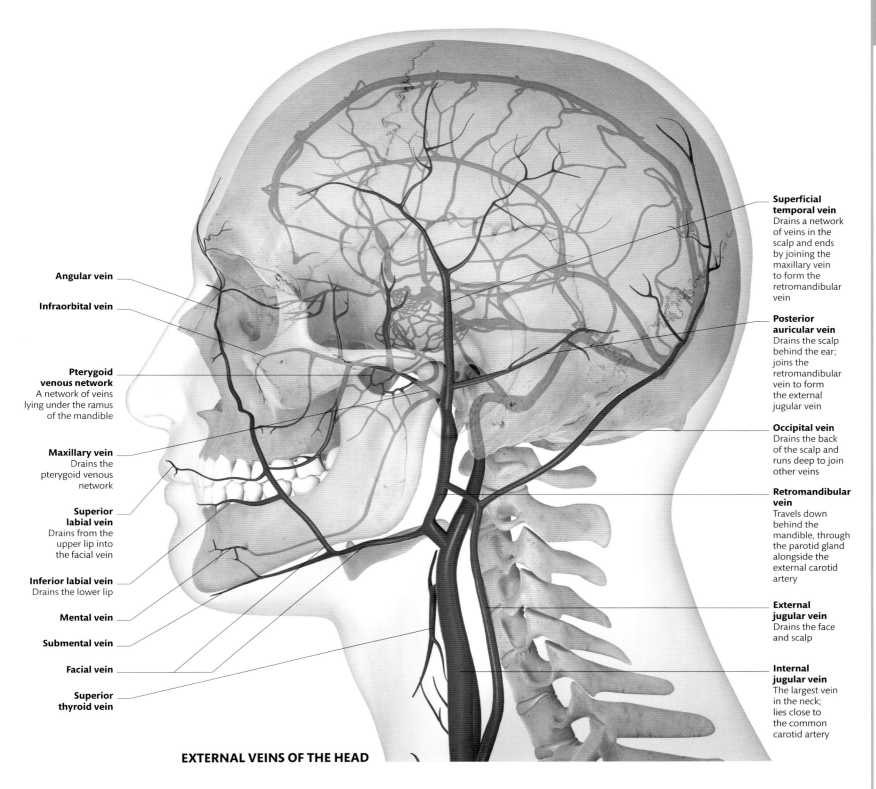

Angular vein

Infraorbital vein

Pterygoid venous network
A network of veins lying under the ramus of the mandible

Maxillary vein
Drains the pterygoid venous network

Superior labial vein
Drains from the upper lip into the facial vein

Inferior labial vein
Drains the lower lip

Mental vein

Submental vein

Facial vein

Superior thyroid vein

Superficial temporal vein
Drains a network of veins in the scalp and ends by joining the maxillary vein to form the retromandibular vein

Posterior auricular vein
Drains the scalp behind the ear; joins the retromandibular vein to form the external jugular vein

Occipital vein
Drains the back of the scalp and runs deep to join other veins

Retromandibular vein
Travels down behind the mandible, through the parotid gland alongside the external carotid artery

External jugular vein
Drains the face and scalp

Internal jugular vein
The largest vein in the neck; lies close to the common carotid artery

EXTERNAL VEINS OF THE HEAD

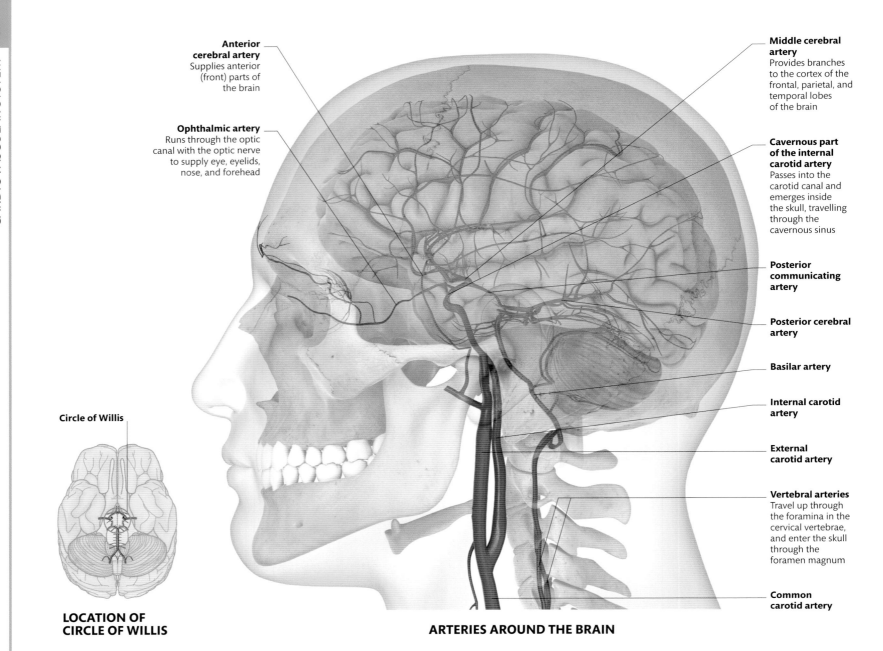

Anterior cerebral artery
Supplies anterior (front) parts of the brain

Ophthalmic artery
Runs through the optic canal with the optic nerve to supply eye, eyelids, nose, and forehead

Middle cerebral artery
Provides branches to the cortex of the frontal, parietal, and temporal lobes of the brain

Cavernous part of the internal carotid artery
Passes into the carotid canal and emerges inside the skull, travelling through the cavernous sinus

Posterior communicating artery

Posterior cerebral artery

Basilar artery

Internal carotid artery

External carotid artery

Vertebral arteries
Travel up through the foramina in the cervical vertebrae, and enter the skull through the foramen magnum

Common carotid artery

ARTERIES AROUND THE BRAIN

Circle of Willis

LOCATION OF CIRCLE OF WILLIS

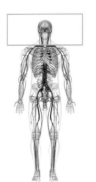

HEAD AND NECK

The brain has a rich blood supply, which arrives via the internal carotid and vertebral arteries. The vertebral arteries join together to form the basilar artery. The internal carotid arteries and basilar artery join up on the undersurface of the brain to form the Circle of Willis. From there, three pairs of cerebral arteries make their way into the brain. The veins of the brain and the skull drain into venous sinuses, which are enclosed within the dura mater (the outermost layer of the meninges) and form grooves on the inner surface of the skull. The sinuses join up and eventually drain out of the base of the skull, into the internal jugular vein.

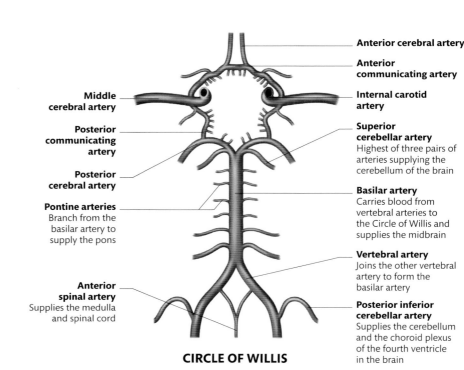

Middle cerebral artery

Posterior communicating artery

Posterior cerebral artery

Pontine arteries
Branch from the basilar artery to supply the pons

Anterior spinal artery
Supplies the medulla and spinal cord

Anterior cerebral artery

Anterior communicating artery

Internal carotid artery

Superior cerebellar artery
Highest of three pairs of arteries supplying the cerebellum of the brain

Basilar artery
Carries blood from vertebral arteries to the Circle of Willis and supplies the midbrain

Vertebral artery
Joins the other vertebral artery to form the basilar artery

Posterior inferior cerebellar artery
Supplies the cerebellum and the choroid plexus of the fourth ventricle in the brain

CIRCLE OF WILLIS

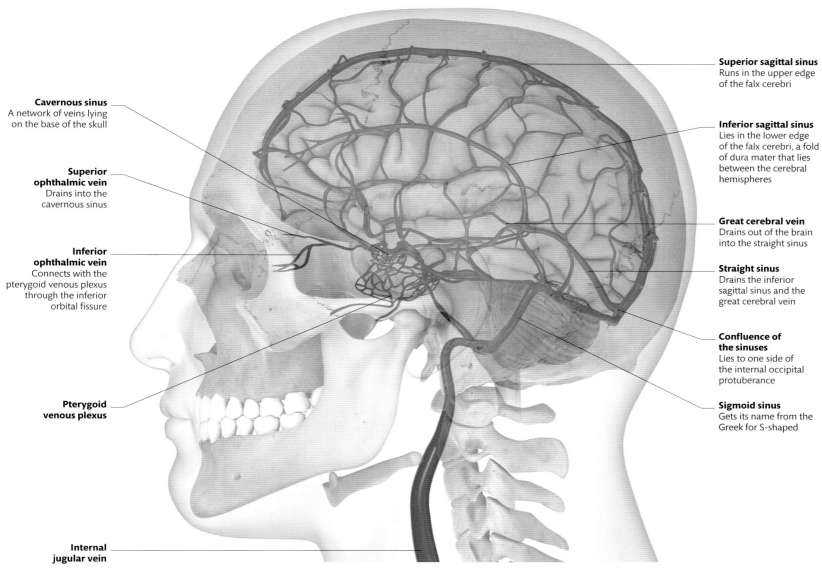

Superior sagittal sinus
Runs in the upper edge
of the falx cerebri

Cavernous sinus
A network of veins lying
on the base of the skull

Inferior sagittal sinus
Lies in the lower edge
of the falx cerebri, a fold
of dura mater that lies
between the cerebral
hemispheres

**Superior
ophthalmic vein**
Drains into the
cavernous sinus

Great cerebral vein
Drains out of the brain
into the straight sinus

**Inferior
ophthalmic vein**
Connects with the
pterygoid venous plexus
through the inferior
orbital fissure

Straight sinus
Drains the inferior
sagittal sinus and the
great cerebral vein

**Confluence of
the sinuses**
Lies to one side of
the internal occipital
protuberance

**Pterygoid
venous plexus**

Sigmoid sinus
Gets its name from the
Greek for S-shaped

**Internal
jugular vein**

VEINS AROUND THE BRAIN

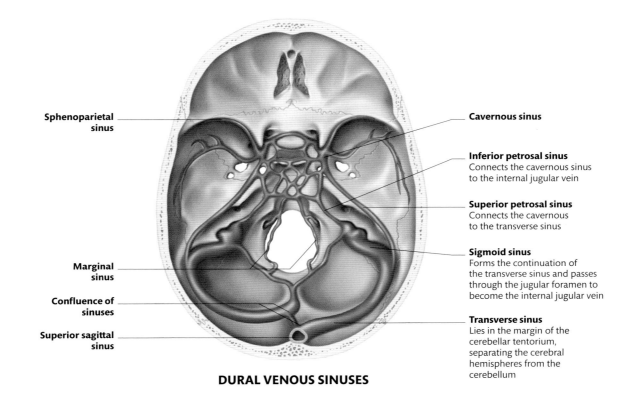

**Sphenoparietal
sinus**

Cavernous sinus

Inferior petrosal sinus
Connects the cavernous sinus
to the internal jugular vein

Superior petrosal sinus
Connects the cavernous
to the transverse sinus

**Marginal
sinus**

Sigmoid sinus
Forms the continuation of
the transverse sinus and passes
through the jugular foramen to
become the internal jugular vein

**Confluence of
sinuses**

**Superior sagittal
sinus**

Transverse sinus
Lies in the margin of the
cerebellar tentorium,
separating the cerebral
hemispheres from the
cerebellum

DURAL VENOUS SINUSES

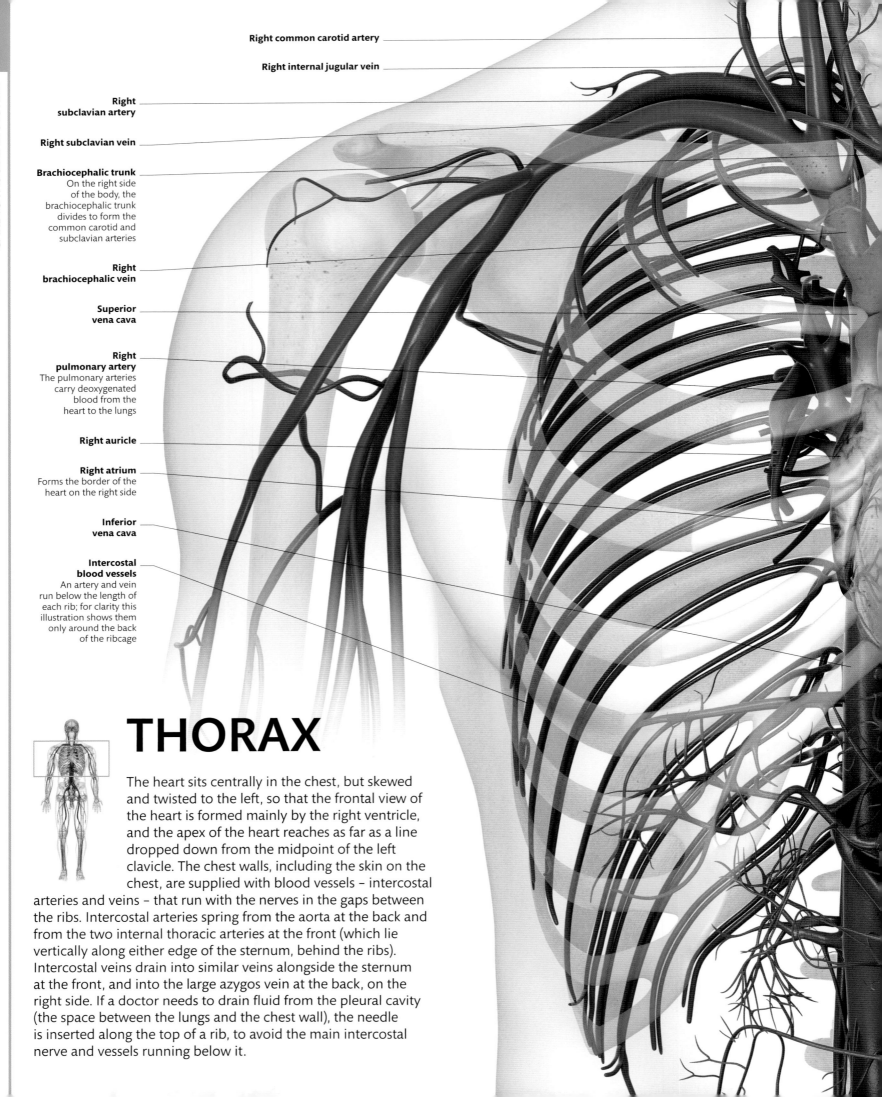

Right common carotid artery

Right internal jugular vein

Right subclavian artery

Right subclavian vein

Brachiocephalic trunk
On the right side of the body, the brachiocephalic trunk divides to form the common carotid and subclavian arteries

Right brachiocephalic vein

Superior vena cava

Right pulmonary artery
The pulmonary arteries carry deoxygenated blood from the heart to the lungs

Right auricle

Right atrium
Forms the border of the heart on the right side

Inferior vena cava

Intercostal blood vessels
An artery and vein run below the length of each rib; for clarity this illustration shows them only around the back of the ribcage

THORAX

The heart sits centrally in the chest, but skewed and twisted to the left, so that the frontal view of the heart is formed mainly by the right ventricle, and the apex of the heart reaches as far as a line dropped down from the midpoint of the left clavicle. The chest walls, including the skin on the chest, are supplied with blood vessels – intercostal arteries and veins – that run with the nerves in the gaps between the ribs. Intercostal arteries spring from the aorta at the back and from the two internal thoracic arteries at the front (which lie vertically along either edge of the sternum, behind the ribs). Intercostal veins drain into similar veins alongside the sternum at the front, and into the large azygos vein at the back, on the right side. If a doctor needs to drain fluid from the pleural cavity (the space between the lungs and the chest wall), the needle is inserted along the top of a rib, to avoid the main intercostal nerve and vessels running below it.

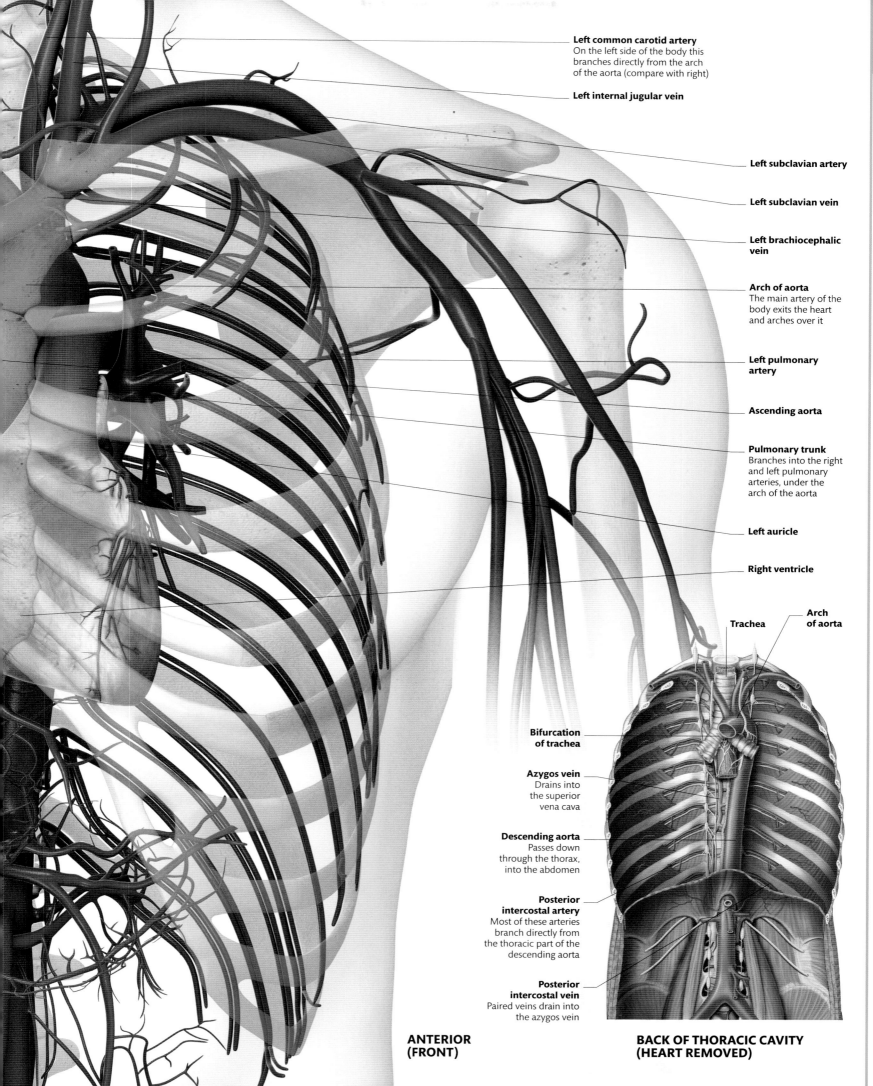

Left common carotid artery
On the left side of the body this branches directly from the arch of the aorta (compare with right)

Left internal jugular vein

Left subclavian artery

Left subclavian vein

Left brachiocephalic vein

Arch of aorta
The main artery of the body exits the heart and arches over it

Left pulmonary artery

Ascending aorta

Pulmonary trunk
Branches into the right and left pulmonary arteries, under the arch of the aorta

Left auricle

Right ventricle

Trachea

Arch of aorta

Bifurcation of trachea

Azygos vein
Drains into the superior vena cava

Descending aorta
Passes down through the thorax, into the abdomen

Posterior intercostal artery
Most of these arteries branch directly from the thoracic part of the descending aorta

Posterior intercostal vein
Paired veins drain into the azygos vein

ANTERIOR (FRONT)

BACK OF THORACIC CAVITY (HEART REMOVED)

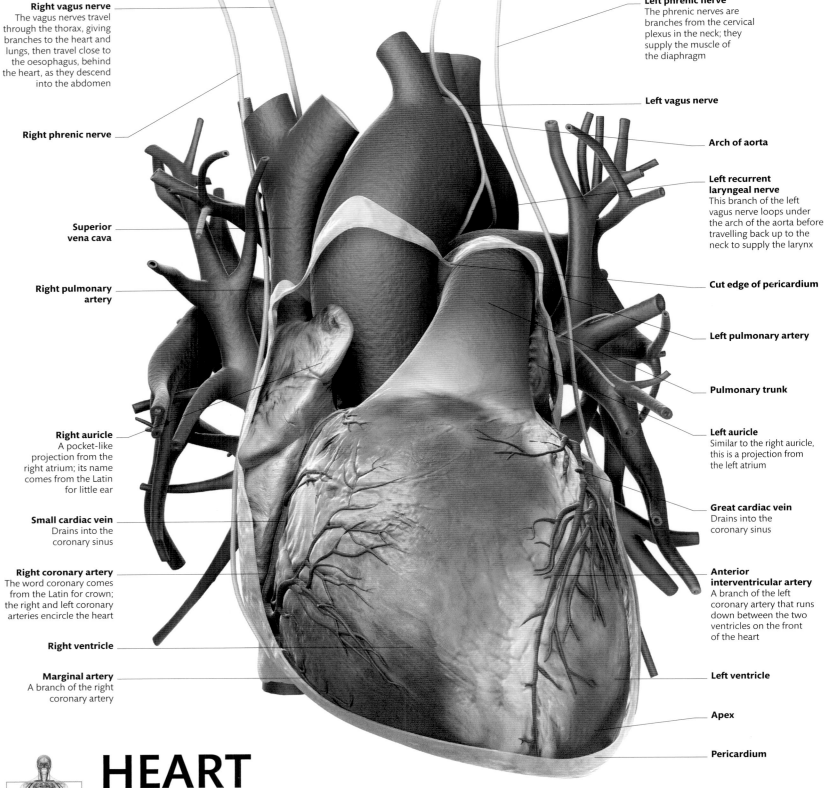

Right vagus nerve
The vagus nerves travel through the thorax, giving branches to the heart and lungs, then travel close to the oesophagus, behind the heart, as they descend into the abdomen

Right phrenic nerve

Superior vena cava

Right pulmonary artery

Right auricle
A pocket-like projection from the right atrium; its name comes from the Latin for little ear

Small cardiac vein
Drains into the coronary sinus

Right coronary artery
The word coronary comes from the Latin for crown; the right and left coronary arteries encircle the heart

Right ventricle

Marginal artery
A branch of the right coronary artery

Left phrenic nerve
The phrenic nerves are branches from the cervical plexus in the neck; they supply the muscle of the diaphragm

Left vagus nerve

Arch of aorta

Left recurrent laryngeal nerve
This branch of the left vagus nerve loops under the arch of the aorta before travelling back up to the neck to supply the larynx

Cut edge of pericardium

Left pulmonary artery

Pulmonary trunk

Left auricle
Similar to the right auricle, this is a projection from the left atrium

Great cardiac vein
Drains into the coronary sinus

Anterior interventricular artery
A branch of the left coronary artery that runs down between the two ventricles on the front of the heart

Left ventricle

Apex

Pericardium

ANTERIOR (FRONT)

HEART

The heart is encased in the pericardium. This has a tough outer layer that is fused to the diaphragm below and to the connective tissue around the large blood vessels above the heart. Lining the inside of this cylinder (and the outer surface of the heart) is a thin membrane called the serous pericardium. Between these two layers is a thin film of fluid that lubricates the movement of the heart as it beats. Inflammation of this membrane, known as pericarditis, can be extremely painful. Branches of the right and left coronary arteries, which spring from the ascending aorta, supply the heart muscle itself. The heart is drained by cardiac veins, most of which drain into the coronary sinus.

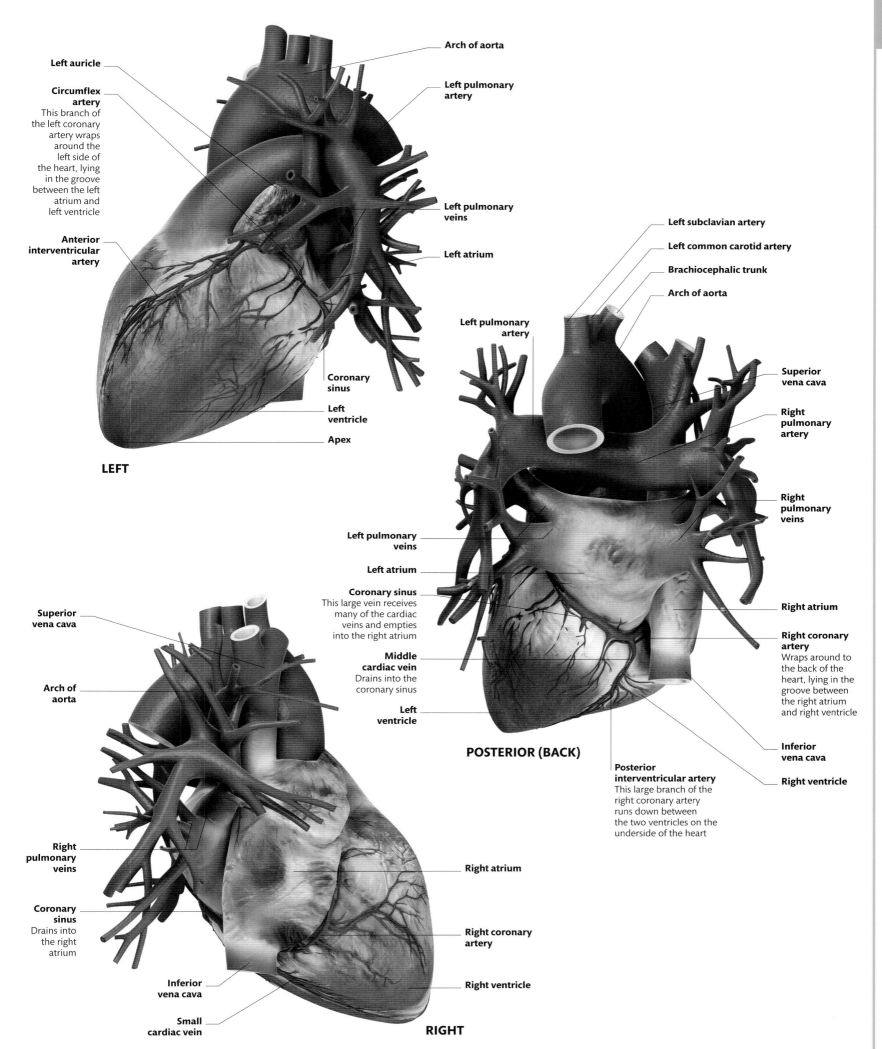

Arch of aorta

Left auricle

Left pulmonary artery

Circumflex artery
This branch of the left coronary artery wraps around the left side of the heart, lying in the groove between the left atrium and left ventricle

Left pulmonary veins

Left atrium

Anterior interventricular artery

Coronary sinus

Left ventricle

Apex

LEFT

Left subclavian artery

Left common carotid artery

Brachiocephalic trunk

Arch of aorta

Left pulmonary artery

Superior vena cava

Right pulmonary artery

Right pulmonary veins

Left pulmonary veins

Left atrium

Coronary sinus
This large vein receives many of the cardiac veins and empties into the right atrium

Right atrium

Middle cardiac vein
Drains into the coronary sinus

Right coronary artery
Wraps around to the back of the heart, lying in the groove between the right atrium and right ventricle

Left ventricle

Inferior vena cava

POSTERIOR (BACK)

Right ventricle

Posterior interventricular artery
This large branch of the right coronary artery runs down between the two ventricles on the underside of the heart

Superior vena cava

Arch of aorta

Right pulmonary veins

Right atrium

Coronary sinus
Drains into the right atrium

Right coronary artery

Inferior vena cava

Right ventricle

Small cardiac vein

RIGHT

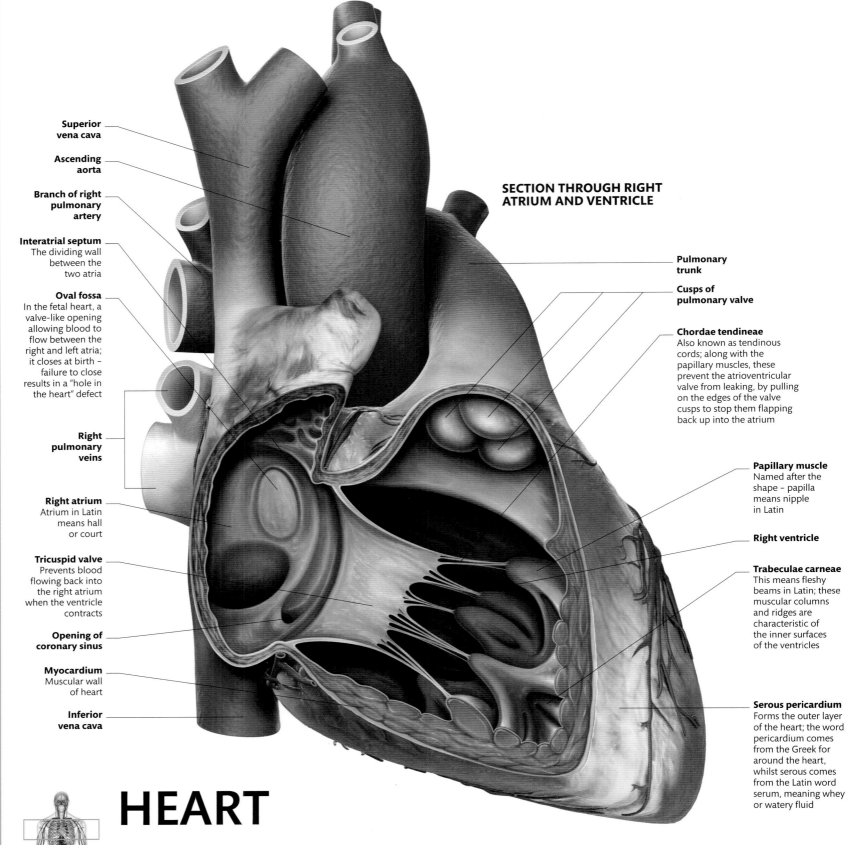

Superior
vena cava

Ascending
aorta

Branch of right
pulmonary
artery

Interatrial septum
The dividing wall
between the
two atria

Oval fossa
In the fetal heart, a
valve-like opening
allowing blood to
flow between the
right and left atria;
it closes at birth –
failure to close
results in a "hole in
the heart" defect

Right
pulmonary
veins

Right atrium
Atrium in Latin
means hall
or court

Tricuspid valve
Prevents blood
flowing back into
the right atrium
when the ventricle
contracts

Opening of
coronary sinus

Myocardium
Muscular wall
of heart

Inferior
vena cava

SECTION THROUGH RIGHT
ATRIUM AND VENTRICLE

Pulmonary
trunk

Cusps of
pulmonary valve

Chordae tendineae
Also known as tendinous
cords; along with the
papillary muscles, these
prevent the atrioventricular
valve from leaking, by pulling
on the edges of the valve
cusps to stop them flapping
back up into the atrium

Papillary muscle
Named after the
shape – papilla
means nipple
in Latin

Right ventricle

Trabeculae carneae
This means fleshy
beams in Latin; these
muscular columns
and ridges are
characteristic of
the inner surfaces
of the ventricles

Serous pericardium
Forms the outer layer
of the heart; the word
pericardium comes
from the Greek for
around the heart,
whilst serous comes
from the Latin word
serum, meaning whey
or watery fluid

HEART

The heart receives blood from veins and pumps it out through
arteries. It has four chambers: two atria and two ventricles. The
heart's left and right sides are separate. The right side receives
deoxygenated blood from the body via the superior and inferior
venae cavae, and pumps it to the lungs through the pulmonary
trunk. The left gets oxygenated blood from the lungs via the
pulmonary veins, and pumps it into the aorta for distribution.
Each atrium opens into its corresponding ventricle via a valve (on the right, the
tricuspid valve, and the bicuspid valve on the left), which shuts when the ventricle
contracts, to stop blood flowing back into the atrium. The aorta and pulmonary
trunk also have valves.

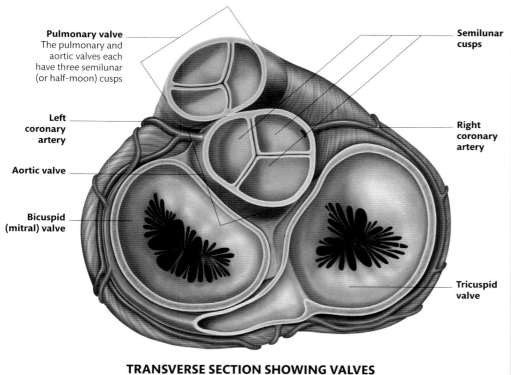

Pulmonary valve
The pulmonary and
aortic valves each
have three semilunar
(or half-moon) cusps

**Left
coronary
artery**

Aortic valve

**Bicuspid
(mitral) valve**

**Semilunar
cusps**

**Right
coronary
artery**

**Tricuspid
valve**

TRANSVERSE SECTION SHOWING VALVES

CARDIAC MUSCLE STRUCTURE

Cardiac muscle (myocardium) is essential to the heart's function as a pump. When the myocardium contracts, blood is squeezed out of the heart. The myocardium is a network of interconnected fibres, which contract rhythmically and spontaneously. Autonomic nerves can adjust the rate of contraction, matching the heart's output to the body's need.

Intercalated disc
These elaborate
junctions firmly
bind cardiac muscle
cells together

**Cardiac
muscle cell**

**Cell
nucleus**

Mitochondrion
Muscle cells are
packed with
energy-producing
mitochondria

Myofibril
The myofibrils of cardiac
muscle are organized in a
similar way to those in
skeletal muscle, giving a
striated appearance under
a light microscope

CARDIAC MUSCLE

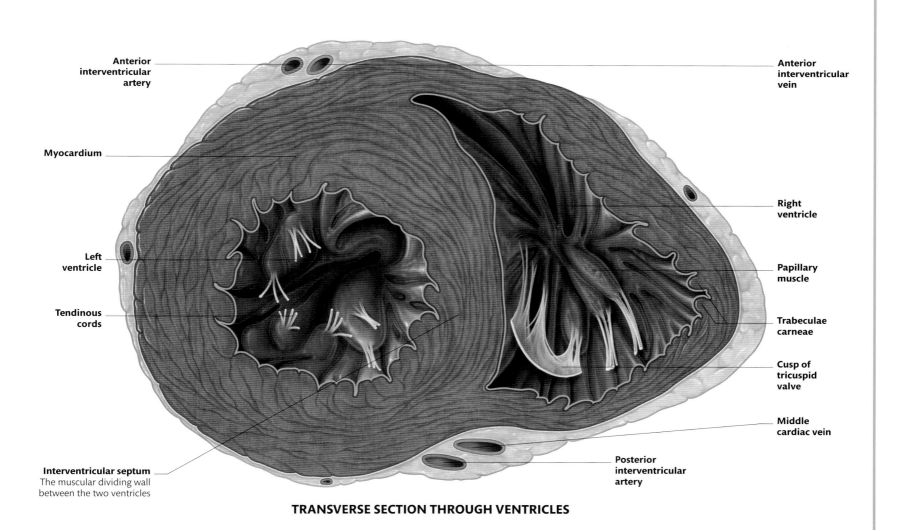

**Anterior
interventricular
artery**

Myocardium

**Left
ventricle**

**Tendinous
cords**

Interventricular septum
The muscular dividing wall
between the two ventricles

**Anterior
interventricular
vein**

**Right
ventricle**

**Papillary
muscle**

**Trabeculae
carneae**

**Cusp of
tricuspid
valve**

**Middle
cardiac vein**

**Posterior
interventricular
artery**

TRANSVERSE SECTION THROUGH VENTRICLES

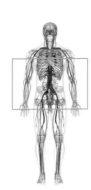

ABDOMEN AND PELVIS

The aorta passes behind the diaphragm, level with the twelfth thoracic vertebra, and enters the abdomen. Pairs of arteries branch from the sides of the aorta to supply the walls of the abdomen, the kidneys, suprarenal glands, and the testes or ovaries with oxygenated blood. A series of branches emerge from the front of the abdominal aorta to supply the abdominal organs: the coeliac trunk gives branches to the liver, stomach, pancreas, and spleen, and the mesenteric arteries provide blood to the gut. The abdominal aorta ends by splitting into two, forming the common iliac arteries. Each of these then divides, in turn, forming an internal iliac artery (which supplies the pelvic organs) and an external iliac artery (which continues into the thigh, becoming the femoral artery). Lying to the right of the aorta is the major vein of the abdomen: the inferior vena cava.

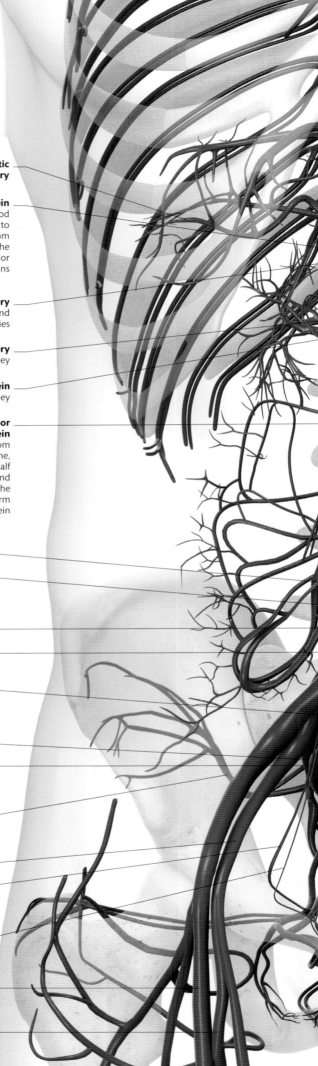

Right hepatic artery

Portal vein
Carries blood from the intestines to the liver; formed from the joining of the splenic and superior mesenteric veins

Common hepatic artery
Branches into right and left hepatic arteries

Right renal artery
Supplies the right kidney

Right renal vein
Drains the right kidney

Superior mesenteric vein
Drains blood from the small intestine, caecum, and half of the colon, and ends by joining the splenic vein to form the portal vein

Inferior vena cava

Ileocolic artery
Branch of the superior mesenteric artery supplying the end of the ileum, the caecum, the start of the ascending colon, and the appendix

Right common iliac vein

Right common iliac artery
Divides into the right external and internal iliac arteries

Right internal iliac artery
Provides branches to the bladder, rectum, perineum, and external genitals, muscles of the inner thigh, bone of the ilium and sacrum, and the buttock, as well as the uterus and vagina in a woman

Right internal iliac vein

Right external iliac artery
Gives a branch to the lower part of the anterior abdominal wall before passing over the pubic bone and under the inguinal ligament to become the femoral artery

Right superior gluteal artery
The largest branch of the internal iliac artery; passes out through the back of the pelvis to supply the upper buttock

Right external iliac vein

Right gonadal artery
In a woman, supplies the ovary on each side; in a man, extends to the scrotum to supply the testis

Right gonadal vein
Drains the ovary or testis and ends by joining the inferior vena cava

Right femoral artery
The main artery of the leg; the continuation of the external iliac artery in the thigh

ANTERIOR (FRONT)

Right femoral vein

Coeliac trunk
Only just over 1cm (⅜in) long, it quickly
branches into the left gastric, splenic, and
common hepatic arteries

Splenic artery
Supplies the spleen, as well as most of the
pancreas, and the upper part of the stomach

Splenic vein
Drains the spleen and receives other veins
from the stomach and pancreas, as well as
the inferior mesenteric vein

Left renal artery
Shorter than the right renal artery, this
supplies the left kidney

Left renal vein
Longer than its counterpart on the
right, this drains the left kidney and
receives the left gonadal vein

Inferior mesenteric vein
Drains blood from the colon and rectum and
ends by emptying into the splenic vein

Superior mesenteric artery
Branches within the mesentery to
supply a great length of intestine,
including all of the jejunum and
ileum and half of the colon

Abdominal aorta
The thoracic aorta becomes
the abdominal aorta as it passes
behind the diaphragm, level with
the twelfth thoracic vertebra

Inferior mesenteric artery
Supplies the last third of the transverse colon, the
descending and sigmoid colon, and the rectum

Bifurcation of aorta
The abdominal aorta divides in front of
the fourth lumbar vertebra

Left common iliac artery

Left common iliac vein
Formed from the union of the external
and internal iliac veins

Left external iliac vein
The continuation of the femoral vein,
after it has passed into the pelvis

Left internal iliac artery

Superior rectal artery
The last branch of the inferior mesenteric artery
passes down into the pelvis to supply the rectum

Left external iliac artery

Left internal iliac vein
Drains the pelvic organs,
perineum, and buttock

Left gonadal artery
Gonadal arteries branch from the
aorta just below the renal arteries

Left gonadal vein
Drains the ovary or testis, and empties
into the left renal vein

Left femoral artery

Left femoral vein
The main vein from the leg;
becomes the external iliac vein

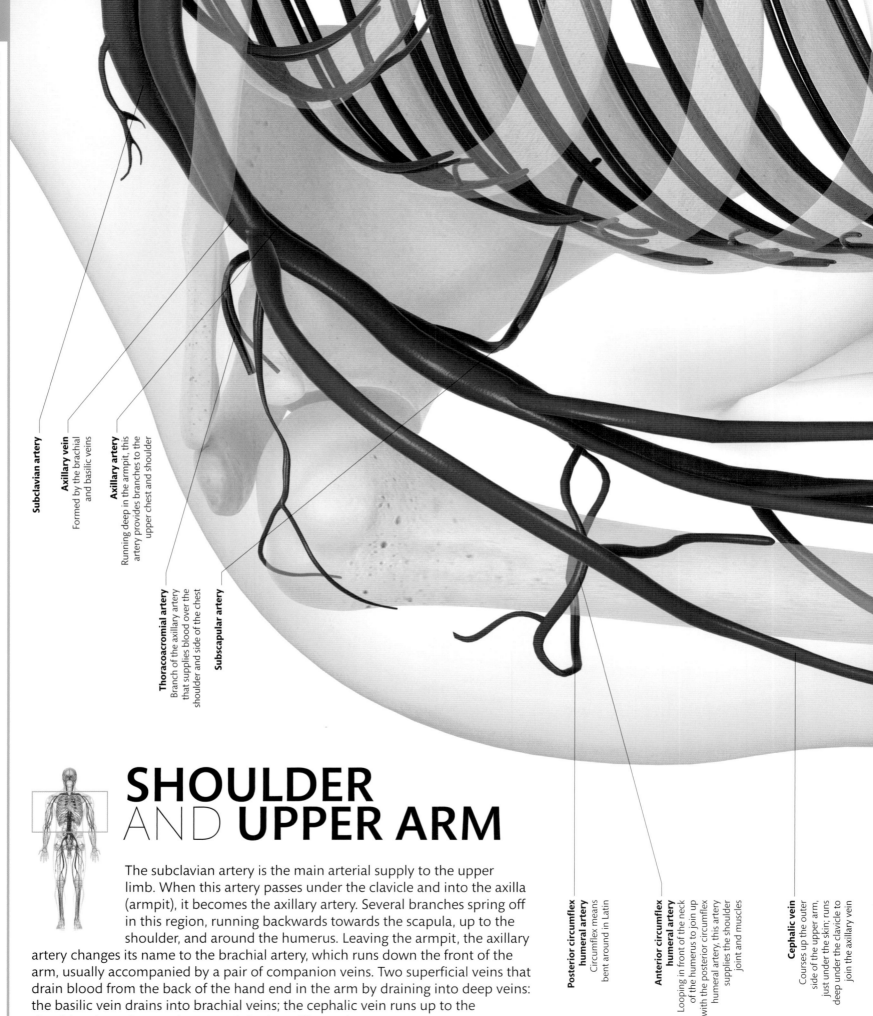

Subclavian artery

Axillary vein
Formed by the brachial
and basilic veins

Axillary artery
Running deep in the armpit, this
artery provides branches to the
upper chest and shoulder

Thoracoacromial artery
Branch of the axillary artery
that supplies blood over the
shoulder and side of the chest

Subscapular artery

**Posterior circumflex
humeral artery**
Circumflex means
bent around in Latin

**Anterior circumflex
humeral artery**
Looping in front of the neck
of the humerus to join up
with the posterior circumflex
humeral artery, this artery
supplies the shoulder
joint and muscles

Cephalic vein
Courses up the outer
side of the upper arm,
just under the skin; runs
deep under the clavicle to
join the axillary vein

SHOULDER
AND **UPPER ARM**

The subclavian artery is the main arterial supply to the upper
limb. When this artery passes under the clavicle and into the axilla
(armpit), it becomes the axillary artery. Several branches spring off
in this region, running backwards towards the scapula, up to the
shoulder, and around the humerus. Leaving the armpit, the axillary
artery changes its name to the brachial artery, which runs down the front of the
arm, usually accompanied by a pair of companion veins. Two superficial veins that
drain blood from the back of the hand end in the arm by draining into deep veins:
the basilic vein drains into brachial veins; the cephalic vein runs up to the
shoulder, then plunges deeper to join the axillary vein.

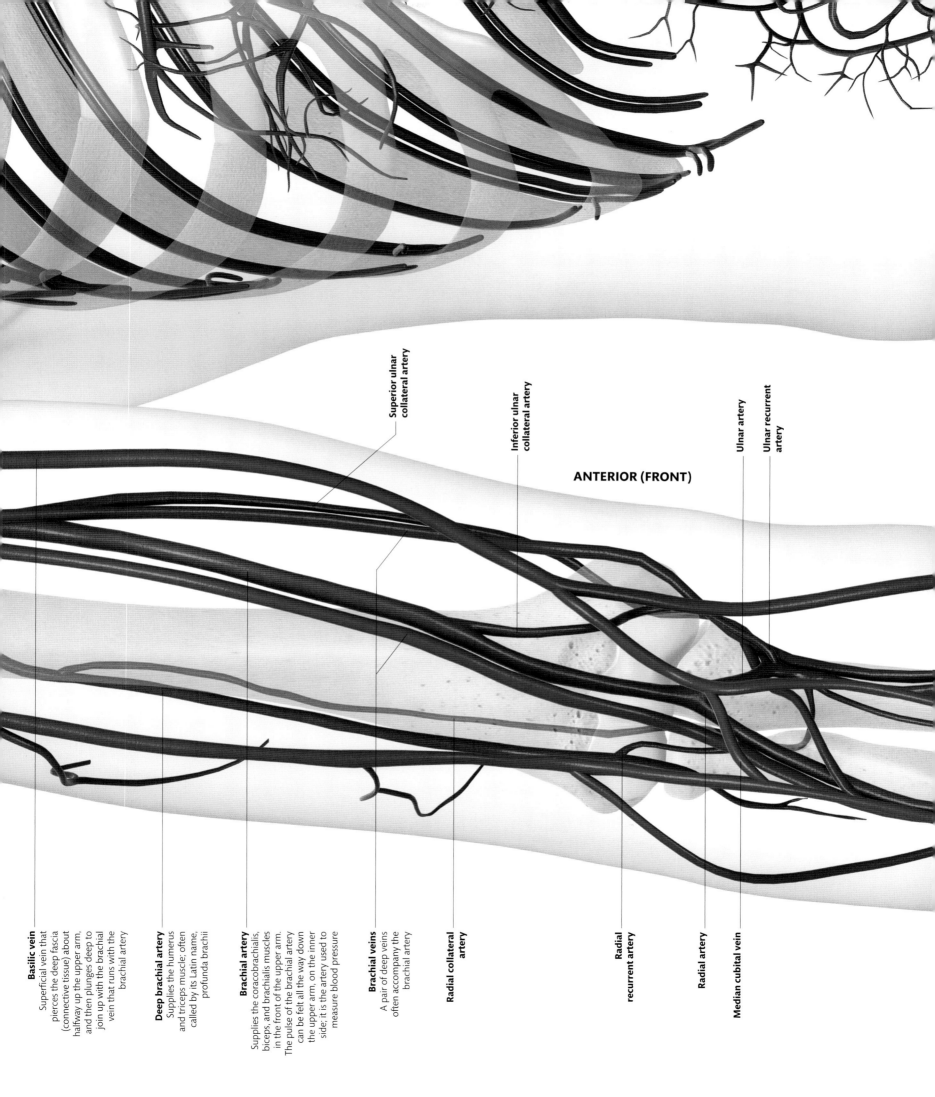

Superior ulnar collateral artery

Inferior ulnar collateral artery

ANTERIOR (FRONT)

Ulnar artery

Ulnar recurrent artery

Basilic vein
Superficial vein that pierces the deep fascia (connective tissue) about halfway up the upper arm, and then plunges deep to join up with the brachial vein that runs with the brachial artery

Deep brachial artery
Supplies the humerus and triceps muscle; often called by its Latin name, profunda brachii

Brachial artery
Supplies the coracobrachialis, biceps, and brachialis muscles in the front of the upper arm. The pulse of the brachial artery can be felt all the way down the upper arm, on the inner side; it is the artery used to measure blood pressure

Brachial veins
A pair of deep veins often accompany the brachial artery

Radial collateral artery

Radial recurrent artery

Radial artery

Median cubital vein

SHOULDER
AND UPPER ARM

Various branches from the axillary and brachial arteries supply the back of the shoulder and upper arm. The posterior circumflex humeral artery, which runs with the axillary nerve, curls around the upper end of the humerus. The deep brachial artery runs with the radial nerve, spiralling around the back of the bone. From this artery, and from the brachial artery itself, collateral branches run down the arm and join up, or anastomose, with recurrent branches running back up from the ulnar and radial arteries of the forearm. There are also anastomoses (links) between branches of the subclavian and axillary arteries around the shoulder. Anastomoses like this, where branches from different regions join up, can provide alternative routes through which blood can flow if the main vessel becomes squashed or blocked.

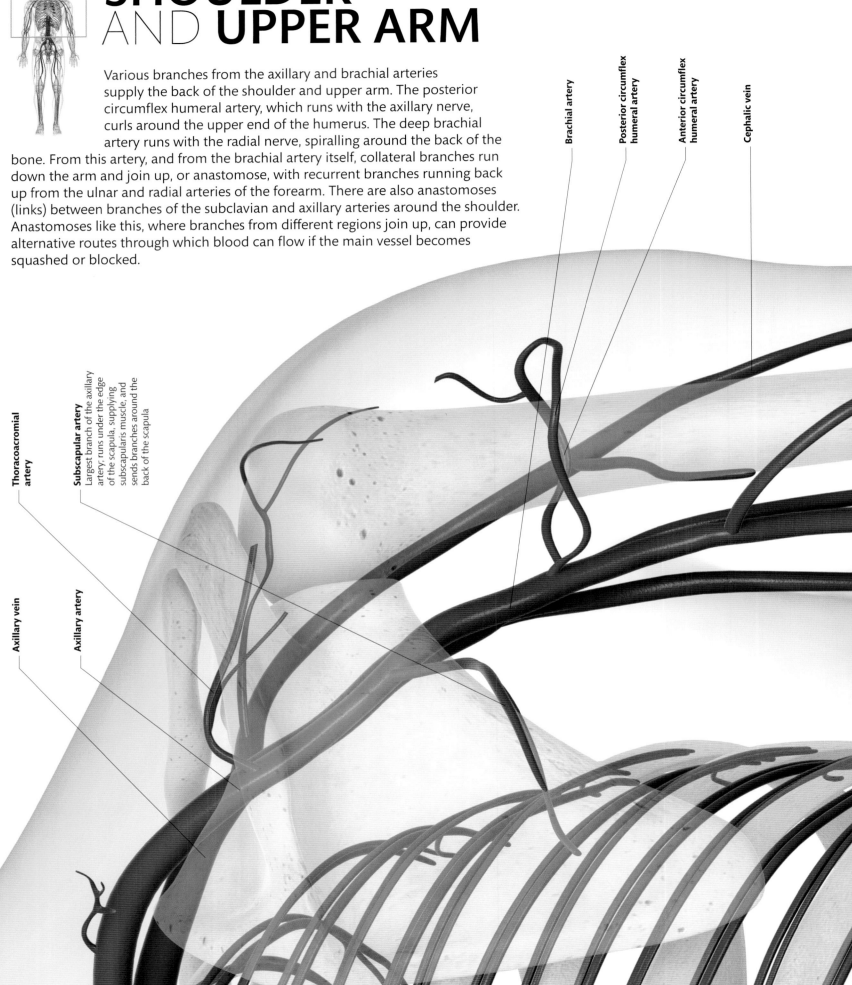

Brachial artery

Posterior circumflex humeral artery

Anterior circumflex humeral artery

Cephalic vein

Thoracoacromial artery

Subscapular artery
Largest branch of the axillary artery; runs under the edge of the scapula, supplying subscapularis muscle, and sends branches around the back of the scapula

Axillary vein

Axillary artery

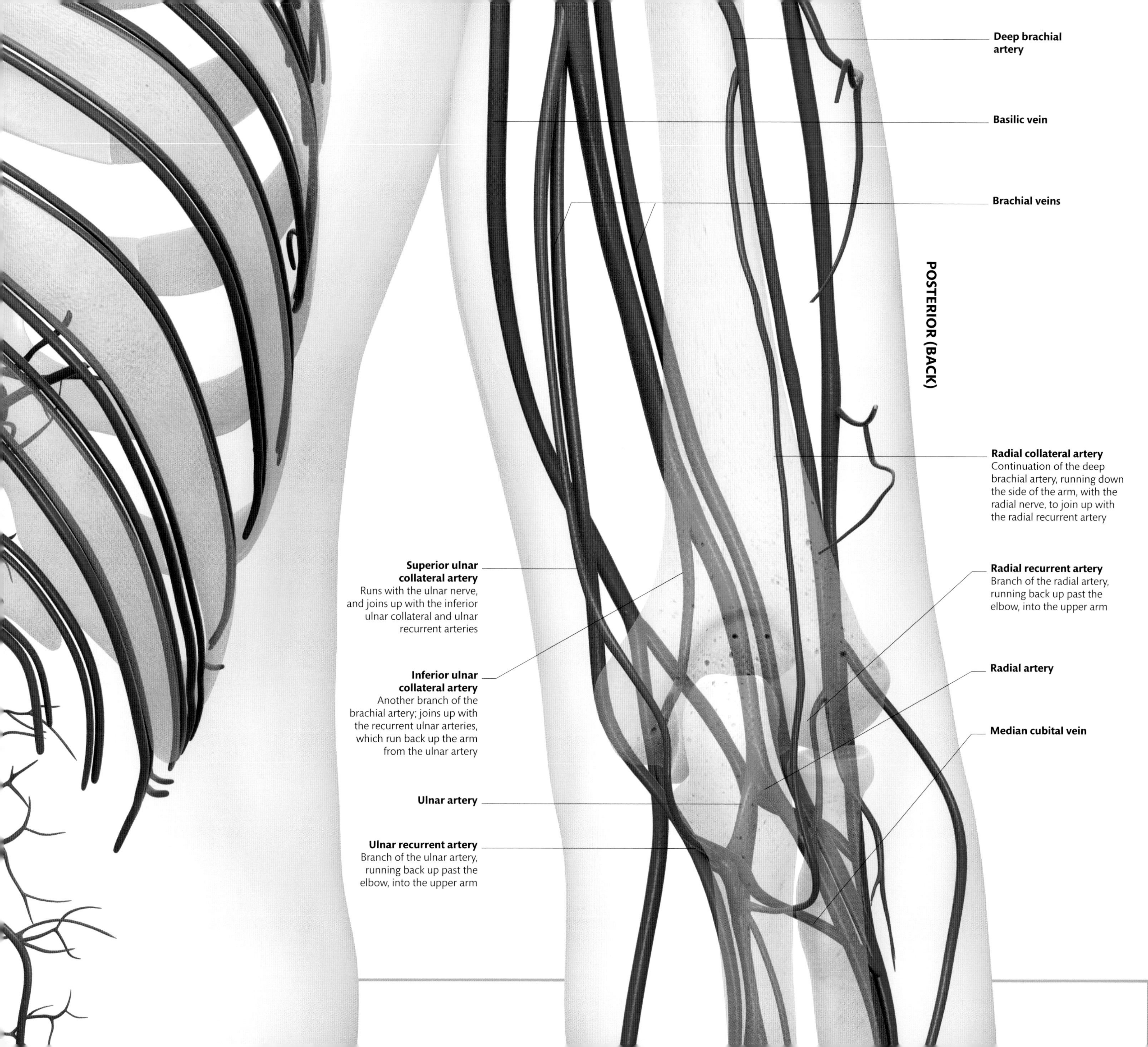

Deep brachial artery

Basilic vein

Brachial veins

POSTERIOR (BACK)

Radial collateral artery
Continuation of the deep brachial artery, running down the side of the arm, with the radial nerve, to join up with the radial recurrent artery

Radial recurrent artery
Branch of the radial artery, running back up past the elbow, into the upper arm

Radial artery

Median cubital vein

Superior ulnar collateral artery
Runs with the ulnar nerve, and joins up with the inferior ulnar collateral and ulnar recurrent arteries

Inferior ulnar collateral artery
Another branch of the brachial artery; joins up with the recurrent ulnar arteries, which run back up the arm from the ulnar artery

Ulnar artery

Ulnar recurrent artery
Branch of the ulnar artery, running back up past the elbow, into the upper arm

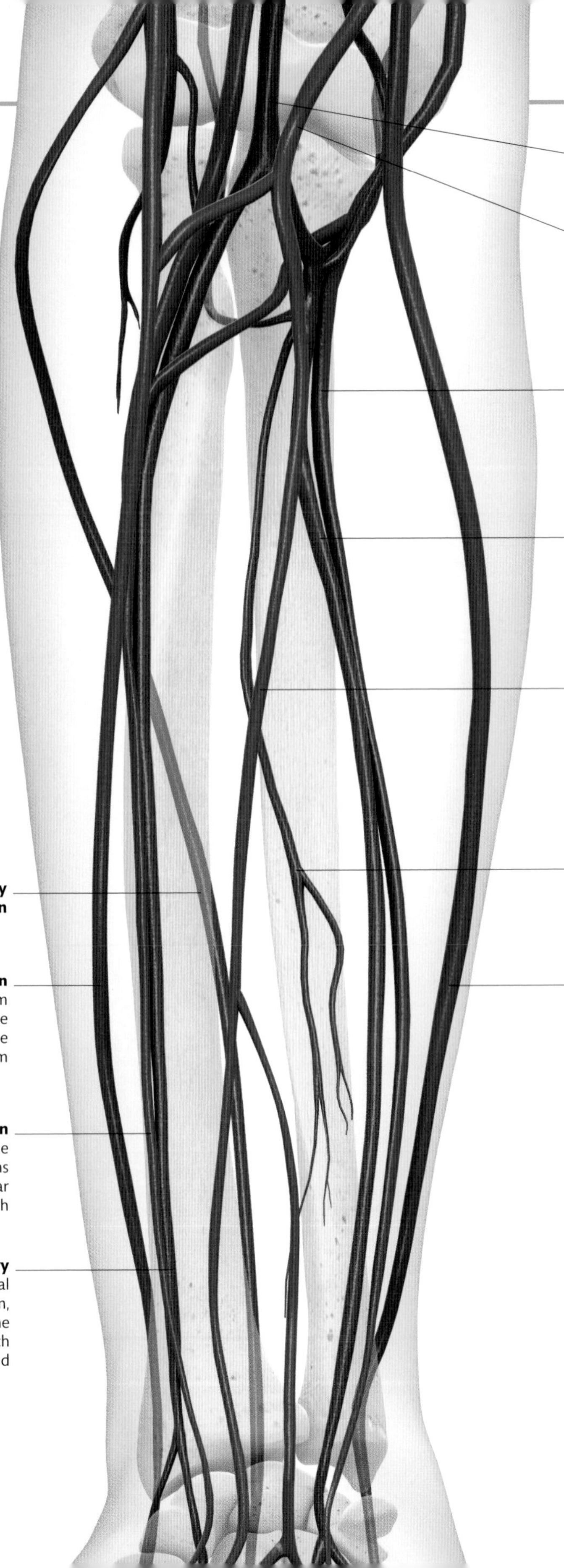

Brachial artery

Median cubital vein
Connects the cephalic and basilic veins; is a favoured site for taking blood

Ulnar vein
Runs with the ulnar artery; drains the deep palmar venous arch

Ulnar artery
Supplies the ulnar side of the forearm; feeds into the superficial palmar arch

Median vein of the forearm
Drains the superficial venous plexus of the palm

Interosseous artery

Basilic vein
Drains blood from the ulnar side of the back of the hand and forearm

ANTERIOR (FRONT)

Accessory cephalic vein

Cephalic vein
Drains blood from the radial side of the back of the hand and forearm

Radial vein
Runs with the radial artery; drains the superficial palmar venous arch

Radial artery
Supplies the radial side of the forearm, and feeds into the deep palmar arch of the hand

Ulna

Basilic vein
The name of this vein means royal and comes from its historical importance in blood-letting

Dorsal digital vein
Drains blood from the sides of the fingers

Dorsal venous network
A plexus of veins visible under the skin that drains blood into the cephalic, accessory cephalic, and basilic veins

Cephalic vein
The name of this vein comes from the Greek for head, because of the historical belief that blood-letting from it could cure headaches

Radius

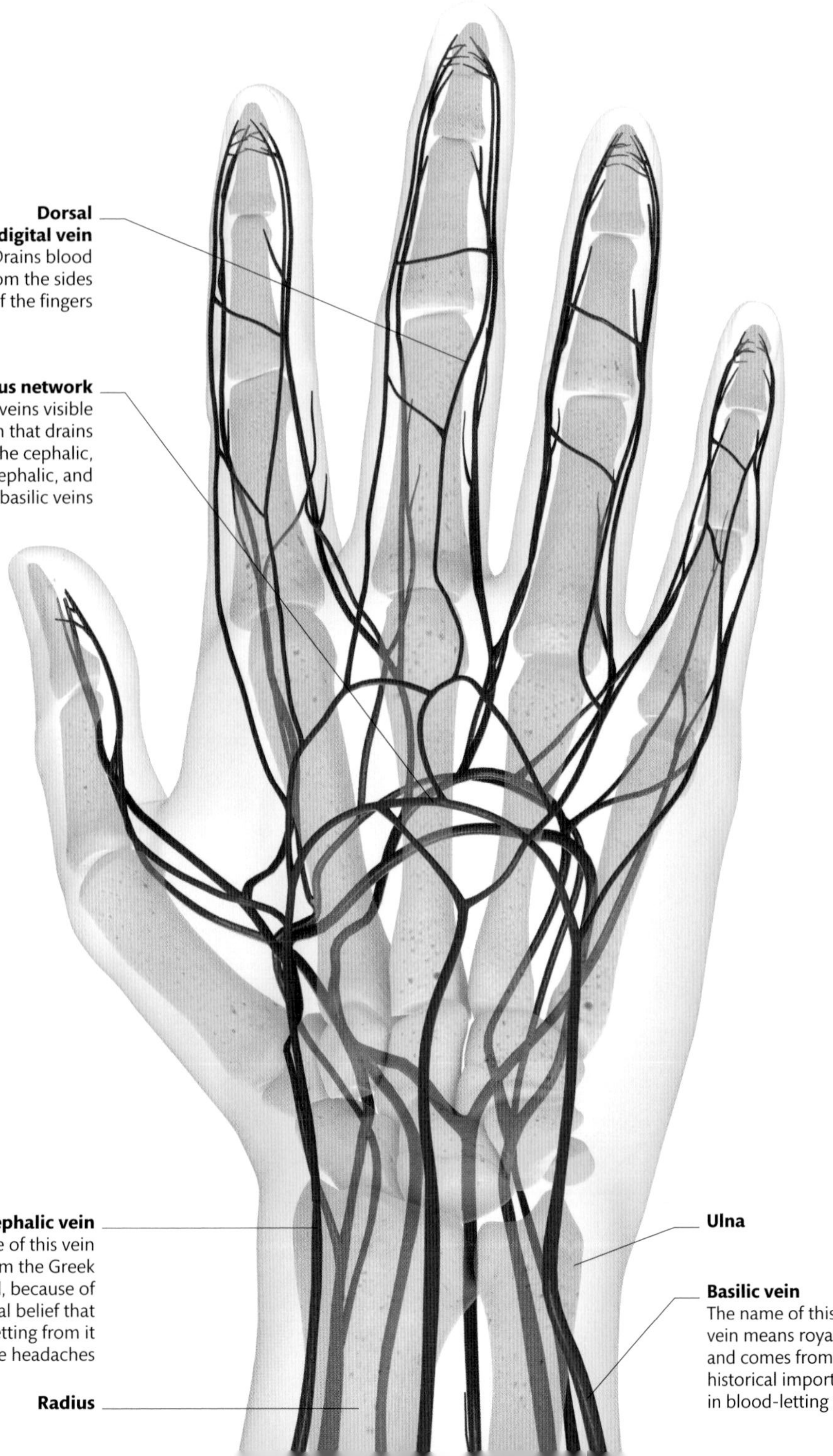

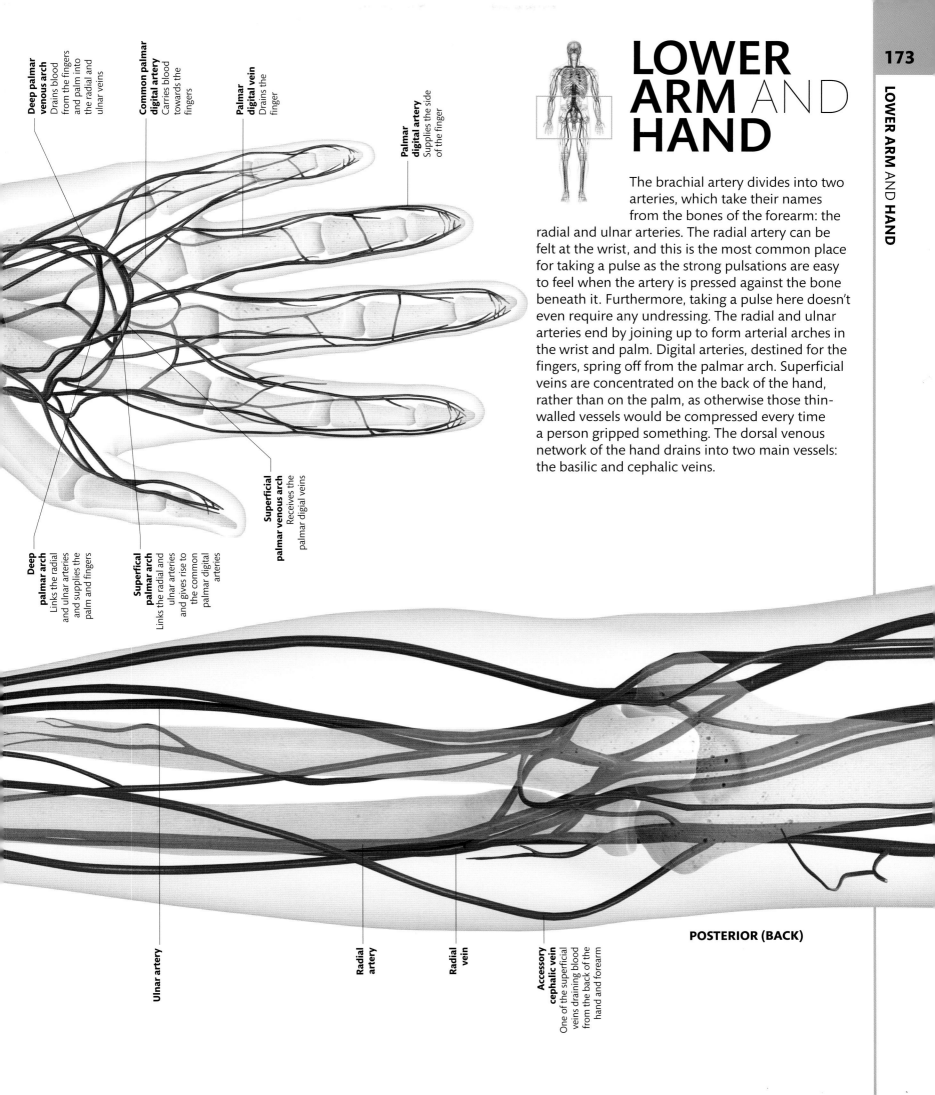

Deep palmar venous arch
Drains blood from the fingers and palm into the radial and ulnar veins

Common palmar digital artery
Carries blood towards the fingers

Palmar digital vein
Drains the finger

Palmar digital artery
Supplies the side of the finger

Superficial palmar venous arch
Receives the palmar digital veins

Deep palmar arch
Links the radial and ulnar arteries and supplies the palm and fingers

Superfical palmar arch
Links the radial and ulnar arteries and gives rise to the common palmar digital arteries

LOWER ARM AND HAND

The brachial artery divides into two arteries, which take their names from the bones of the forearm: the radial and ulnar arteries. The radial artery can be felt at the wrist, and this is the most common place for taking a pulse as the strong pulsations are easy to feel when the artery is pressed against the bone beneath it. Furthermore, taking a pulse here doesn't even require any undressing. The radial and ulnar arteries end by joining up to form arterial arches in the wrist and palm. Digital arteries, destined for the fingers, spring off from the palmar arch. Superficial veins are concentrated on the back of the hand, rather than on the palm, as otherwise those thin-walled vessels would be compressed every time a person gripped something. The dorsal venous network of the hand drains into two main vessels: the basilic and cephalic veins.

Ulnar artery

Radial artery

Radial vein

Accessory cephalic vein
One of the superficial veins draining blood from the back of the hand and forearm

POSTERIOR (BACK)

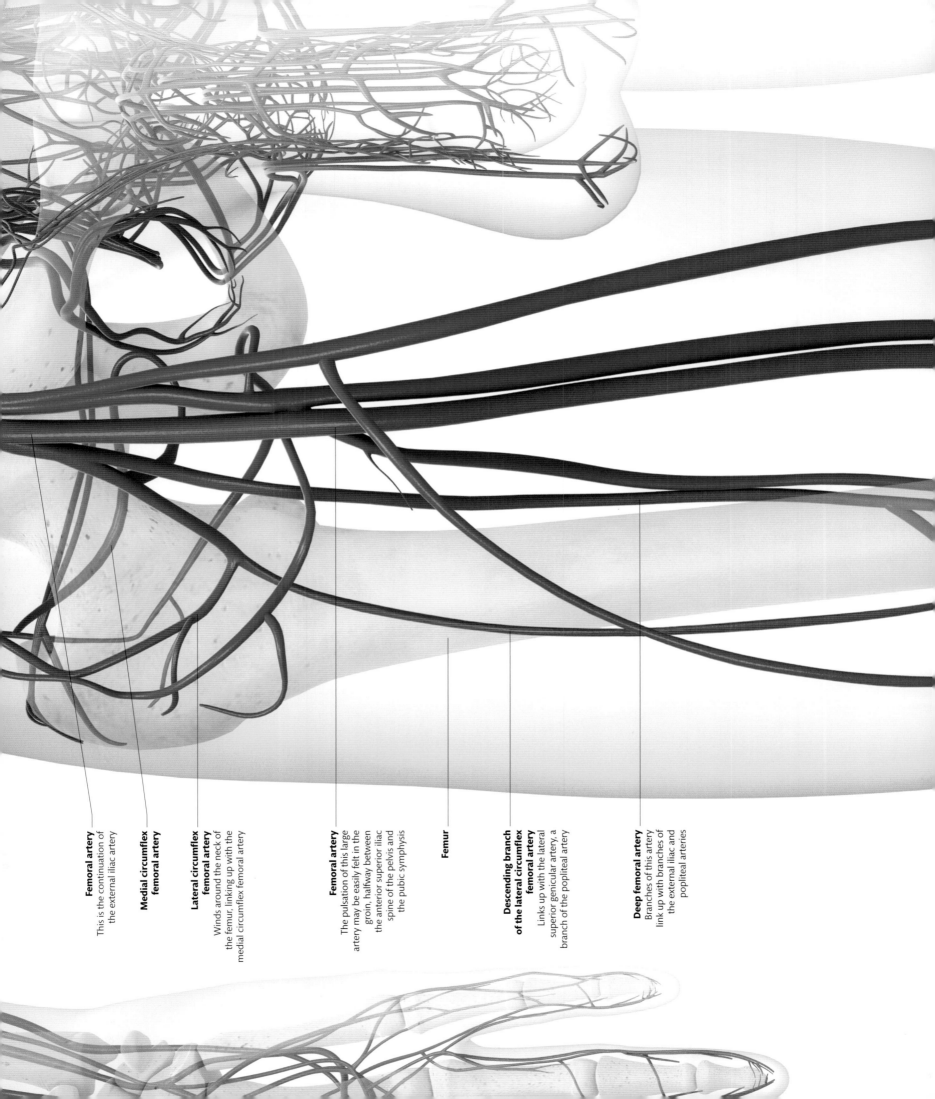

Femoral artery
This is the continuation of the external iliac artery

Medial circumflex femoral artery

Lateral circumflex femoral artery
Winds around the neck of the femur, linking up with the medial circumflex femoral artery

Femoral artery
The pulsation of this large artery may be easily felt in the groin, halfway between the anterior superior iliac spine of the pelvis and the pubic symphysis

Femur

Descending branch of the lateral circumflex femoral artery
Links up with the lateral superior genicular artery, a branch of the popliteal artery

Deep femoral artery
Branches of this artery link up with branches of the external iliac and popliteal arteries

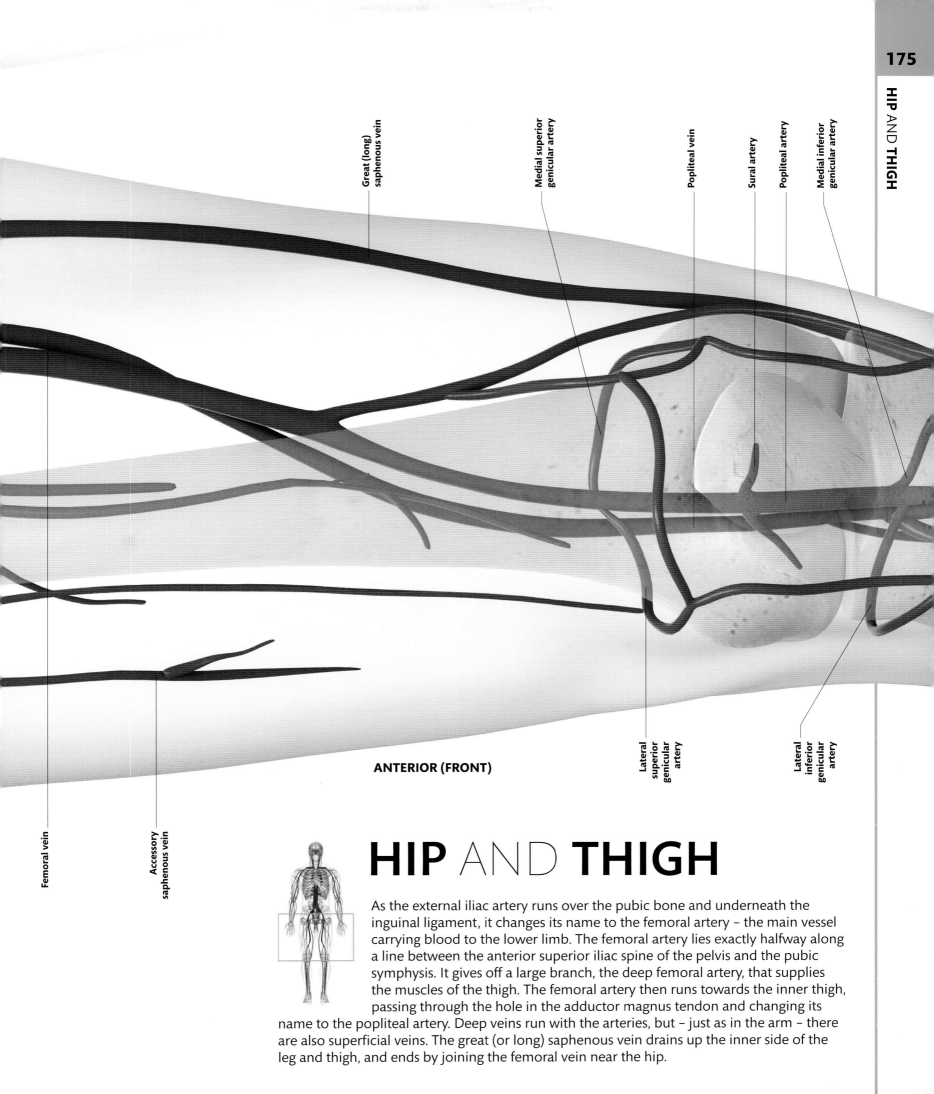

Great (long) saphenous vein

Medial superior genicular artery

Popliteal vein

Sural artery

Popliteal artery

Medial inferior genicular artery

Lateral superior genicular artery

Lateral inferior genicular artery

ANTERIOR (FRONT)

Femoral vein

Accessory saphenous vein

HIP AND **THIGH**

As the external iliac artery runs over the pubic bone and underneath the inguinal ligament, it changes its name to the femoral artery – the main vessel carrying blood to the lower limb. The femoral artery lies exactly halfway along a line between the anterior superior iliac spine of the pelvis and the pubic symphysis. It gives off a large branch, the deep femoral artery, that supplies the muscles of the thigh. The femoral artery then runs towards the inner thigh, passing through the hole in the adductor magnus tendon and changing its name to the popliteal artery. Deep veins run with the arteries, but – just as in the arm – there are also superficial veins. The great (or long) saphenous vein drains up the inner side of the leg and thigh, and ends by joining the femoral vein near the hip.

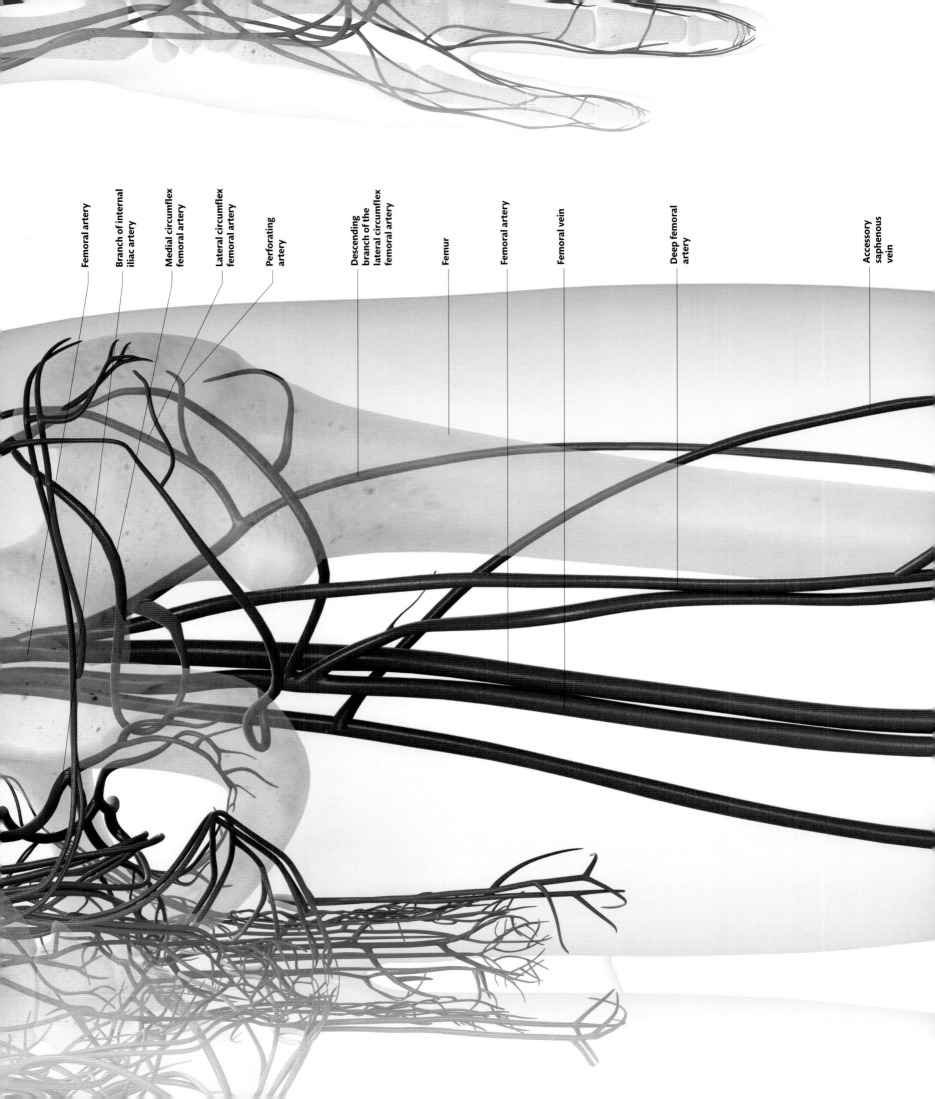

Femoral artery

Branch of internal iliac artery

Medial circumflex femoral artery

Lateral circumflex femoral artery

Perforating artery

Descending branch of the lateral circumflex femoral artery

Femur

Femoral artery

Femoral vein

Deep femoral artery

Accessory saphenous vein

HIP AND THIGH

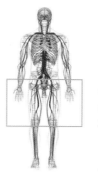

In this back view, gluteal branches of the internal iliac artery can be clearly seen, emerging through the greater sciatic foramen to supply the buttock. The muscles and skin of the inner part and back of the thigh are supplied by branches of the deep femoral artery. These are the perforating arteries, so-called because they pierce through the adductor magnus muscle. Higher up, the circumflex femoral arteries encircle the femur. The popliteal artery, formed after the femoral artery passes through the hiatus (gap) in adductor magnus, lies on the back of the femur, deep to the popliteal vein.

POSTERIOR (BACK)

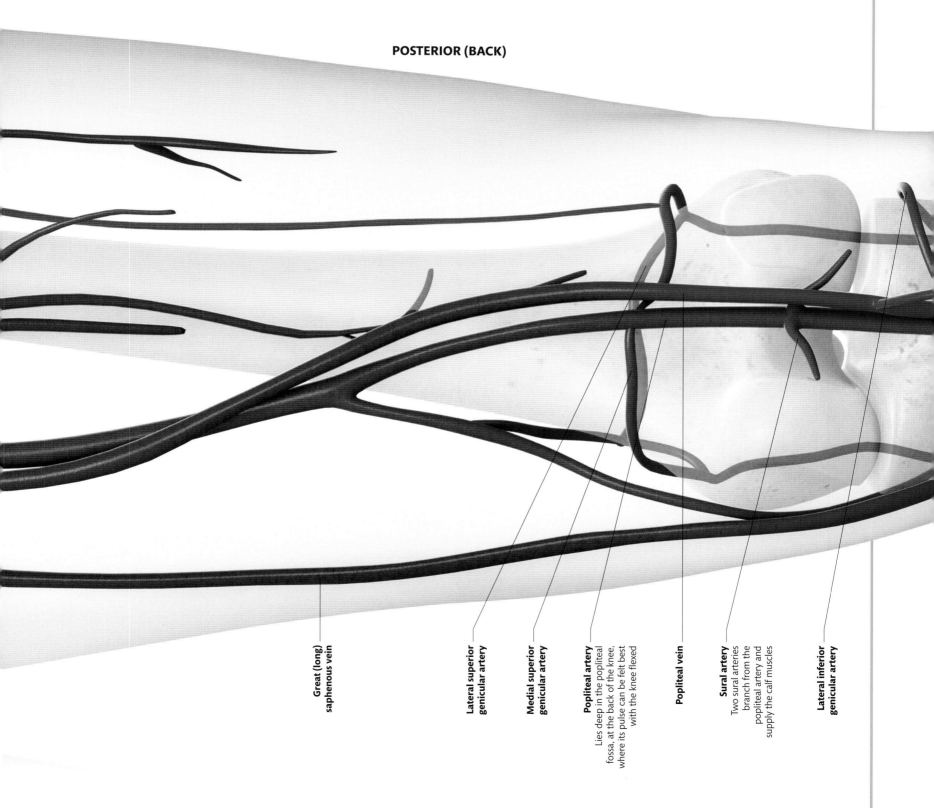

Great (long) saphenous vein

Lateral superior genicular artery

Medial superior genicular artery

Popliteal artery
Lies deep in the popliteal fossa, at the back of the knee, where its pulse can be felt best with the knee flexed

Popliteal vein

Sural artery
Two sural arteries branch from the popliteal artery and supply the calf muscles

Lateral inferior genicular artery

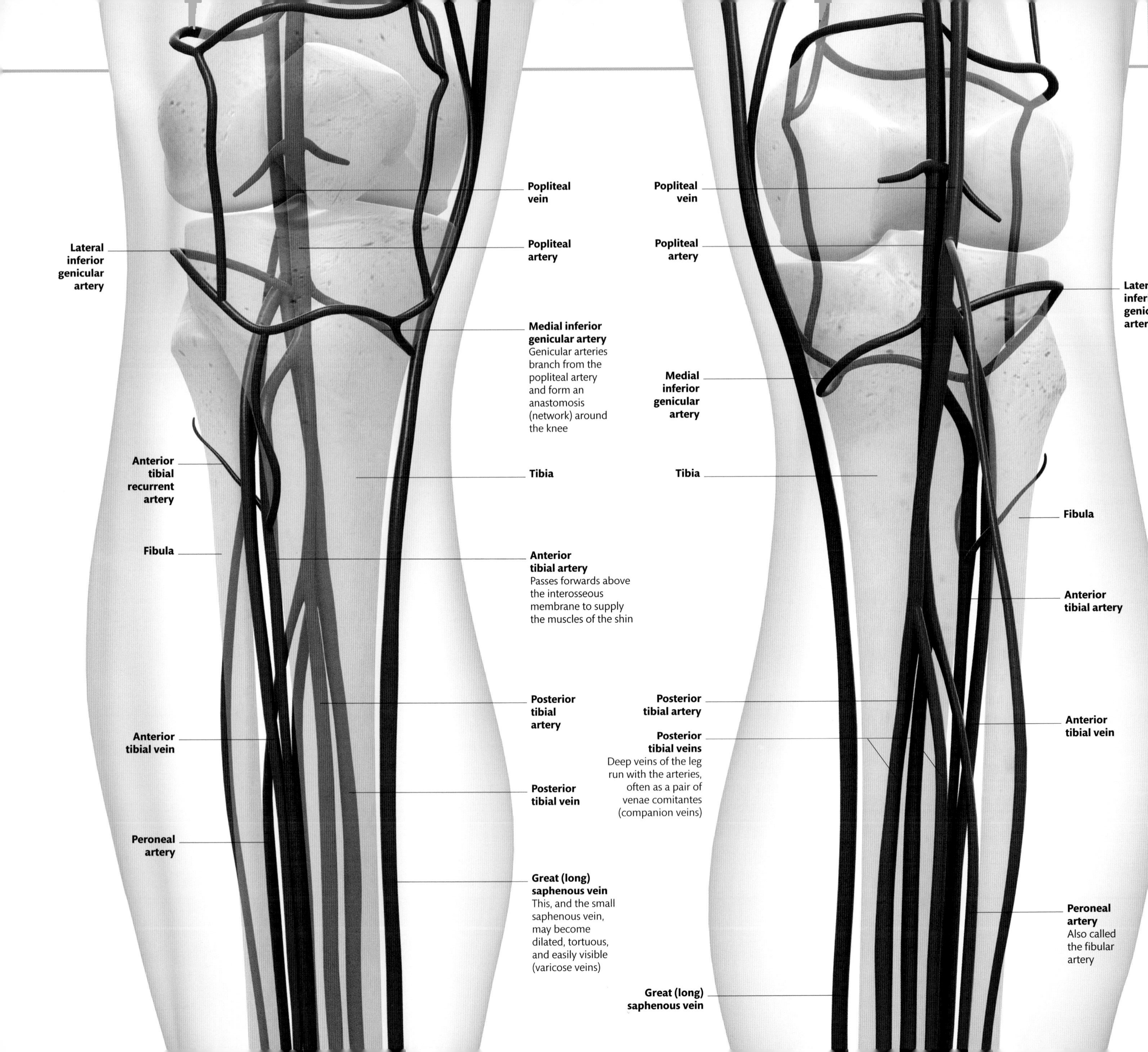

Lateral inferior genicular artery

Fibula

Anterior tibial artery

Anterior tibial vein

Peroneal artery
Also called the fibular artery

Popliteal vein

Popliteal artery

Medial inferior genicular artery

Tibia

Posterior tibial artery

Posterior tibial veins
Deep veins of the leg run with the arteries, often as a pair of venae comitantes (companion veins)

Great (long) saphenous vein

Popliteal vein

Popliteal artery

Medial inferior genicular artery
Genicular arteries branch from the popliteal artery and form an anastomosis (network) around the knee

Tibia

Anterior tibial artery
Passes forwards above the interosseous membrane to supply the muscles of the shin

Posterior tibial artery

Posterior tibial vein

Great (long) saphenous vein
This, and the small saphenous vein, may become dilated, tortuous, and easily visible (varicose veins)

Anterior tibial recurrent artery

Fibula

Anterior tibial vein

Peroneal artery

Lateral inferior genicular artery

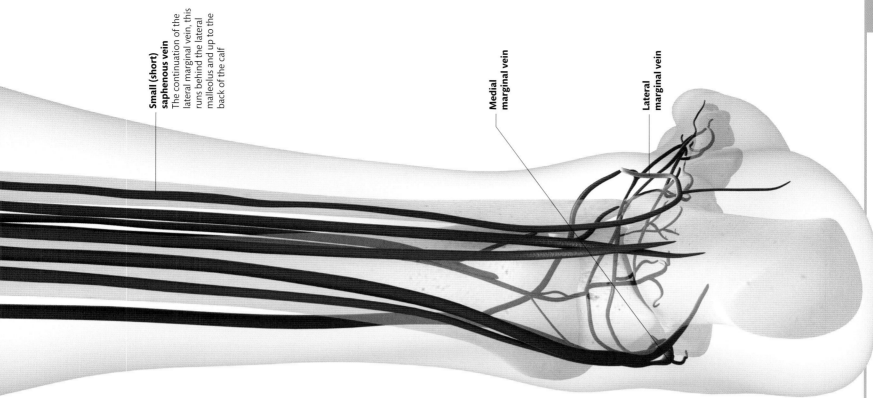

Small (short) saphenous vein
The continuation of the lateral marginal vein, this runs behind the lateral malleolus and up to the back of the calf

Medial marginal vein

Lateral marginal vein

POSTERIOR (BACK)

LOWER LEG AND FOOT

The popliteal artery runs deep across the back of the knee, dividing into two branches: the anterior and posterior tibial arteries. The former runs forwards, piercing the interosseous membrane between the tibia and fibula, to supply the extensor muscles of the shin. It runs down past the ankle, onto the top of the foot, as the dorsalis pedis artery. The latter gives off a peroneal branch, supplying the muscles and skin on the leg's outer side. The posterior tibial artery itself continues in the calf, running with the tibial nerve and, like the nerve, divides into plantar branches to supply the sole of the foot. There is a network of superficial veins on the back of the foot, which is drained by the saphenous veins.

ANTERIOR (FRONT)

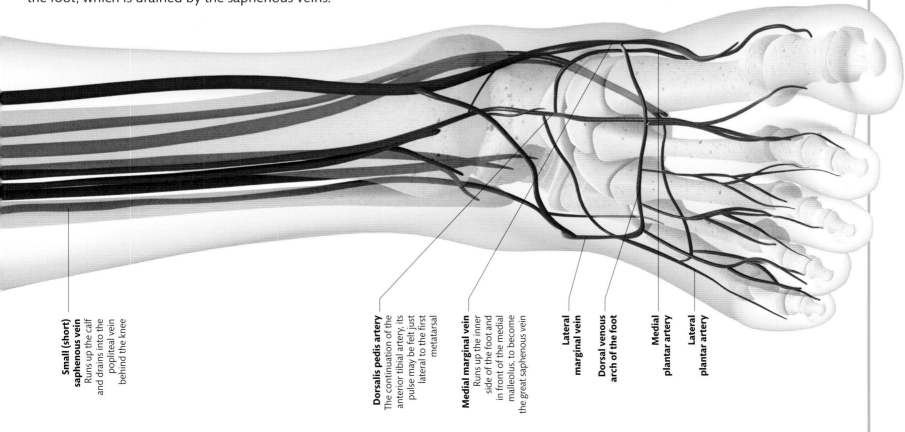

Small (short) saphenous vein
Runs up the calf and drains into the popliteal vein behind the knee

Dorsalis pedis artery
The continuation of the anterior tibial artery, its pulse may be felt just lateral to the first metatarsal

Medial marginal vein
Runs up the inner side of the foot and in front of the medial malleolus, to become the great saphenous vein

Lateral marginal vein

Dorsal venous arch of the foot

Medial plantar artery

Lateral plantar artery

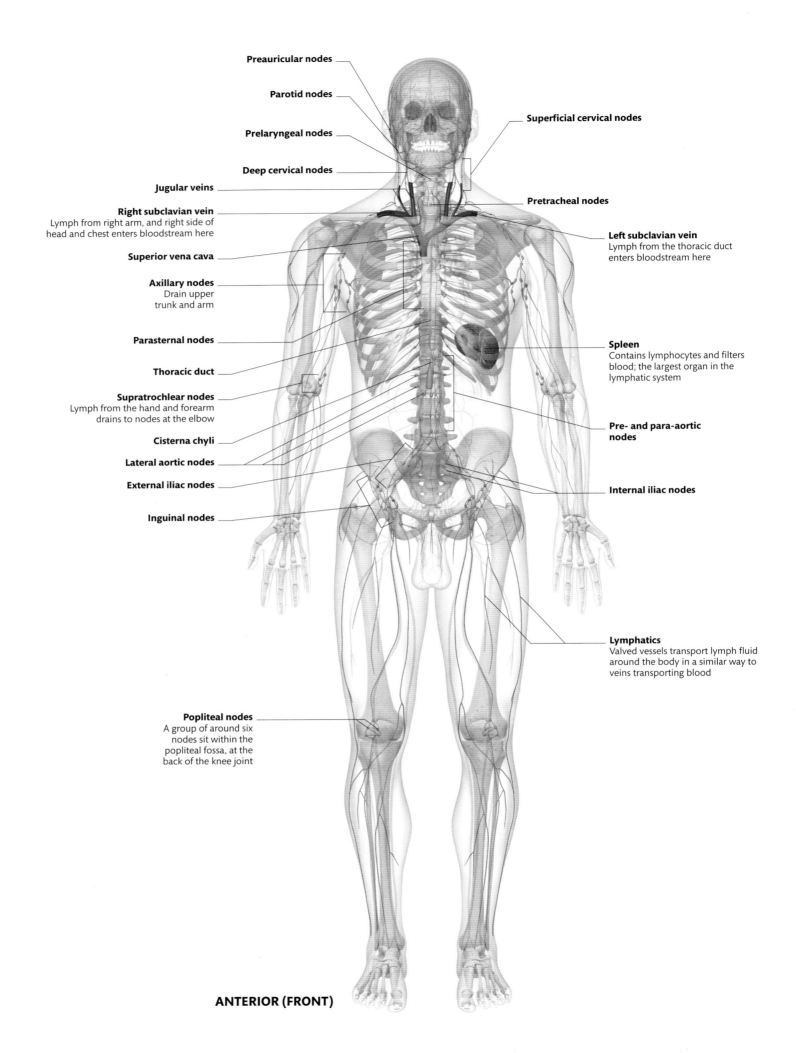

Preauricular nodes

Parotid nodes

Prelaryngeal nodes

Deep cervical nodes

Jugular veins

Right subclavian vein
Lymph from right arm, and right side of
head and chest enters bloodstream here

Superior vena cava

Axillary nodes
Drain upper
trunk and arm

Parasternal nodes

Thoracic duct

Supratrochlear nodes
Lymph from the hand and forearm
drains to nodes at the elbow

Cisterna chyli

Lateral aortic nodes

External iliac nodes

Inguinal nodes

Popliteal nodes
A group of around six
nodes sit within the
popliteal fossa, at the
back of the knee joint

Superficial cervical nodes

Pretracheal nodes

Left subclavian vein
Lymph from the thoracic duct
enters bloodstream here

Spleen
Contains lymphocytes and filters
blood; the largest organ in the
lymphatic system

**Pre- and para-aortic
nodes**

Internal iliac nodes

Lymphatics
Valved vessels transport lymph fluid
around the body in a similar way to
veins transporting blood

ANTERIOR (FRONT)

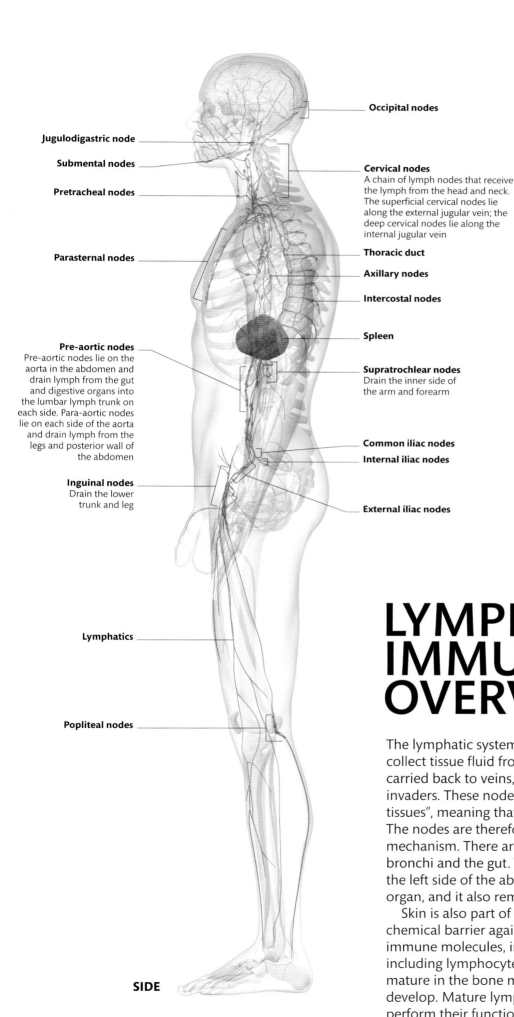

Occipital nodes

Jugulodigastric node

Submental nodes

Pretracheal nodes

Cervical nodes
A chain of lymph nodes that receive the lymph from the head and neck. The superficial cervical nodes lie along the external jugular vein; the deep cervical nodes lie along the internal jugular vein

Parasternal nodes

Thoracic duct

Axillary nodes

Intercostal nodes

Spleen

Pre-aortic nodes
Pre-aortic nodes lie on the aorta in the abdomen and drain lymph from the gut and digestive organs into the lumbar lymph trunk on each side. Para-aortic nodes lie on each side of the aorta and drain lymph from the legs and posterior wall of the abdomen

Supratrochlear nodes
Drain the inner side of the arm and forearm

Common iliac nodes

Internal iliac nodes

Inguinal nodes
Drain the lower trunk and leg

External iliac nodes

Lymphatics

Popliteal nodes

SIDE

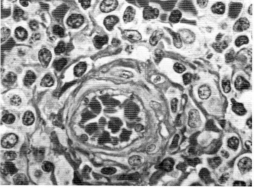

Lymphoid tissue
At a high magnification, individual lymphocytes (purple) can be seen in a section of lymphoid tissue. The blue circle in the image is an arteriole, packed full of blood cells (stained pink).

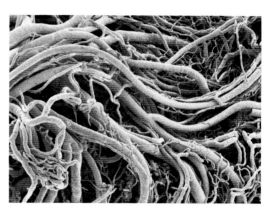

Blood vessels of lymph node
This image, produced using a scanning electron microscope, shows a resin cast of the dense network of tiny blood vessels inside a lymph node.

LYMPHATIC AND IMMUNE SYSTEM OVERVIEW

The lymphatic system consists of a network of lymphatic vessels that collect tissue fluid from the spaces between cells. Before this fluid is carried back to veins, it is delivered to lymph nodes to check for potential invaders. These nodes, like the tonsils, spleen, and thymus, are "lymphoid tissues", meaning that they contain immune cells known as lymphocytes. The nodes are therefore part of the immune system, the body's defence mechanism. There are also patches of lymphoid tissue in the walls of the bronchi and the gut. The spleen, which lies tucked up under the ribs on the left side of the abdomen, has two important roles: it is a lymphoid organ, and it also removes old red blood cells from the circulation.

Skin is also part of the immune system as it forms a physical and chemical barrier against infections. The formation of some important immune molecules, including antibodies, and a range of immune cells, including lymphocytes, happens in the bone marrow. Some lymphocytes mature in the bone marrow, whereas others move to the thymus to develop. Mature lymphocytes stay in the lymph nodes, where they perform their function.

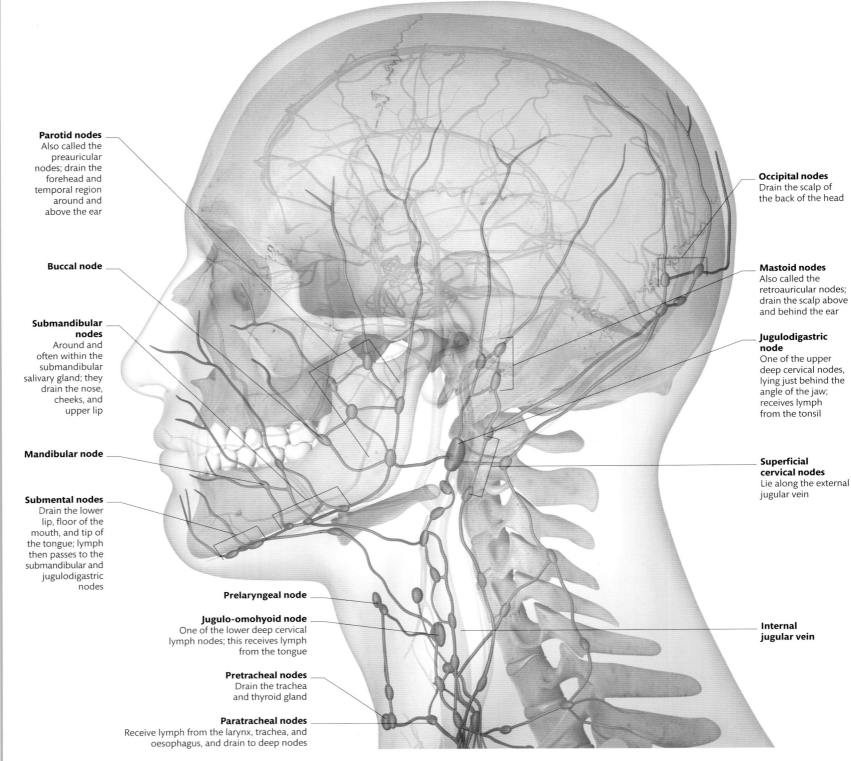

Parotid nodes
Also called the preauricular nodes; drain the forehead and temporal region around and above the ear

Buccal node

Submandibular nodes
Around and often within the submandibular salivary gland; they drain the nose, cheeks, and upper lip

Mandibular node

Submental nodes
Drain the lower lip, floor of the mouth, and tip of the tongue; lymph then passes to the submandibular and jugulodigastric nodes

Occipital nodes
Drain the scalp of the back of the head

Mastoid nodes
Also called the retroauricular nodes; drain the scalp above and behind the ear

Jugulodigastric node
One of the upper deep cervical nodes, lying just behind the angle of the jaw; receives lymph from the tonsil

Superficial cervical nodes
Lie along the external jugular vein

Internal jugular vein

Prelaryngeal node

Jugulo-omohyoid node
One of the lower deep cervical lymph nodes; this receives lymph from the tongue

Pretracheal nodes
Drain the trachea and thyroid gland

Paratracheal nodes
Receive lymph from the larynx, trachea, and oesophagus, and drain to deep nodes

LYMPH NODES OF HEAD

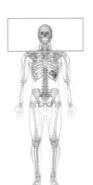

HEAD AND NECK

There is a ring of lymph nodes lying close to the skin where the head meets the neck, from the occipital nodes (against the skull at the back) to the submandibular and submental nodes (which are tucked under the jaw). Superficial nodes lie along the sides and front of the neck, and deep nodes are clustered around the internal jugular vein, under cover of sternocleidomastoid muscle. Lymph from all other nodes passes to these deep ones, then into the jugular lymphatic trunk before draining back into veins in the base of the neck. Lymphoid tissue, in the form of the palatine, pharyngeal, and lingual tonsils, forms a protective ring around the upper parts of the respiratory and digestive tracts.

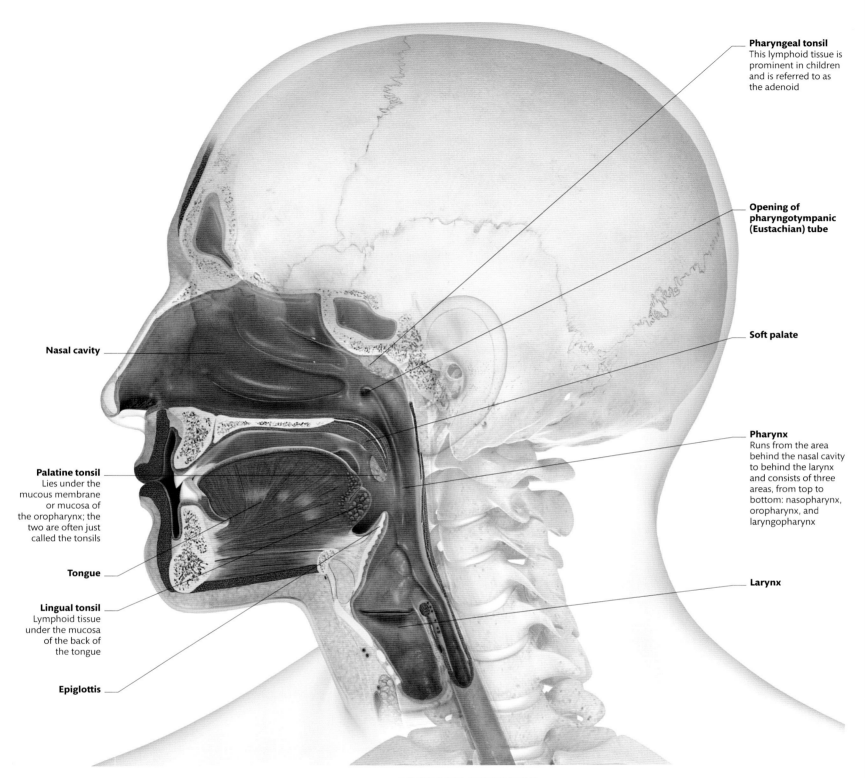

Pharyngeal tonsil
This lymphoid tissue is prominent in children and is referred to as the adenoid

Opening of pharyngotympanic (Eustachian) tube

Soft palate

Nasal cavity

Pharynx
Runs from the area behind the nasal cavity to behind the larynx and consists of three areas, from top to bottom: nasopharynx, oropharynx, and laryngopharynx

Palatine tonsil
Lies under the mucous membrane or mucosa of the oropharynx; the two are often just called the tonsils

Tongue

Larynx

Lingual tonsil
Lymphoid tissue under the mucosa of the back of the tongue

Epiglottis

LOCATION OF TONSILS

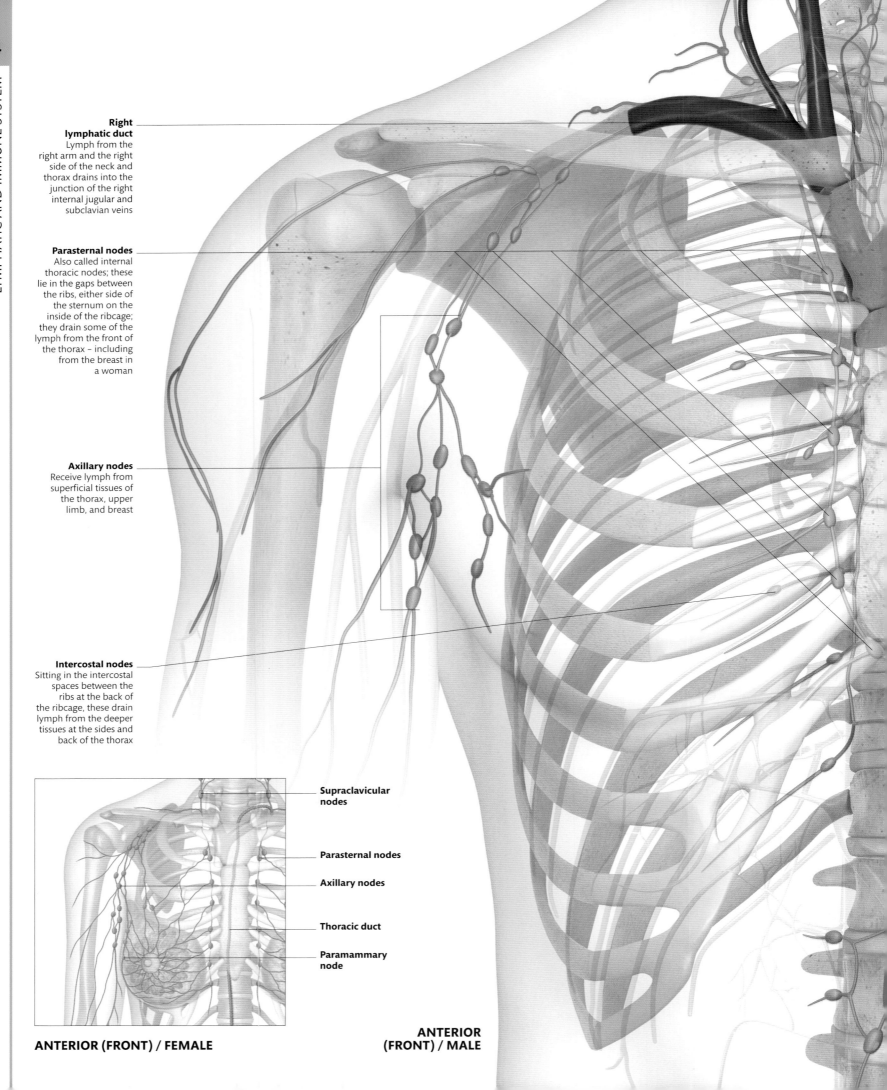

Right lymphatic duct
Lymph from the right arm and the right side of the neck and thorax drains into the junction of the right internal jugular and subclavian veins

Parasternal nodes
Also called internal thoracic nodes; these lie in the gaps between the ribs, either side of the sternum on the inside of the ribcage; they drain some of the lymph from the front of the thorax – including from the breast in a woman

Axillary nodes
Receive lymph from superficial tissues of the thorax, upper limb, and breast

Intercostal nodes
Sitting in the intercostal spaces between the ribs at the back of the ribcage, these drain lymph from the deeper tissues at the sides and back of the thorax

Supraclavicular nodes

Parasternal nodes

Axillary nodes

Thoracic duct

Paramammary node

ANTERIOR (FRONT) / FEMALE

ANTERIOR (FRONT) / MALE

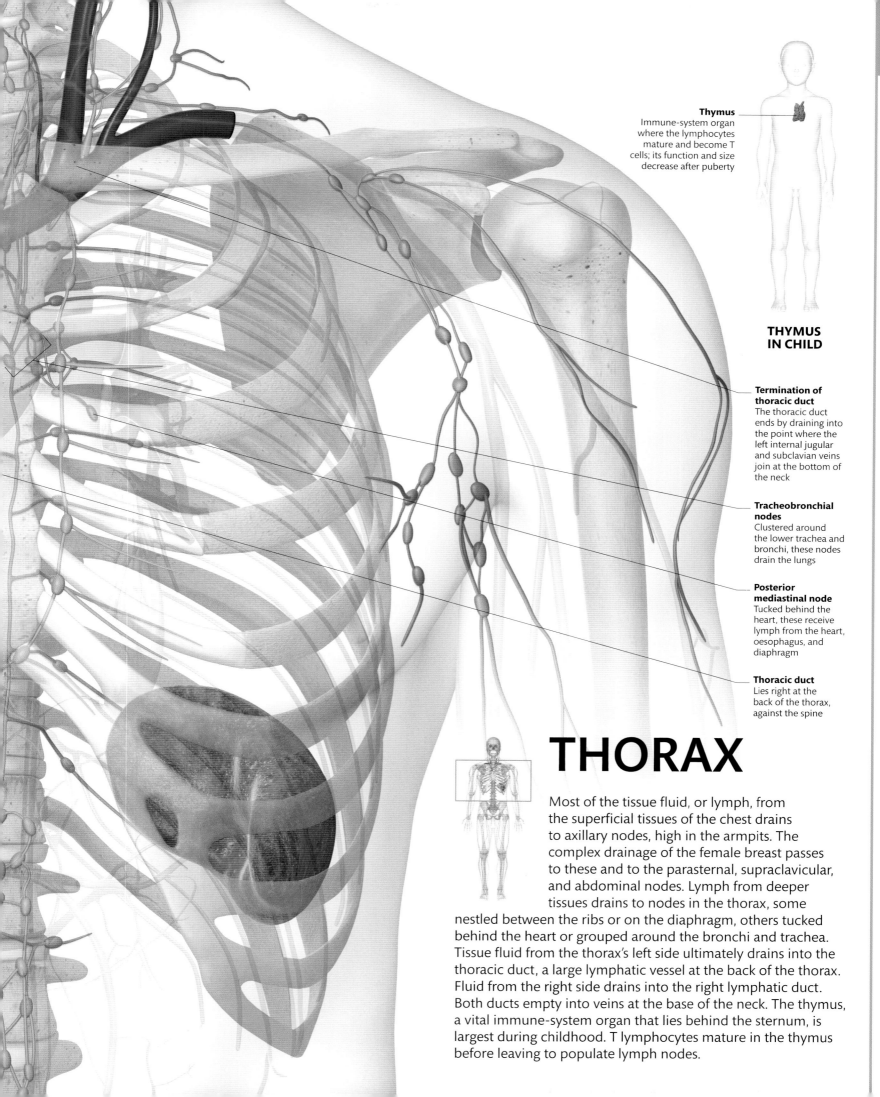

Thymus
Immune-system organ where the lymphocytes mature and become T cells; its function and size decrease after puberty

THYMUS IN CHILD

Termination of thoracic duct
The thoracic duct ends by draining into the point where the left internal jugular and subclavian veins join at the bottom of the neck

Tracheobronchial nodes
Clustered around the lower trachea and bronchi, these nodes drain the lungs

Posterior mediastinal node
Tucked behind the heart, these receive lymph from the heart, oesophagus, and diaphragm

Thoracic duct
Lies right at the back of the thorax, against the spine

THORAX

Most of the tissue fluid, or lymph, from the superficial tissues of the chest drains to axillary nodes, high in the armpits. The complex drainage of the female breast passes to these and to the parasternal, supraclavicular, and abdominal nodes. Lymph from deeper tissues drains to nodes in the thorax, some nestled between the ribs or on the diaphragm, others tucked behind the heart or grouped around the bronchi and trachea. Tissue fluid from the thorax's left side ultimately drains into the thoracic duct, a large lymphatic vessel at the back of the thorax. Fluid from the right side drains into the right lymphatic duct. Both ducts empty into veins at the base of the neck. The thymus, a vital immune-system organ that lies behind the sternum, is largest during childhood. T lymphocytes mature in the thymus before leaving to populate lymph nodes.

ABDOMEN AND PELVIS

The deep lymph nodes of the abdomen are clustered around arteries. Nodes lying along each side of the aorta receive lymph from paired structures, such as the muscles of the abdominal wall, the kidneys and adrenal glands, and the testes or ovaries. Iliac nodes collect lymph returning from the legs and pelvis. Nodes clustered around the branches on the front of the aorta collect lymph from the gut and abdominal organs. Eventually, all this lymph from the legs, pelvis, and abdomen passes into a swollen lymphatic vessel called the cisterna chyli; this narrows down to become the thoracic duct, which runs up into the chest. Most lymph nodes are small, bean-sized structures, but the abdomen also contains a large and important organ of the immune system – the spleen.

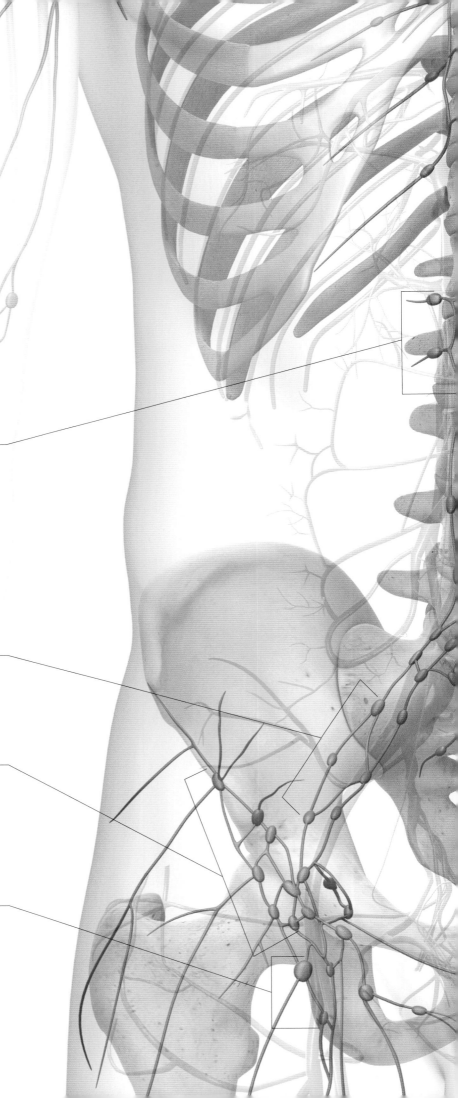

Lateral aortic nodes
Lying along each side of the aorta, these collect lymph from the kidneys, posterior abdominal wall, and pelvic viscera; they drain into the right and left intestinal trunks

External iliac nodes
Collect lymph from the inguinal nodes in the groin, the perineum, and the inner thigh

Proximal superficial inguinal nodes
Lying just below the inguinal ligament, this upper group of superficial inguinal nodes receives lymph from the lower abdominal wall, below the umbilicus, as well as from the external genitalia

Distal superficial inguinal nodes
The lower nodes in the groin drain most of the superficial lymphatics of the thigh and leg

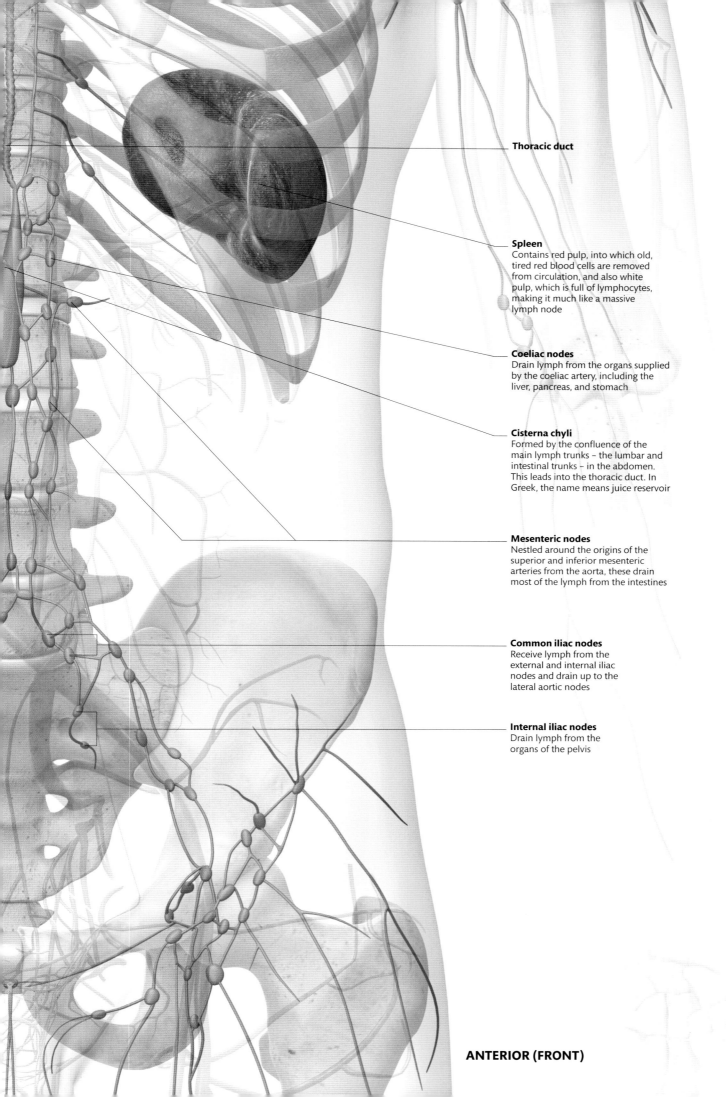

Thoracic duct

Spleen
Contains red pulp, into which old, tired red blood cells are removed from circulation, and also white pulp, which is full of lymphocytes, making it much like a massive lymph node

Coeliac nodes
Drain lymph from the organs supplied by the coeliac artery, including the liver, pancreas, and stomach

Cisterna chyli
Formed by the confluence of the main lymph trunks – the lumbar and intestinal trunks – in the abdomen. This leads into the thoracic duct. In Greek, the name means juice reservoir

Mesenteric nodes
Nestled around the origins of the superior and inferior mesenteric arteries from the aorta, these drain most of the lymph from the intestines

Common iliac nodes
Receive lymph from the external and internal iliac nodes and drain up to the lateral aortic nodes

Internal iliac nodes
Drain lymph from the organs of the pelvis

ANTERIOR (FRONT)

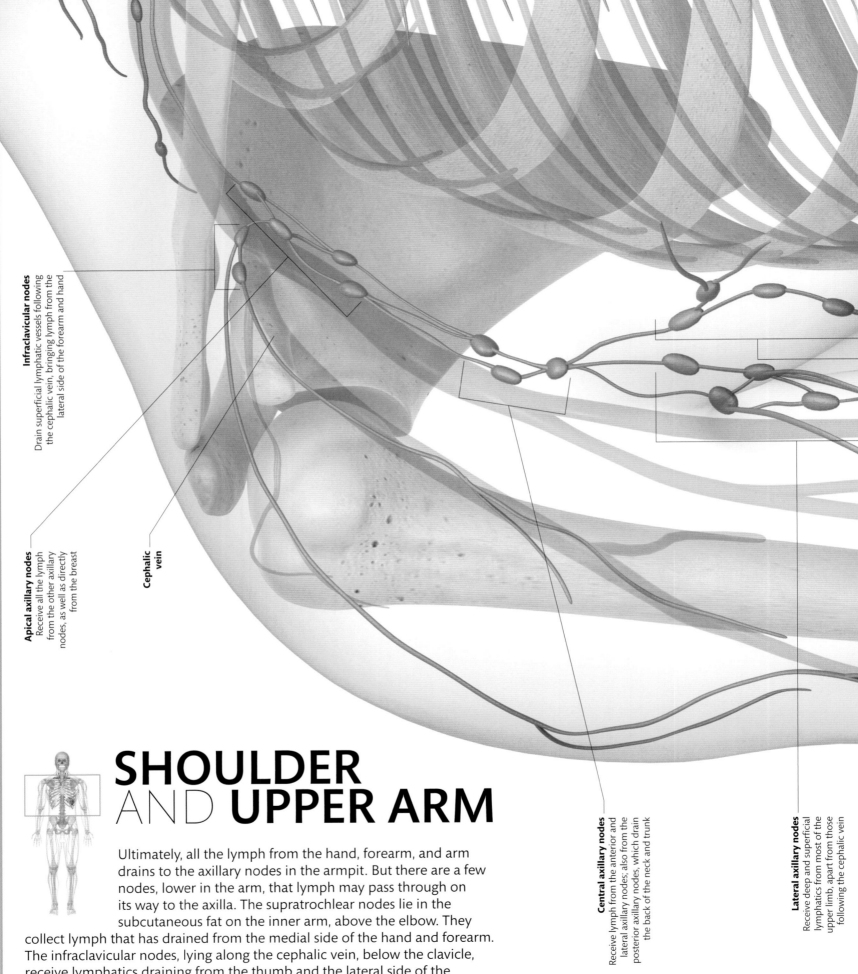

Infraclavicular nodes
Drain superficial lymphatic vessels following the cephalic vein, bringing lymph from the lateral side of the forearm and hand

Apical axillary nodes
Receive all the lymph from the other axillary nodes, as well as directly from the breast

Cephalic vein

Central axillary nodes
Receive lymph from the anterior and lateral axillary nodes; also from the posterior axillary nodes, which drain the back of the neck and trunk

Lateral axillary nodes
Receive deep and superficial lymphatics from most of the upper limb, apart from those following the cephalic vein

SHOULDER
AND **UPPER ARM**

Ultimately, all the lymph from the hand, forearm, and arm drains to the axillary nodes in the armpit. But there are a few nodes, lower in the arm, that lymph may pass through on its way to the axilla. The supratrochlear nodes lie in the subcutaneous fat on the inner arm, above the elbow. They collect lymph that has drained from the medial side of the hand and forearm. The infraclavicular nodes, lying along the cephalic vein, below the clavicle, receive lymphatics draining from the thumb and the lateral side of the forearm and arm. Axillary nodes drain lymph from the arm and receive it from the chest wall. They may become infiltrated with cancerous cells spreading from a tumour in the breast.

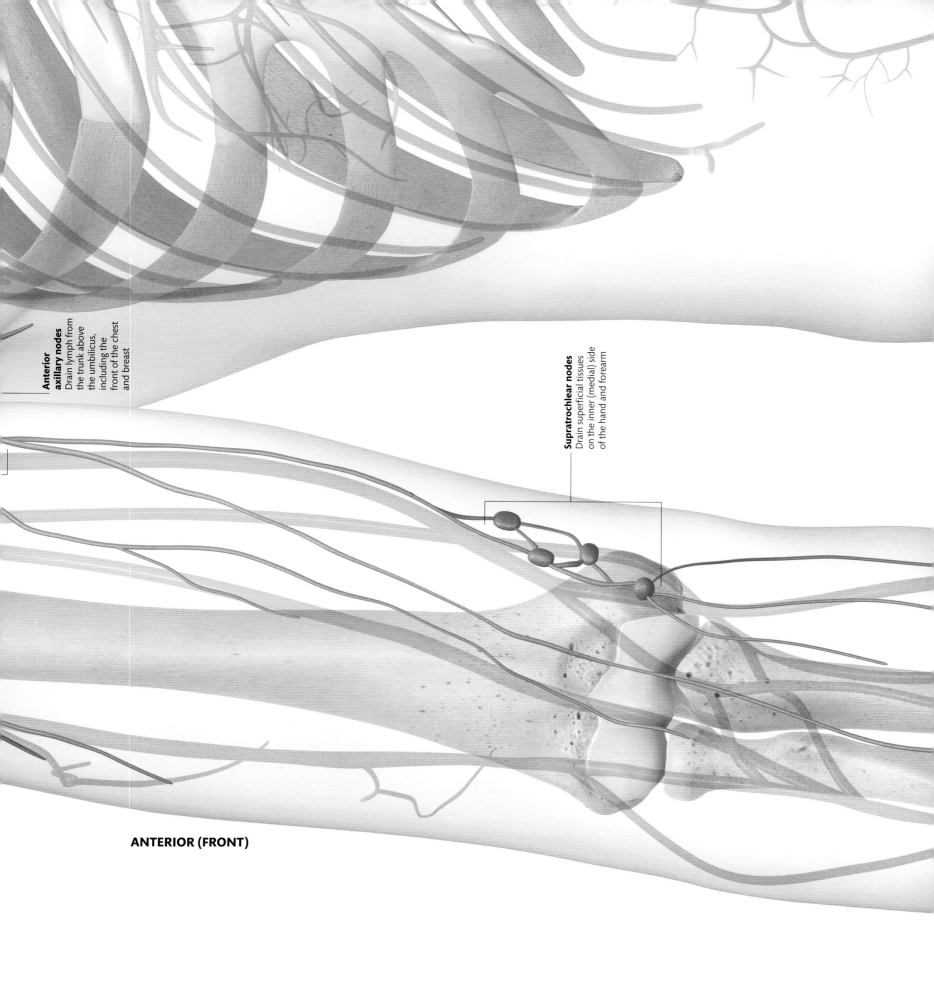

Anterior axillary nodes
Drain lymph from the trunk above the umbilicus, including the front of the chest and breast

Supratrochlear nodes
Drain superficial tissues on the inner (medial) side of the hand and forearm

ANTERIOR (FRONT)

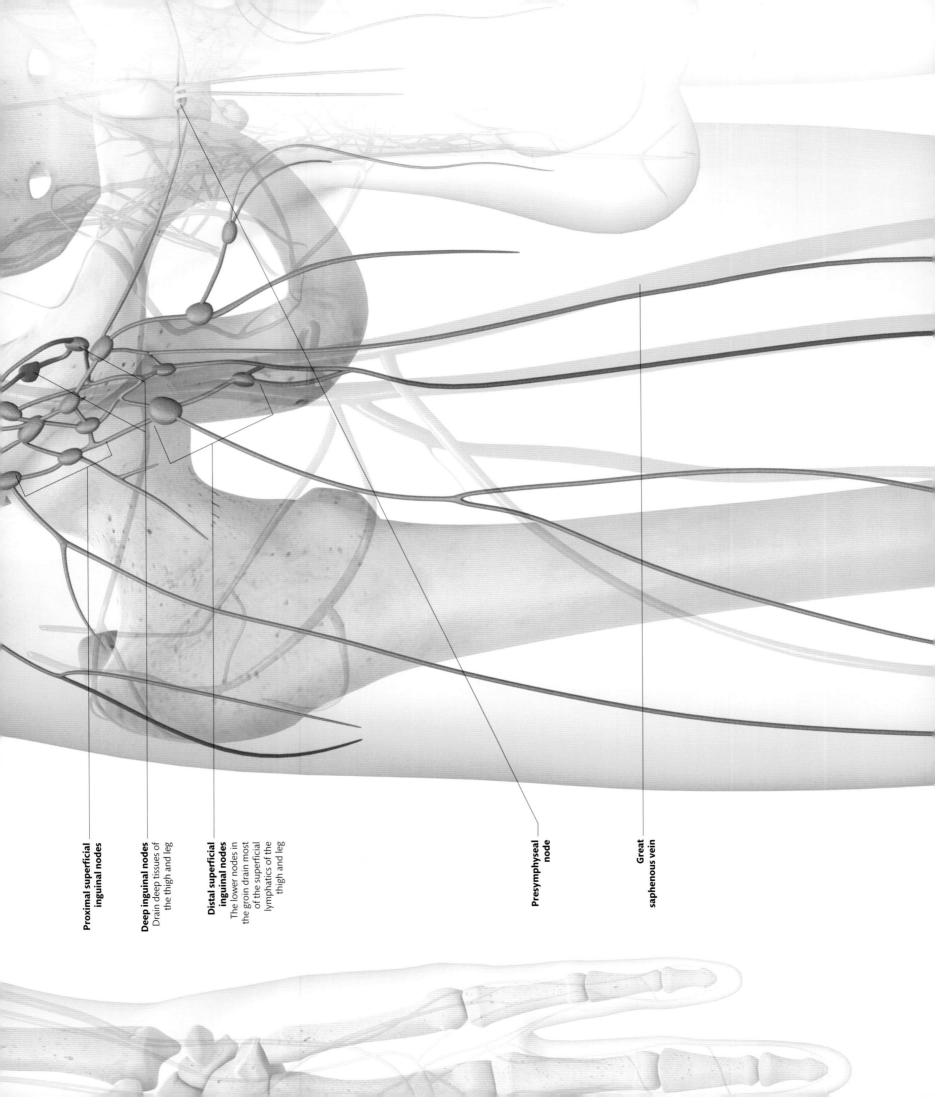

Proximal superficial inguinal nodes

Deep inguinal nodes
Drain deep tissues of the thigh and leg

Distal superficial inguinal nodes
The lower nodes in the groin drain most of the superficial lymphatics of the thigh and leg

Presymphyseal node

Great saphenous vein

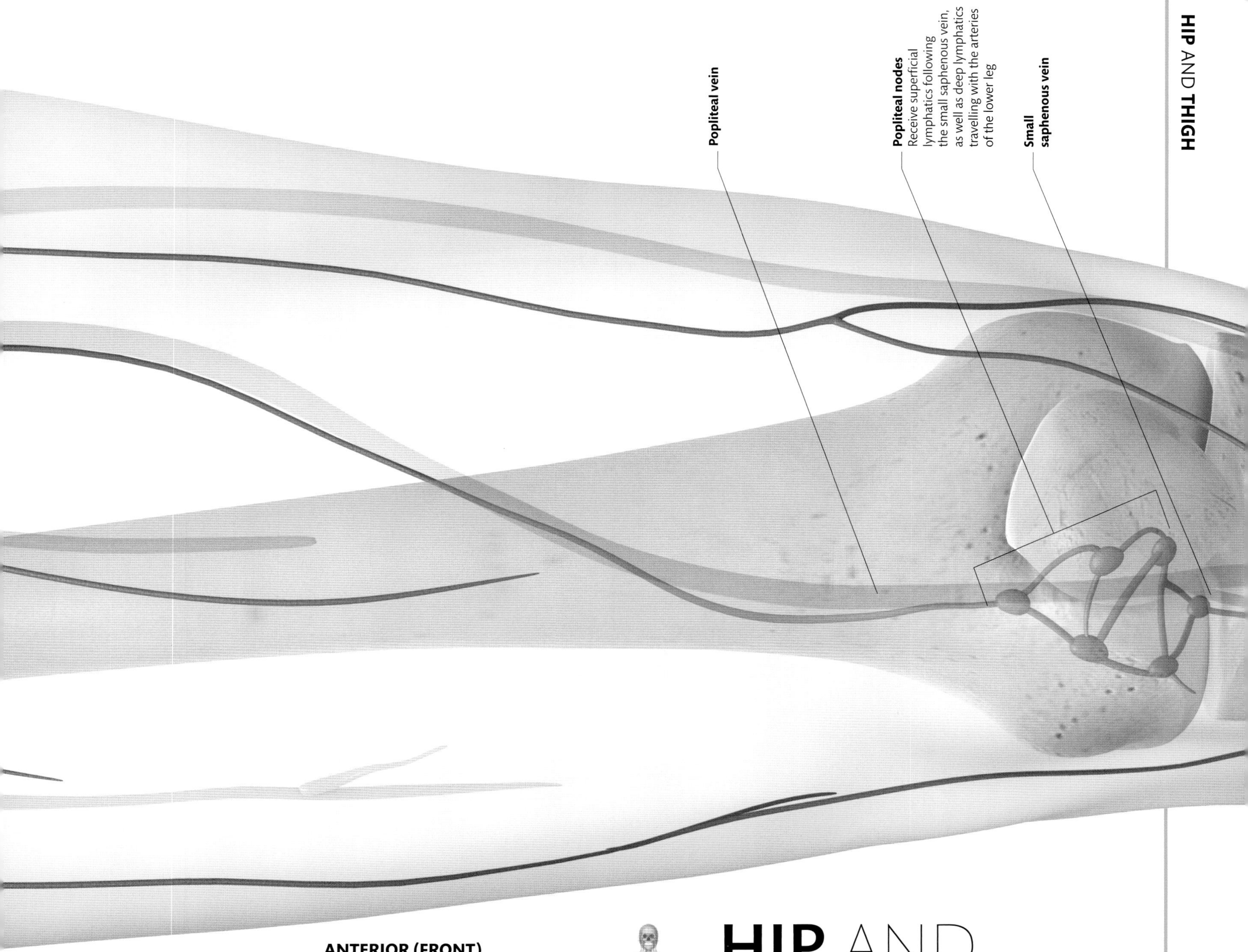

Popliteal vein

Popliteal nodes
Receive superficial lymphatics following the small saphenous vein, as well as deep lymphatics travelling with the arteries of the lower leg

Small saphenous vein

ANTERIOR (FRONT)

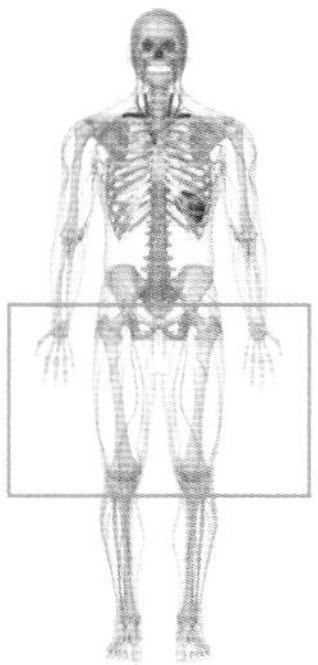

HIP AND THIGH

Most lymph from the thigh, leg, and foot passes through the inguinal group of lymph nodes, which are in the groin. But lymph from the deep tissues of the buttock passes straight to nodes inside the pelvis (see pp.184–85), along the internal and common iliac arteries. Eventually, all the lymph from the leg reaches the lateral aortic nodes, on the back wall of the abdomen. As in the arm, there are groups of nodes clustered around points at which superficial veins drain into deep veins. Popliteal nodes are close to the drainage of the small saphenous vein into the popliteal vein, while the superficial inguinal nodes lie close to the great saphenous vein, just before it empties into the femoral vein.

DIGESTIVE SYSTEM **OVERVIEW**

The digestive system comprises the organs that enable us to take in food, break it down physically and chemically, extract useful nutrients from it, and excrete what we don't need. This process begins in the mouth, where the teeth, tongue, and saliva work together to form a food into a moist ball that can be swallowed. The mouth, pharynx, stomach, intestines, rectum, and anal canal form a long tube that is referred to as the digestive tract. It usually takes between one and two days for ingested food to travel all the way from the mouth to the anus. Other organs – including the salivary glands, liver, gallbladder, and pancreas – complete the digestive system.

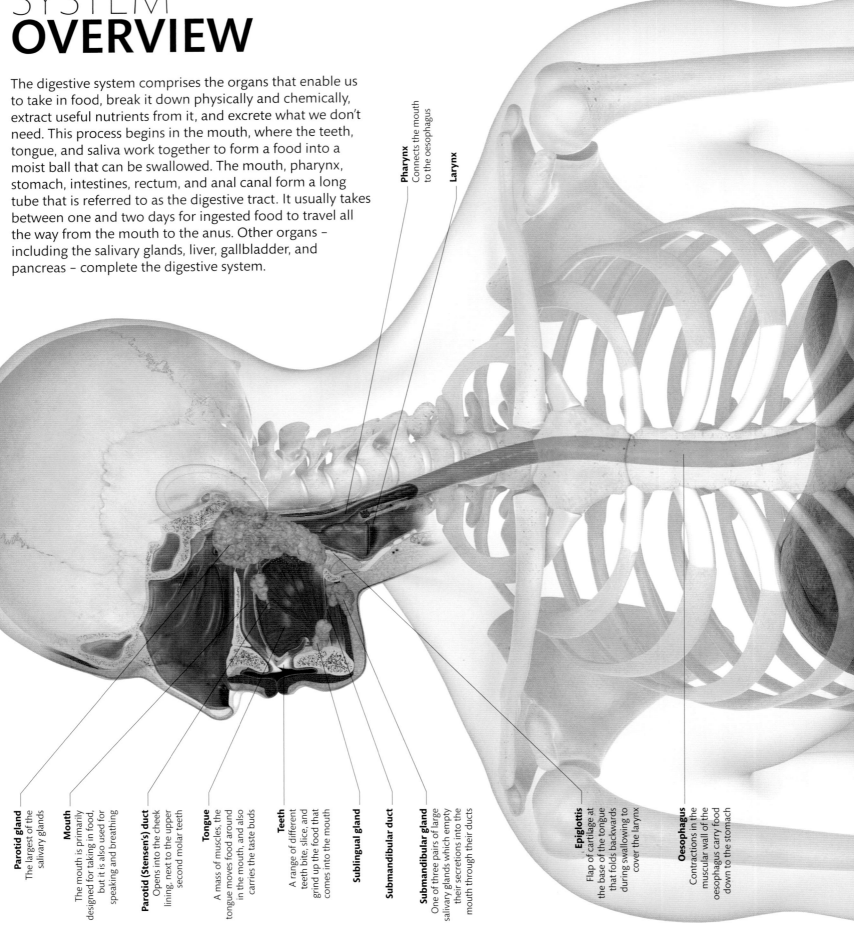

Pharynx
Connects the mouth to the oesophagus

Larynx

Parotid gland
The largest of the salivary glands

Mouth
The mouth is primarily designed for taking in food, but it is also used for speaking and breathing

Parotid (Stensen's) duct
Opens into the cheek lining, next to the upper second molar teeth

Tongue
A mass of muscles, the tongue moves food around in the mouth, and also carries the taste buds

Teeth
A range of different teeth bite, slice, and grind up the food that comes into the mouth

Sublingual gland

Submandibular duct

Submandibular gland
One of three pairs of large salivary glands which empty their secretions into the mouth through their ducts

Epiglottis
Flap of cartilage at the base of the tongue that folds backwards during swallowing to cover the larynx

Oesophagus
Contractions in the muscular wall of the oesophagus carry food down to the stomach

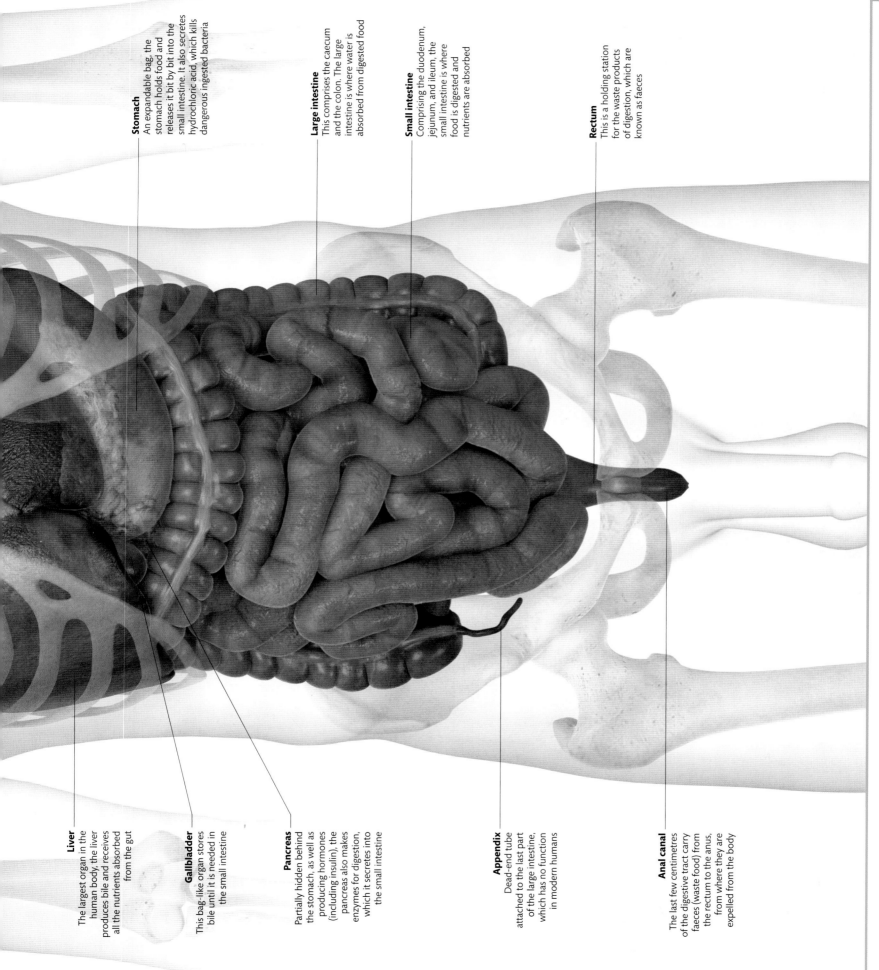

Stomach
An expandable bag, the stomach holds food and releases it bit by bit into the small intestine. It also secretes hydrochloric acid, which kills dangerous ingested bacteria

Large intestine
This comprises the caecum and the colon. The large intestine is where water is absorbed from digested food

Small intestine
Comprising the duodenum, jejunum, and ileum, the small intestine is where food is digested and nutrients are absorbed

Rectum
This is a holding station for the waste products of digestion, which are known as faeces

Liver
The largest organ in the human body, the liver produces bile and receives all the nutrients absorbed from the gut

Gallbladder
This bag-like organ stores bile until it is needed in the small intestine

Pancreas
Partially hidden behind the stomach, as well as producing hormones (including insulin), the pancreas also makes enzymes for digestion, which it secretes into the small intestine

Appendix
Dead-end tube attached to the last part of the large intestine, which has no function in modern humans

Anal canal
The last few centimetres of the digestive tract carry faeces (waste food) from the rectum to the anus, from where they are expelled from the body

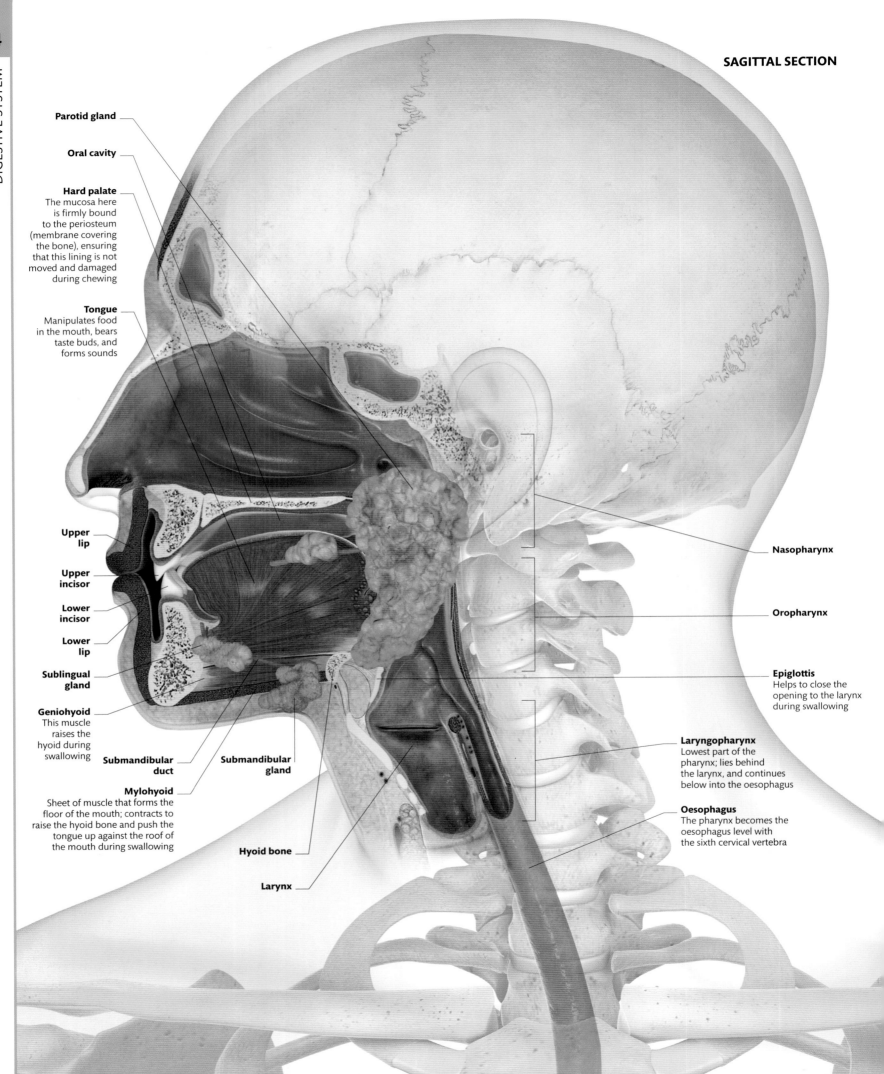

Parotid gland

Oral cavity

Hard palate
The mucosa here is firmly bound to the periosteum (membrane covering the bone), ensuring that this lining is not moved and damaged during chewing

Tongue
Manipulates food in the mouth, bears taste buds, and forms sounds

Upper lip

Upper incisor

Lower incisor

Lower lip

Sublingual gland

Geniohyoid
This muscle raises the hyoid during swallowing

Submandibular duct

Mylohyoid
Sheet of muscle that forms the floor of the mouth; contracts to raise the hyoid bone and push the tongue up against the roof of the mouth during swallowing

Submandibular gland

Hyoid bone

Larynx

Nasopharynx

Oropharynx

Epiglottis
Helps to close the opening to the larynx during swallowing

Laryngopharynx
Lowest part of the pharynx; lies behind the larynx, and continues below into the oesophagus

Oesophagus
The pharynx becomes the oesophagus level with the sixth cervical vertebra

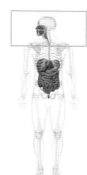

HEAD AND NECK

The mouth is the first part of the digestive tract, and it is here that the processes of mechanical and chemical digestion get underway. Your teeth grind each mouthful, and you have three pairs of major salivary glands – parotid, submandibular, and sublingual – that secrete saliva through ducts into the mouth. Saliva contains digestive enzymes that begin to chemically break down the food in your mouth. The tongue manipulates the food, and also has taste buds that allow you to quickly make the important distinction between delicious food and potentially harmful toxins. As you swallow, the tongue pushes up against the hard palate, the soft palate seals off the airway, and the muscular tube of the pharynx contracts in a wave to push the ball of food down into the oesophagus, ready for the next stage of its journey.

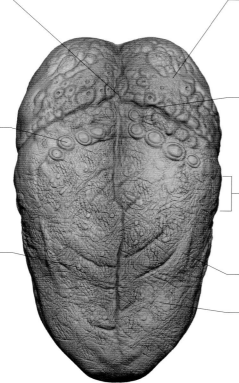

Foramen caecum
This small, blind hole at the back of the tongue is a remnant of where the thyroid gland started to develop in the embryo, before it dropped down into the neck

Vallate papillae
There are around a dozen of these large papillae at the back of the tongue; each one is surrounded by a circular furrow that contains taste buds

Fungiform papilla
Literally means mushroom-shaped; these are scattered over the tongue like mushrooms across the lawn of filiform papillae; fungiform papillae also bear taste buds

Pharyngeal part of tongue
Lymphoid tissue underlies the mucosa here, forming the lingual tonsil

Sulcus terminalis
Border between the pharyngeal and oral parts of the tongue, lying in the oropharynx and oral cavity respectively

Foliate papillae
Leaf-shaped papillae that form a series of ridges on each side of the back of the tongue

Oral part of the tongue

Filiform papilla
Tiny, hair-shaped papillae that give the tongue a velvety texture

TONGUE

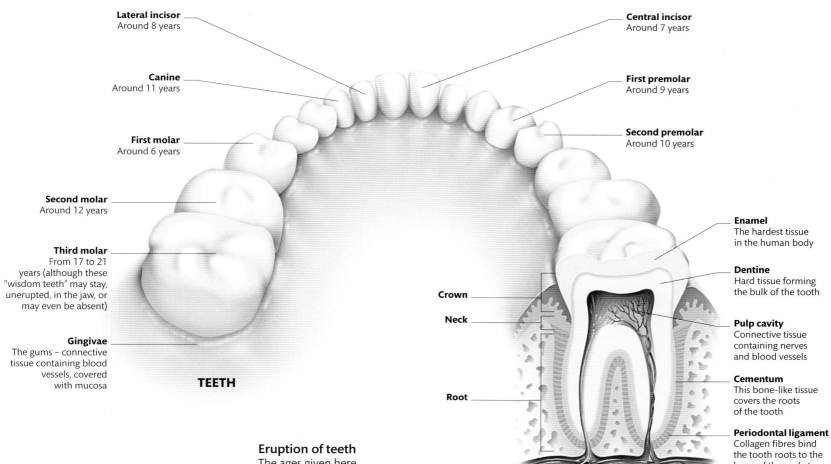

Lateral incisor
Around 8 years

Canine
Around 11 years

First molar
Around 6 years

Second molar
Around 12 years

Third molar
From 17 to 21 years (although these "wisdom teeth" may stay, unerupted, in the jaw, or may even be absent)

Gingivae
The gums – connective tissue containing blood vessels, covered with mucosa

TEETH

Central incisor
Around 7 years

First premolar
Around 9 years

Second premolar
Around 10 years

Enamel
The hardest tissue in the human body

Dentine
Hard tissue forming the bulk of the tooth

Crown

Neck

Pulp cavity
Connective tissue containing nerves and blood vessels

Cementum
This bone-like tissue covers the roots of the tooth

Root

Periodontal ligament
Collagen fibres bind the tooth roots to the bone of the socket

Eruption of teeth
The ages given here are the approximate times of eruption of the permanent teeth.

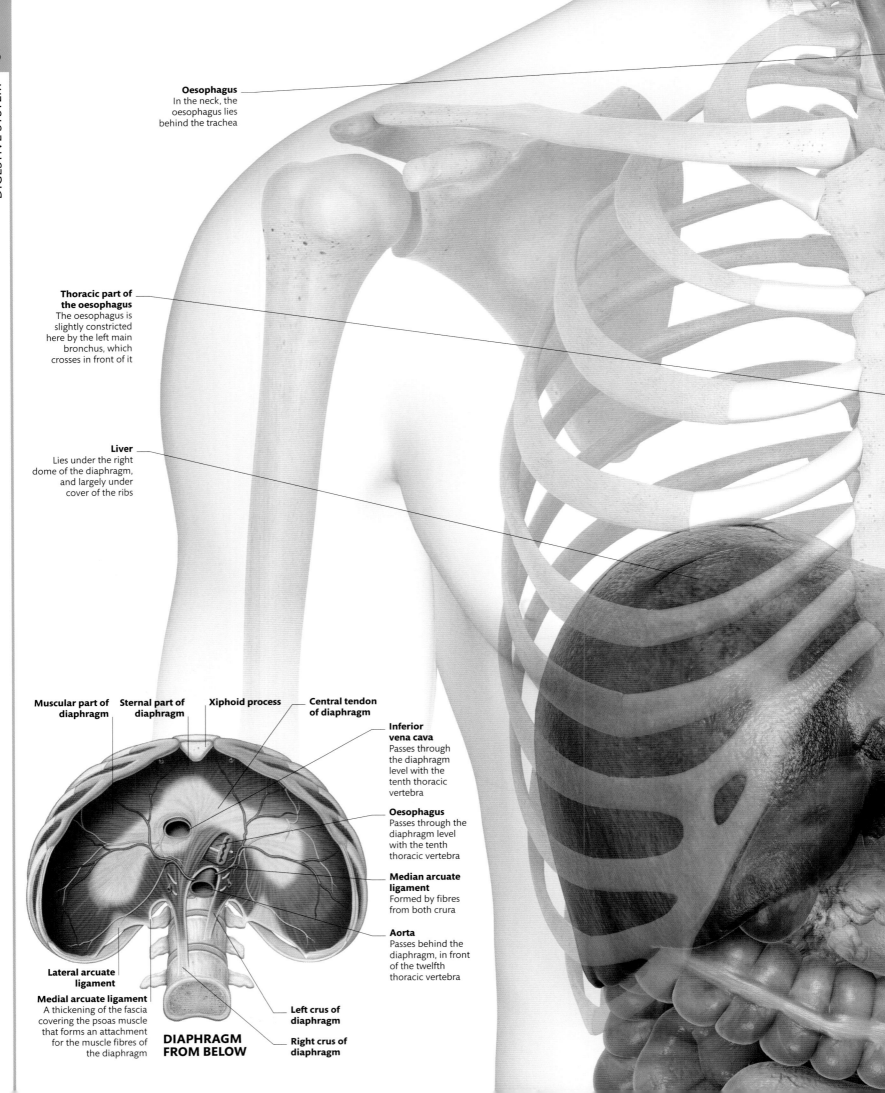

Oesophagus
In the neck, the oesophagus lies behind the trachea

Thoracic part of the oesophagus
The oesophagus is slightly constricted here by the left main bronchus, which crosses in front of it

Liver
Lies under the right dome of the diaphragm, and largely under cover of the ribs

Muscular part of diaphragm

Sternal part of diaphragm

Xiphoid process

Central tendon of diaphragm

Inferior vena cava
Passes through the diaphragm level with the tenth thoracic vertebra

Oesophagus
Passes through the diaphragm level with the tenth thoracic vertebra

Median arcuate ligament
Formed by fibres from both crura

Aorta
Passes behind the diaphragm, in front of the twelfth thoracic vertebra

Lateral arcuate ligament

Medial arcuate ligament
A thickening of the fascia covering the psoas muscle that forms an attachment for the muscle fibres of the diaphragm

DIAPHRAGM FROM BELOW

Left crus of diaphragm

Right crus of diaphragm

ANTERIOR (FRONT)

Fundus of stomach
The upper part of the stomach lies below the left dome of the diaphragm, under the ribs

THORAX

There are several large tubes crammed into the space behind the heart. These include the descending aorta, the azygos vein, and the lymphatic duct, but also a part of the digestive tract – the oesophagus. This tube of smooth muscle starts in the neck as a continuation of the pharynx. It runs down through the thorax, slightly to the left of centre, and pierces through the diaphragm level with the tenth thoracic vertebra. A couple of centimetres below this, it empties into the stomach and ends. The oesophagus, like much of the digestive tract, has an outer layer of longitudinal muscle and an inner layer of circular muscle within its wall. During swallowing, a wave of constriction passes downwards to push food or fluid down into the stomach.

Right lobe of liver

Fundus of gallbladder
Bottom of the bag-like
gallbladder, which just
sticks out under the liver

Transverse colon
Hanging down below the liver
and stomach, this part of the
colon has a mesentery (fold of
the peritoneum that connects the
intestines to the dorsal abdominal
wall) through which its blood
vessels and nerves travel

**Hepatic flexure
of colon**
Junction between
the ascending and
transverse colon, tucked
up under the liver

Ascending colon
This part of the large
intestine is firmly bound
down to the back wall
of the abdomen

Ileum
Lying mainly in the suprapubic
region of the abdomen, this
part of the small intestine is
about 4m (13ft) long; ileum
simply means entrails in Latin

Caecum
First part of the large
intestine, lying in the right
iliac fossa of the abdomen

Appendix
Properly known as the vermiform
(worm-like) appendix; usually a
few centimetres long, it is full of
lymphoid tissue, and thus forms
part of the gut's immune system

Rectum
About 12cm (4¾in) long, this
penultimate part of the gut is
stretchy; it can expand to store
faeces, until a convenient time
for emptying presents itself

Anal canal
Muscular sphincters in and around the anal
canal keep it closed; the sphincters relax during
defaecation, as the diaphragm and abdominal
wall muscles contract to raise pressure in the
abdomen and force the faeces out

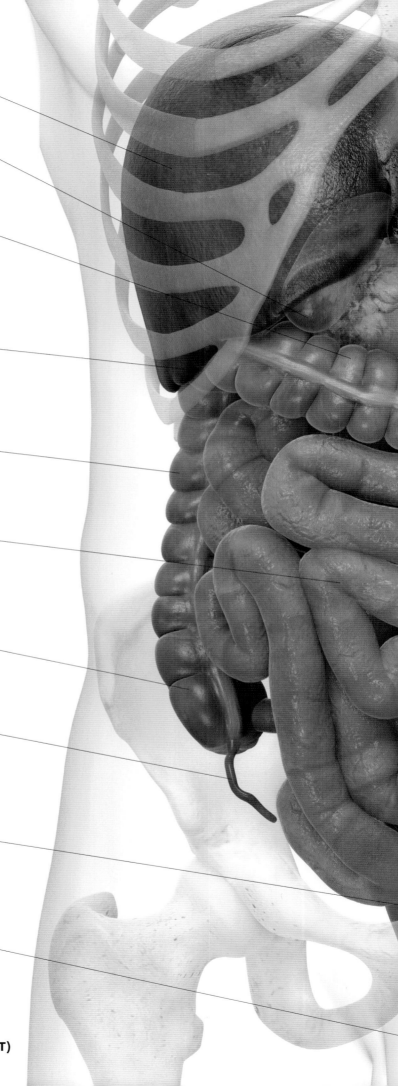

ANTERIOR (FRONT)

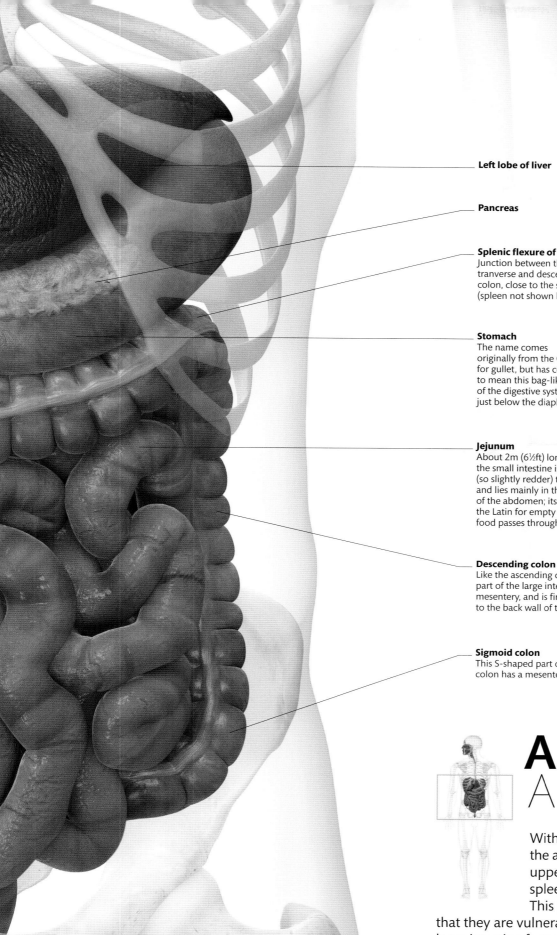

Left lobe of liver

Pancreas

Splenic flexure of colon
Junction between the
tranverse and descending
colon, close to the spleen
(spleen not shown here)

Stomach
The name comes
originally from the Greek
for gullet, but has come
to mean this bag-like part
of the digestive system,
just below the diaphragm

Jejunum
About 2m (6½ft) long, this part of
the small intestine is more vascular
(so slightly redder) than the ileum,
and lies mainly in the umbilical region
of the abdomen; its name comes from
the Latin for empty – perhaps because
food passes through here quickly

Descending colon
Like the ascending colon, this
part of the large intestine has no
mesentery, and is firmly bound
to the back wall of the abdomen

Sigmoid colon
This S-shaped part of
colon has a mesentery

ABDOMEN AND PELVIS

With the organs in situ, it is clear how much
the abdominal cavity extends up under the ribs. The
upper abdominal organs – the liver, stomach, and
spleen – are largely under cover of the ribcage.
This gives them some protection, but it also means
that they are vulnerable to injury if a lower rib is fractured. The
large intestine forms an M shape in the abdomen, starting with
the caecum low down on the right, and the ascending colon
running up the right flank and tucking under the liver. The
transverse colon hangs down below the liver and stomach, and
the descending colon runs down the left side of the abdomen.
This becomes the S-shaped sigmoid colon, which runs down into
the pelvis to become the rectum. The coils of the small intestine
occupy the middle of the abdomen.

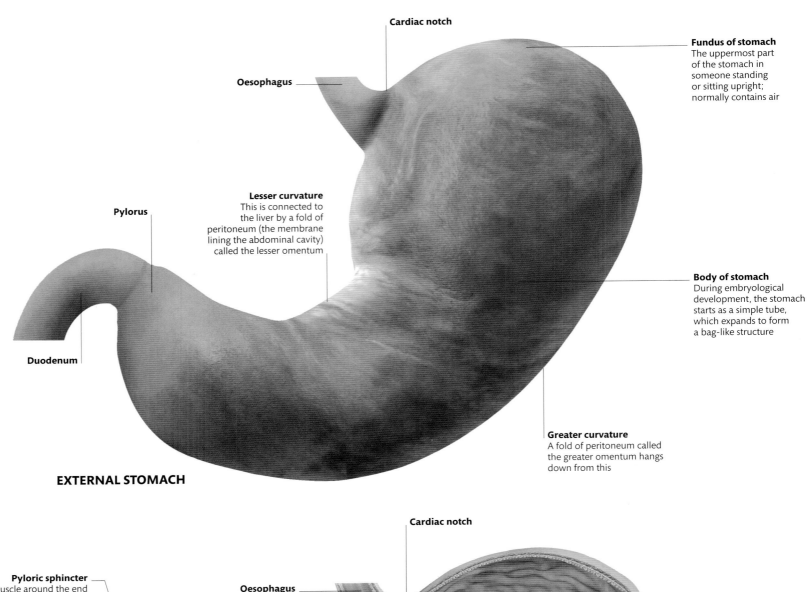

Cardiac notch

Oesophagus

Fundus of stomach
The uppermost part of the stomach in someone standing or sitting upright; normally contains air

Pylorus

Lesser curvature
This is connected to the liver by a fold of peritoneum (the membrane lining the abdominal cavity) called the lesser omentum

Body of stomach
During embryological development, the stomach starts as a simple tube, which expands to form a bag-like structure

Duodenum

Greater curvature
A fold of peritoneum called the greater omentum hangs down from this

EXTERNAL STOMACH

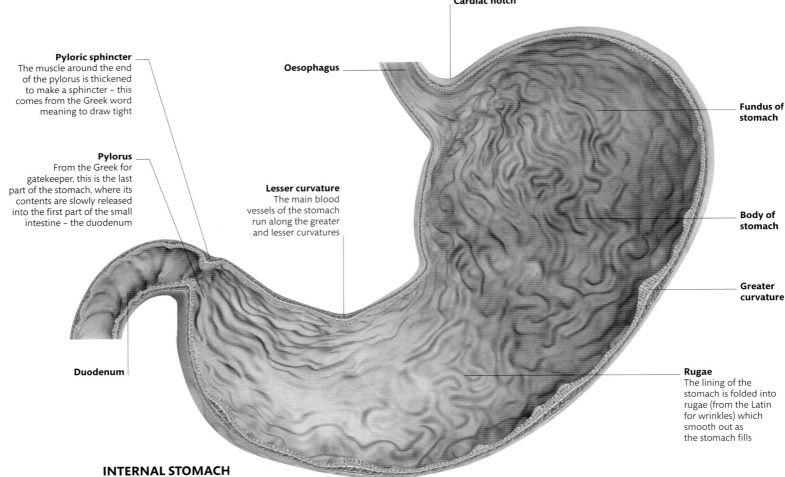

Cardiac notch

Pyloric sphincter
The muscle around the end of the pylorus is thickened to make a sphincter – this comes from the Greek word meaning to draw tight

Oesophagus

Fundus of stomach

Pylorus
From the Greek for gatekeeper, this is the last part of the stomach, where its contents are slowly released into the first part of the small intestine – the duodenum

Lesser curvature
The main blood vessels of the stomach run along the greater and lesser curvatures

Body of stomach

Greater curvature

Duodenum

Rugae
The lining of the stomach is folded into rugae (from the Latin for wrinkles) which smooth out as the stomach fills

INTERNAL STOMACH

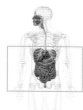

STOMACH AND INTESTINES

The stomach is a muscular bag, where food is held before moving on to the intestines. Inside the stomach, food is exposed to a cocktail of hydrochloric acid, which kills off bacteria, and protein-digesting enzymes. The layered muscle of the stomach wall contracts to churn up its contents. Semi-digested food is released from the stomach into the first part of the small intestine, the duodenum, where bile and pancreatic juices are added. Contractions in the intestine wall then push the liquid food into the jejunum and ileum, where digestion continues. What is left passes into the caecum, the beginning of the large intestine. In the colon, the next part of the large intestine, water is absorbed so that the gut contents become more solid. The resulting faeces pass into the rectum where they are stored until excretion.

SMOOTH MUSCLE STRUCTURE

Functions of the gut, blood vessels, and respiratory tract are carried out involuntarily, at a subconscious level, with the help of a special type of muscle called smooth muscle. This is supplied by autonomic motor nerves.

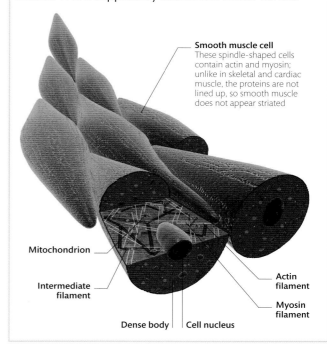

Smooth muscle cell
These spindle-shaped cells contain actin and myosin; unlike in skeletal and cardiac muscle, the proteins are not lined up, so smooth muscle does not appear striated

Mitochondrion

Intermediate filament

Dense body | Cell nucleus

Actin filament

Myosin filament

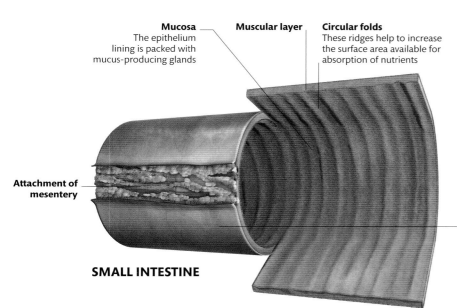

Mucosa
The epithelium lining is packed with mucus-producing glands

Muscular layer

Circular folds
These ridges help to increase the surface area available for absorption of nutrients

Attachment of mesentery

Serous lining of the small intestine
This is formed by the mesentery (membranous folds) enveloping the gut tube

SMALL INTESTINE

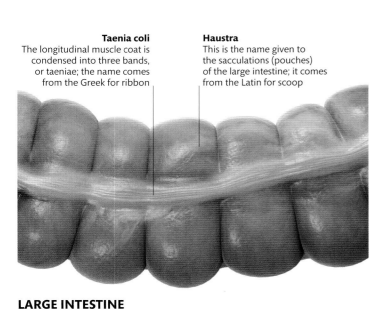

Taenia coli
The longitudinal muscle coat is condensed into three bands, or taeniae; the name comes from the Greek for ribbon

Haustra
This is the name given to the sacculations (pouches) of the large intestine; it comes from the Latin for scoop

LARGE INTESTINE

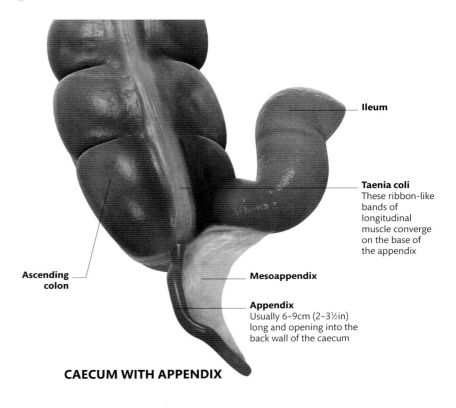

Ileum

Taenia coli
These ribbon-like bands of longitudinal muscle converge on the base of the appendix

Ascending colon

Mesoappendix

Appendix
Usually 6–9cm (2–3½in) long and opening into the back wall of the caecum

CAECUM WITH APPENDIX

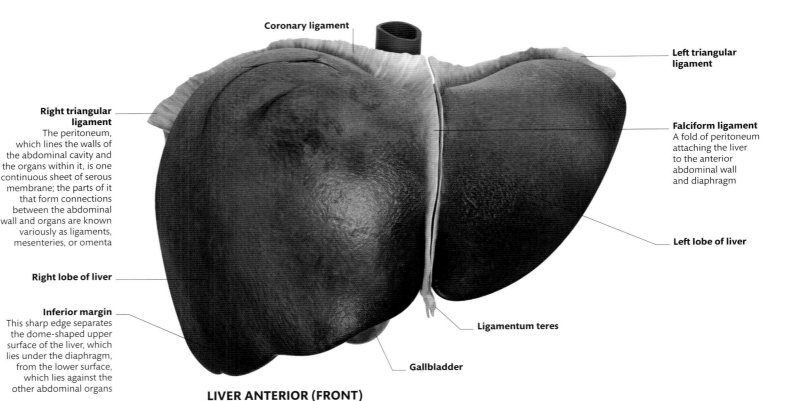

Coronary ligament

Left triangular ligament

Right triangular ligament
The peritoneum, which lines the walls of the abdominal cavity and the organs within it, is one continuous sheet of serous membrane; the parts of it that form connections between the abdominal wall and organs are known variously as ligaments, mesenteries, or omenta

Falciform ligament
A fold of peritoneum attaching the liver to the anterior abdominal wall and diaphragm

Right lobe of liver

Left lobe of liver

Inferior margin
This sharp edge separates the dome-shaped upper surface of the liver, which lies under the diaphragm, from the lower surface, which lies against the other abdominal organs

Ligamentum teres

Gallbladder

LIVER ANTERIOR (FRONT)

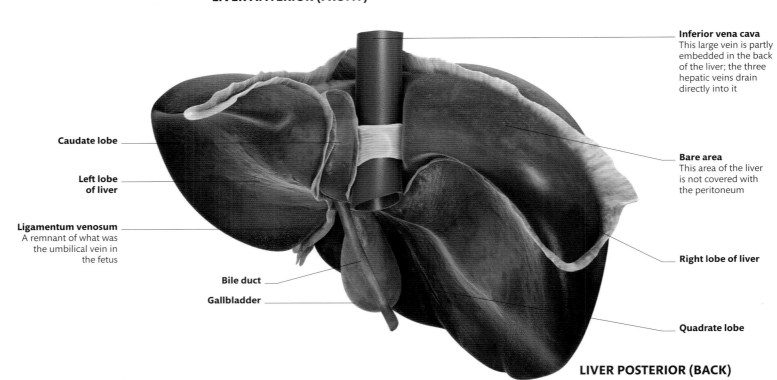

Inferior vena cava
This large vein is partly embedded in the back of the liver; the three hepatic veins drain directly into it

Caudate lobe

Left lobe of liver

Bare area
This area of the liver is not covered with the peritoneum

Ligamentum venosum
A remnant of what was the umbilical vein in the fetus

Right lobe of liver

Bile duct

Gallbladder

Quadrate lobe

LIVER POSTERIOR (BACK)

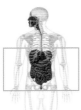

LIVER, PANCREAS, AND GALLBLADDER

The liver, the largest internal organ, can weigh up to 3kg (6lb). It does hundreds of jobs simultaneously, many of them related to digestion. It produces bile, which is stored in the gallbladder and helps to digest fats. It also receives nutrients from the gut via the portal vein and processes them. It breaks down or builds up proteins, carbohydrates, and fats according to need; detoxifies or deactivates substances such as alcohol and drugs; and plays a role in the immune system. The pancreas, a long, thin, leaf-shaped gland lying under the liver and behind the stomach, produces hormones that are secreted into the blood, and makes pancreatic juice, full of digestive enzymes, which it empties into the duodenum.

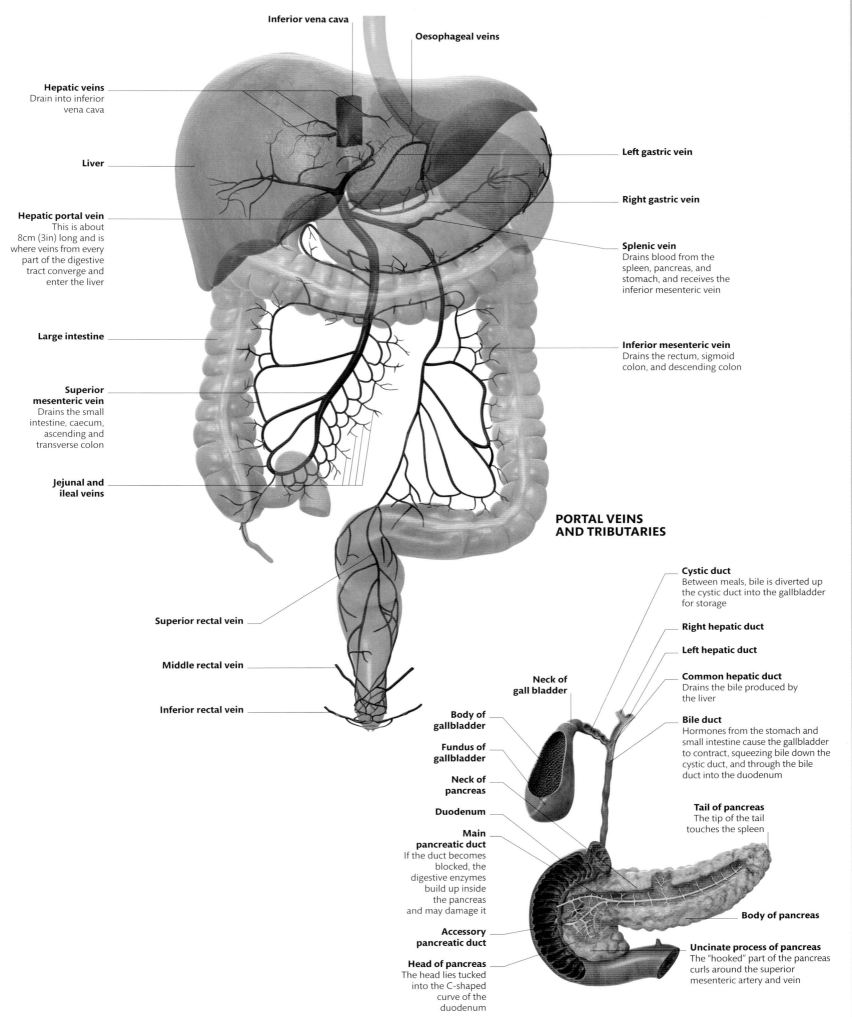

Inferior vena cava

Oesophageal veins

Hepatic veins
Drain into inferior
vena cava

Liver

Left gastric vein

Right gastric vein

Hepatic portal vein
This is about
8cm (3in) long and is
where veins from every
part of the digestive
tract converge and
enter the liver

Splenic vein
Drains blood from the
spleen, pancreas, and
stomach, and receives the
inferior mesenteric vein

Large intestine

Inferior mesenteric vein
Drains the rectum, sigmoid
colon, and descending colon

**Superior
mesenteric vein**
Drains the small
intestine, caecum,
ascending and
transverse colon

**Jejunal and
ileal veins**

**PORTAL VEINS
AND TRIBUTARIES**

Superior rectal vein

Middle rectal vein

Inferior rectal vein

Cystic duct
Between meals, bile is diverted up
the cystic duct into the gallbladder
for storage

Right hepatic duct

Left hepatic duct

Common hepatic duct
Drains the bile produced by
the liver

**Neck of
gall bladder**

**Body of
gallbladder**

Bile duct
Hormones from the stomach and
small intestine cause the gallbladder
to contract, squeezing bile down the
cystic duct, and through the bile
duct into the duodenum

**Fundus of
gallbladder**

**Neck of
pancreas**

Tail of pancreas
The tip of the tail
touches the spleen

Duodenum

**Main
pancreatic duct**
If the duct becomes
blocked, the
digestive enzymes
build up inside
the pancreas
and may damage it

Body of pancreas

**Accessory
pancreatic duct**

Uncinate process of pancreas
The "hooked" part of the pancreas
curls around the superior
mesenteric artery and vein

Head of pancreas
The head lies tucked
into the C-shaped
curve of the
duodenum

GALLBLADDER AND PANCREAS

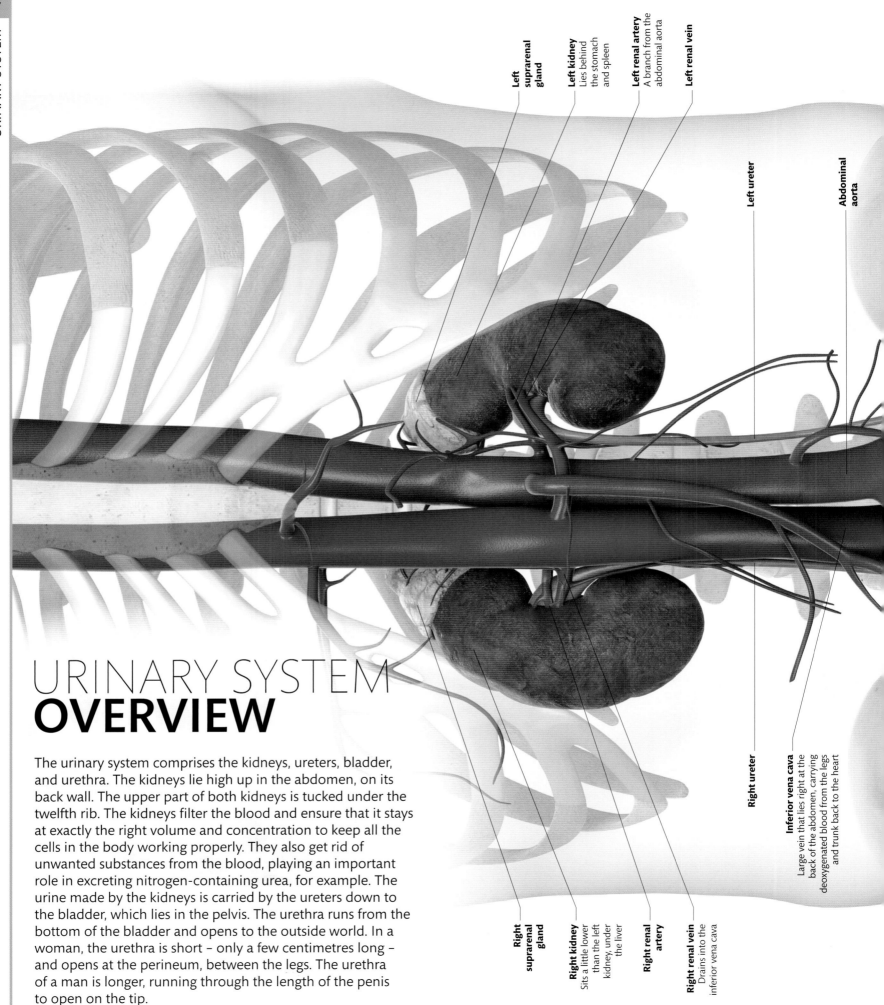

Left suprarenal gland

Left kidney
Lies behind the stomach and spleen

Left renal artery
A branch from the abdominal aorta

Left renal vein

Left ureter

Abdominal aorta

Right ureter

Inferior vena cava
Large vein that lies right at the back of the abdomen, carrying deoxygenated blood from the legs and trunk back to the heart

Right suprarenal gland

Right kidney
Sits a little lower than the left kidney, under the liver

Right renal artery

Right renal vein
Drains into the inferior vena cava

URINARY SYSTEM
OVERVIEW

The urinary system comprises the kidneys, ureters, bladder, and urethra. The kidneys lie high up in the abdomen, on its back wall. The upper part of both kidneys is tucked under the twelfth rib. The kidneys filter the blood and ensure that it stays at exactly the right volume and concentration to keep all the cells in the body working properly. They also get rid of unwanted substances from the blood, playing an important role in excreting nitrogen-containing urea, for example. The urine made by the kidneys is carried by the ureters down to the bladder, which lies in the pelvis. The urethra runs from the bottom of the bladder and opens to the outside world. In a woman, the urethra is short – only a few centimetres long – and opens at the perineum, between the legs. The urethra of a man is longer, running through the length of the penis to open on the tip.

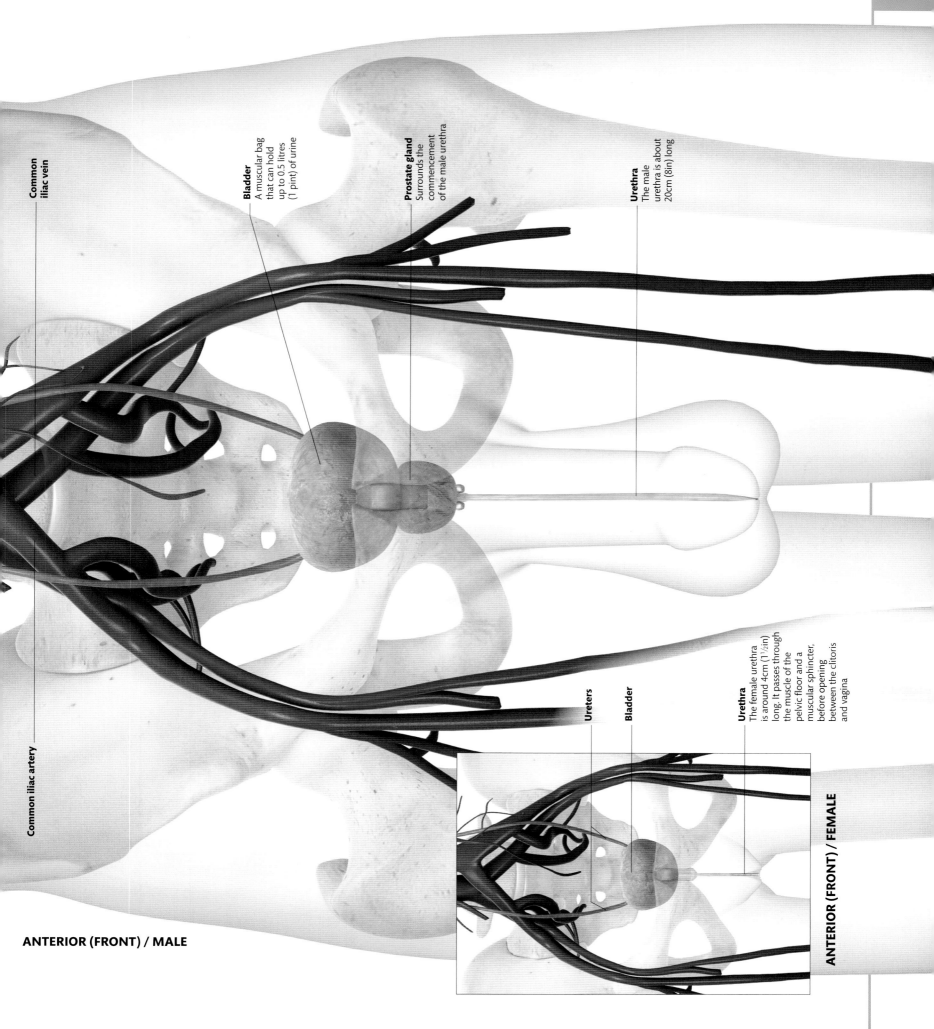

Common
iliac vein

Bladder
A muscular bag
that can hold
up to 0.5 litres
(1 pint) of urine

Prostate gland
Surrounds the
commencement
of the male urethra

Urethra
The male
urethra is about
20cm (8in) long

Common iliac artery

Ureters

Bladder

Urethra
The female urethra
is around 4cm (1½in)
long. It passes through
the muscle of the
pelvic floor and a
muscular sphincter,
before opening
between the clitoris
and vagina

ANTERIOR (FRONT) / MALE

ANTERIOR (FRONT) / FEMALE

Suprarenal gland

Upper pole

Right kidney

Right renal artery
Renal comes from
the Latin for kidney

Hilum
Where the artery enters
and the vein and ureter
exit the kidney; the word
just means small thing
in Latin, but is used in
botany to describe the
area on a seed where
the seed-vessel attaches,
such as the eye of a bean

Right renal vein

Lower pole

Inferior vena cava

Right common iliac vein

Right internal iliac vein
Veins from the bladder
eventually drain into the
internal iliac veins

Right internal iliac artery
Vesical branches of
the internal iliac artery
supply the bladder

Right external iliac vein

Right external iliac artery

Right ureter
The two ureters are muscular tubes:
peristaltic (wave-like) contractions
pump urine down into the bladder,
even if you stand on your head; each
ureter is about 25cm (10in) long

**ANTERIOR
(FRONT)**

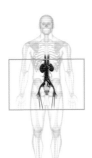

ABDOMEN
AND PELVIS

The kidneys lie high up on the back wall of the abdomen,
tucked up under the twelfth ribs. A thick layer of perinephric
fat surrounds and protects each kidney. The kidneys filter the
blood, which is carried to them via the renal arteries. They
remove waste from the blood, and keep a tight check on
blood volume and concentration. The urine they produce collects first in
cup-shaped calyces, which join to form the renal pelvis. The urine then flows
out of the kidneys and down narrow, muscular tubes called ureters to the
bladder in the pelvis. The bladder is a muscular bag that can expand to hold
up to about 0.5 litres (1 pint) of urine, and empties itself when the individual
decides it is convenient. Urine travels through the urethra before leaving
the body.

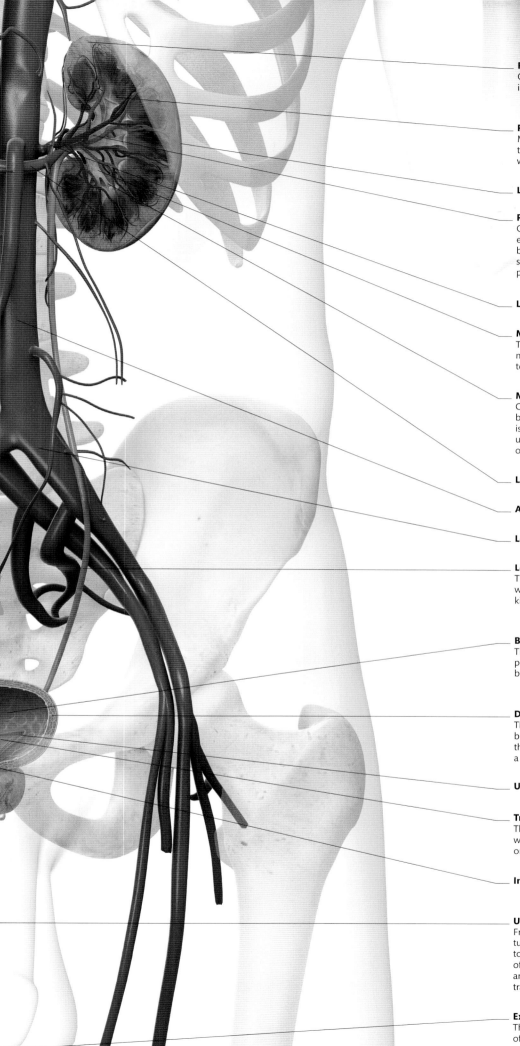

Renal cortex
Cortex means rind or bark; this
is the outer tissue of the kidney

Renal medullary pyramid
Medulla means marrow or pith; this core
tissue of the kidney is arranged as pyramids,
which look triangular in cross-section

Left kidney

Renal pelvis
Collects all urine from the kidney, and
empties into the ureter; pelvis means
basin in Latin, and the renal pelvis
should not be confused with the bony
pelvis – also shaped like a large basin

Left renal artery

Major calyx
The major calyces collect urine from the
minor calyces, then themselves join together
to form the renal pelvis

Minor calyx
Calyx originally meant flower-covering in Greek,
but because it is similar to the Latin word for cup it
is used to describe cup-shaped structures in biology;
urine from the microscopic collecting tubules
of the kidney flows out into the minor calyces

Left renal vein

Abdominal aorta

Left common iliac artery

Left ureter
This name comes from the Greek for to make
water; the two ureters carry urine from the
kidneys to the bladder

Bladder
The empty bladder lies low down, in the true
pelvis, behind the pubic symphysis; as the
bladder fills, it expands up into the abdomen

Detrusor muscle
The criss-crossing smooth muscle
bundles of the bladder wall give
the inner surface of the bladder
a net-like appearance

Ureteric orifice

Trigone
The three-cornered region of the back
wall of the bladder, between the ureteric
orifices and the internal urethral orifice

Internal urethral orifice

Urethra
From the Greek for urinate; this
tube carries urine from the bladder
to the outside world, a distance
of around 4cm (1½in) in women,
and about 20cm (8in) in men (as it
travels the length of the penis)

External urethral orifice
The male urethra opens at the tip
of the glans penis

REPRODUCTIVE SYSTEM
OVERVIEW

Most organs in the body are similar in men and women. However, when it comes to the reproductive organs, there is a world of difference. In a woman, the ovaries, which produce eggs and female sex hormones, are tucked away, deep inside the pelvis. Also located within the pelvis are the vagina, uterus, and paired oviducts, or fallopian tubes, in which eggs are conveyed from the ovaries to the uterus. The woman's reproductive system also includes the mammary glands, which are important in providing milk for the newborn.

In a man, the testes, which produce sperm and sex hormones, hang well outside the pelvis, in the scrotum. The rest of the male reproductive system consists of a pair of tubes called the vasa deferentia (singular, vas deferens), the accessory sex glands (the seminal vesicles and the prostate), and the urethra.

Lactiferous duct
A series of 15 to 20 ducts each drain a lobe of the breast

Nipple
Lactiferous ducts open on highest point (apex) of the nipple, which extends from centre of the breast

Secretory lobule containing alveoli
One of several small compartments housed within each lobe of the breast. A lobule is composed of grape-like clusters of milk-secreting glands called alveoli

ANTERIOR (FRONT) / FEMALE

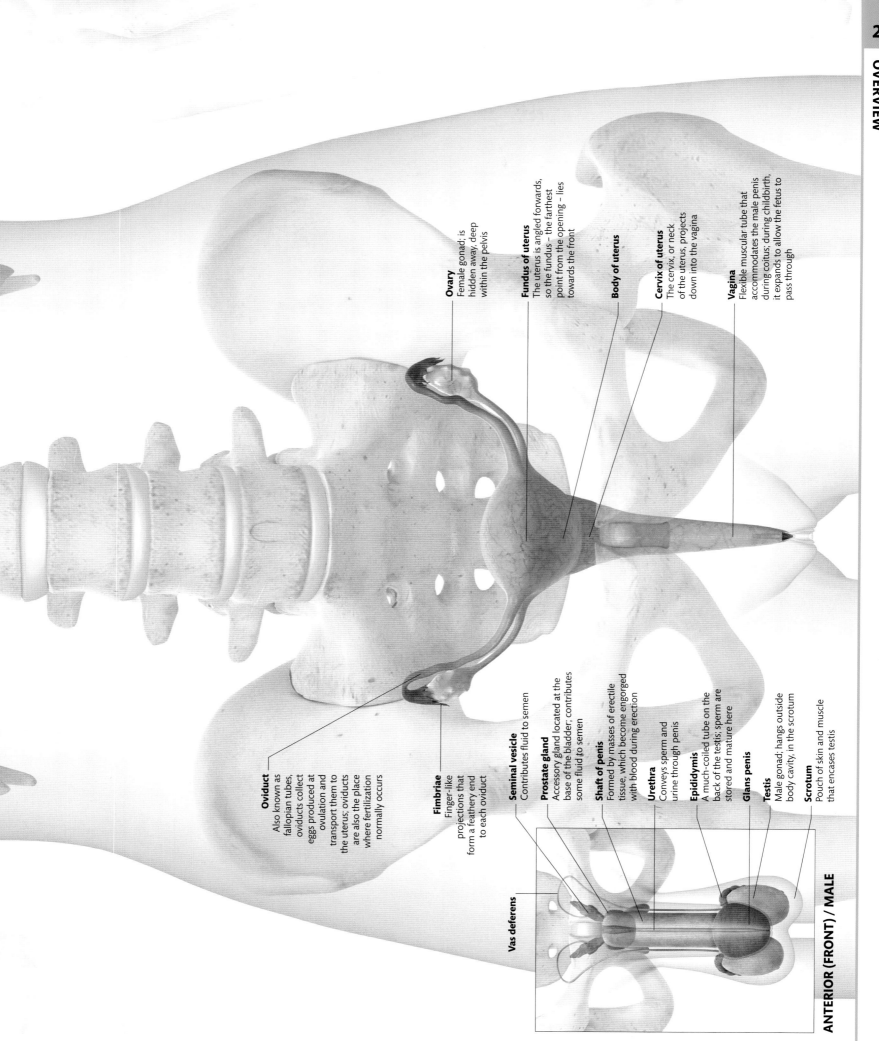

Ovary
Female gonad; is hidden away, deep within the pelvis

Fundus of uterus
The uterus is angled forwards, so the fundus – the farthest point from the opening – lies towards the front

Body of uterus

Cervix of uterus
The cervix, or neck of the uterus, projects down into the vagina

Vagina
Flexible muscular tube that accommodates the male penis during coitus; during childbirth, it expands to allow the fetus to pass through

Oviduct
Also known as fallopian tubes, oviducts collect eggs produced at ovulation and transport them to the uterus; oviducts are also the place where fertilization normally occurs

Fimbriae
Finger-like projections that form a feathery end to each oviduct

Seminal vesicle
Contributes fluid to semen

Prostate gland
Accessory gland located at the base of the bladder; contributes some fluid to semen

Shaft of penis
Formed by masses of erectile tissue, which become engorged with blood during erection

Urethra
Conveys sperm and urine through penis

Epididymis
A much-coiled tube on the back of the testis; sperm are stored and mature here

Glans penis

Testis
Male gonad; hangs outside body cavity, in the scrotum

Scrotum
Pouch of skin and muscle that encases testis

Vas deferens

ANTERIOR (FRONT) / MALE

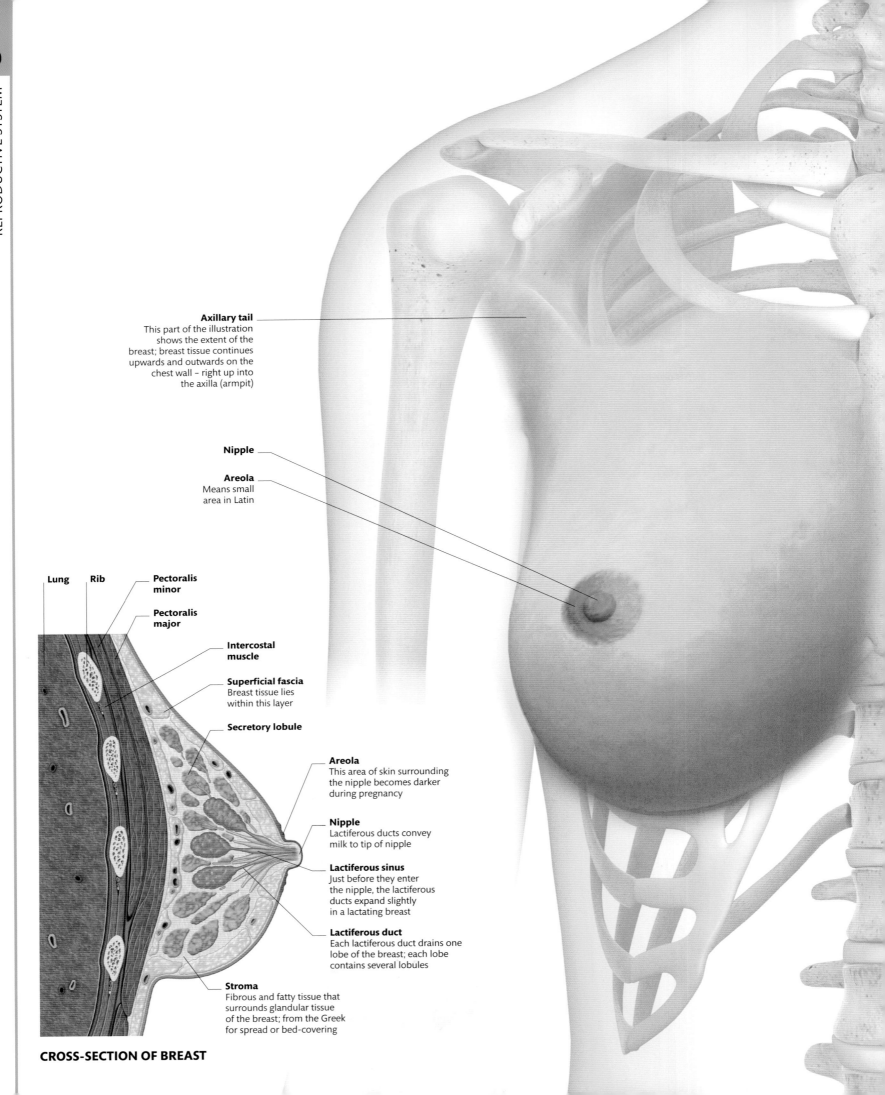

Axillary tail
This part of the illustration shows the extent of the breast; breast tissue continues upwards and outwards on the chest wall – right up into the axilla (armpit)

Nipple

Areola
Means small area in Latin

Lung   **Rib**   **Pectoralis minor**

Pectoralis major

Intercostal muscle

Superficial fascia
Breast tissue lies within this layer

Secretory lobule

Areola
This area of skin surrounding the nipple becomes darker during pregnancy

Nipple
Lactiferous ducts convey milk to tip of nipple

Lactiferous sinus
Just before they enter the nipple, the lactiferous ducts expand slightly in a lactating breast

Lactiferous duct
Each lactiferous duct drains one lobe of the breast; each lobe contains several lobules

Stroma
Fibrous and fatty tissue that surrounds glandular tissue of the breast; from the Greek for spread or bed-covering

CROSS-SECTION OF BREAST

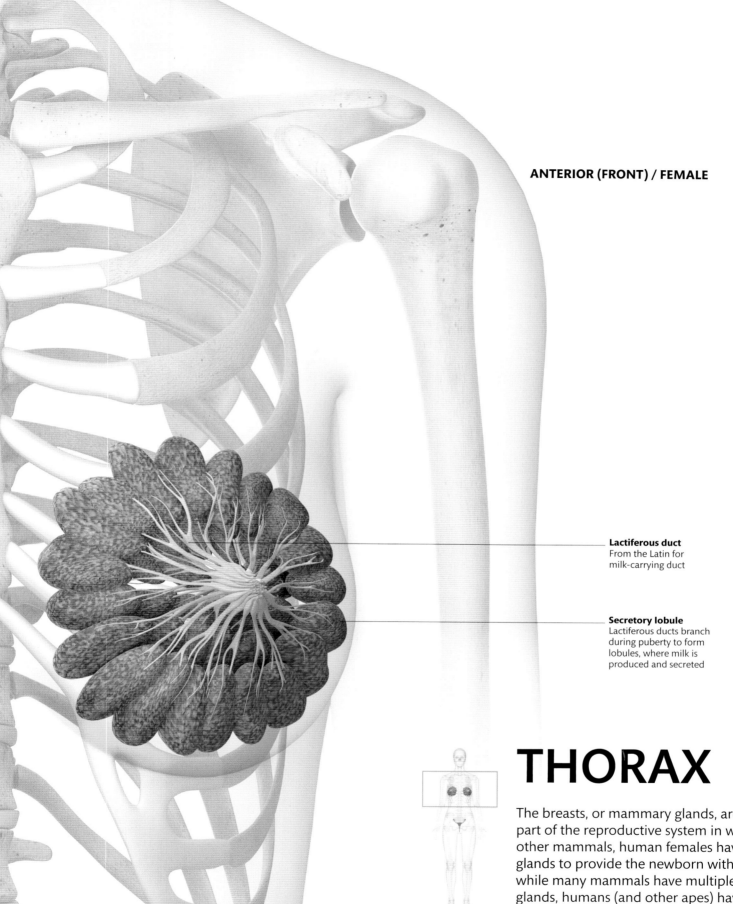

ANTERIOR (FRONT) / FEMALE

Lactiferous duct
From the Latin for
milk-carrying duct

Secretory lobule
Lactiferous ducts branch
during puberty to form
lobules, where milk is
produced and secreted

THORAX

The breasts, or mammary glands, are an important
part of the reproductive system in women. Like all
other mammals, human females have mammary
glands to provide the newborn with milk. But
while many mammals have multiple mammary
glands, humans (and other apes) have just two,
on the front of the chest. The breasts develop at
puberty, when they grow due to the increased production of
glandular tissue and fat. The breasts lie on the pectoralis major
muscle on each side. Each breast contains 15 to 20 lobes, which
are connected to the nipple by lactiferous ducts. There seems to
be a basic plan in the developing embryo, so that male nipples
appear, although the breast does not form.

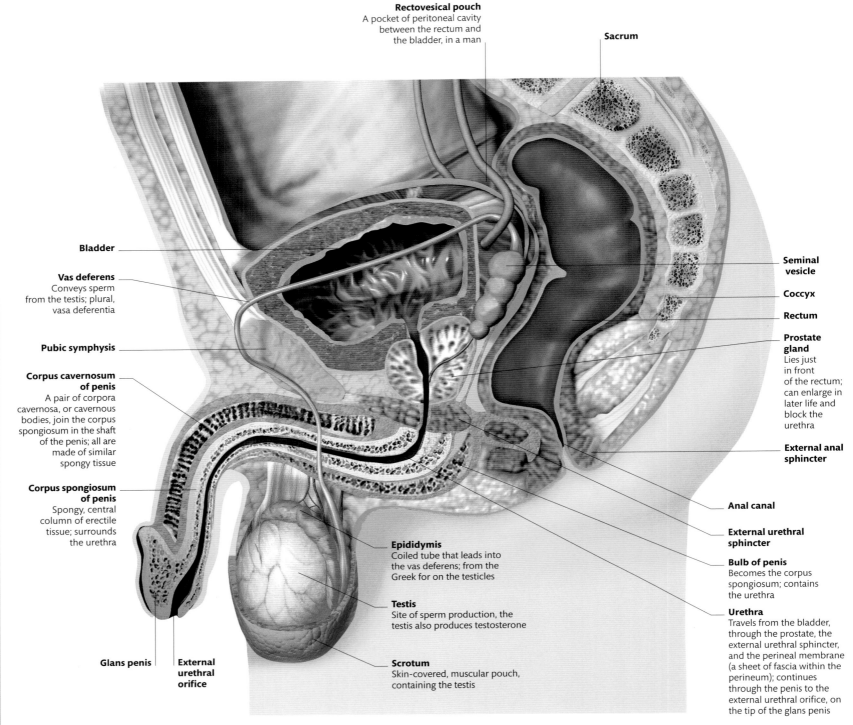

Rectovesical pouch
A pocket of peritoneal cavity between the rectum and the bladder, in a man

Sacrum

Bladder

Vas deferens
Conveys sperm from the testis; plural, vasa deferentia

Pubic symphysis

Corpus cavernosum of penis
A pair of corpora cavernosa, or cavernous bodies, join the corpus spongiosum in the shaft of the penis; all are made of similar spongy tissue

Corpus spongiosum of penis
Spongy, central column of erectile tissue; surrounds the urethra

Glans penis

External urethral orifice

Epididymis
Coiled tube that leads into the vas deferens; from the Greek for on the testicles

Testis
Site of sperm production, the testis also produces testosterone

Scrotum
Skin-covered, muscular pouch, containing the testis

Seminal vesicle

Coccyx

Rectum

Prostate gland
Lies just in front of the rectum; can enlarge in later life and block the urethra

External anal sphincter

Anal canal

External urethral sphincter

Bulb of penis
Becomes the corpus spongiosum; contains the urethra

Urethra
Travels from the bladder, through the prostate, the external urethral sphincter, and the perineal membrane (a sheet of fascia within the perineum); continues through the penis to the external urethral orifice, on the tip of the glans penis

SAGITTAL SECTION / MALE

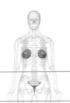

ABDOMEN AND PELVIS

The male and female reproductive systems are both comprised of a series of internal and external organs, although structurally these are very different. It is true that both sexes possess gonads (ovaries in women and testes in men) and a tract, or set of tubes, but the similarity ends there. When we look in detail at the anatomy of the pelvis in each sex, the differences are obvious. The pelvis of a man contains only part of the reproductive tract, as well as the lower parts of the digestive and urinary tracts, including the rectum and bladder. Beneath the bladder is the prostate gland; this is where the vasa deferentia, which bring sperm from the testis, empty into the urethra. A woman's pelvic cavity contains more of the reproductive tract than a man's. The vagina and uterus are situated between the bladder and rectum in the pelvis.

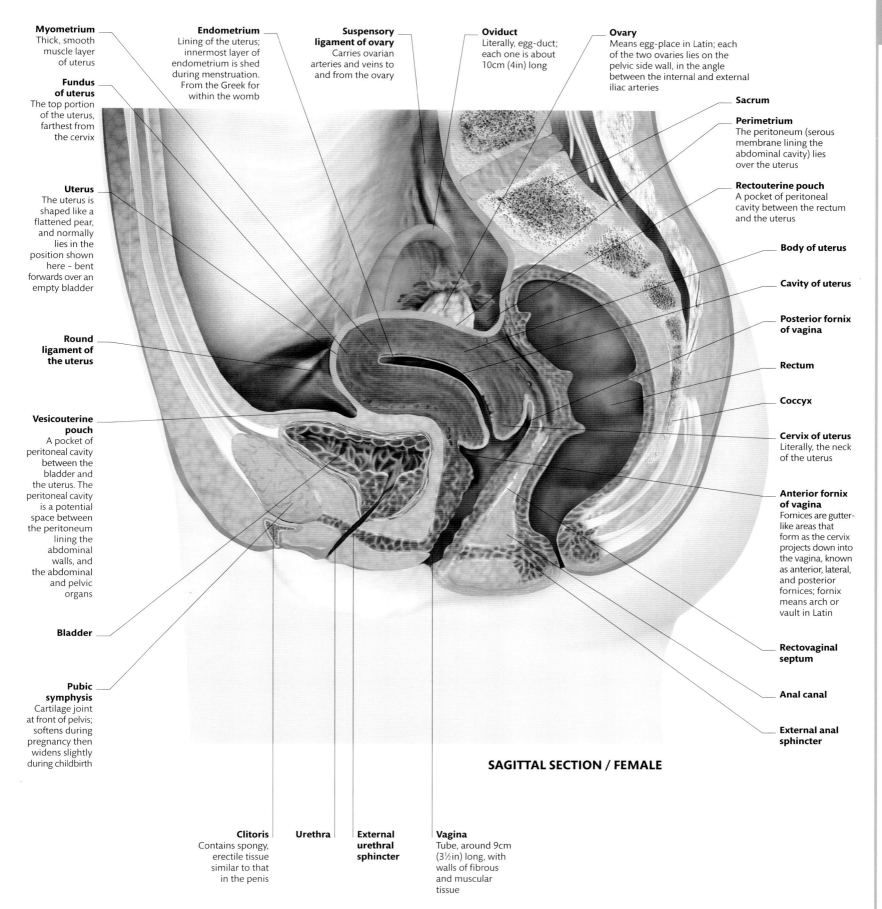

Myometrium
Thick, smooth muscle layer of uterus

Fundus of uterus
The top portion of the uterus, farthest from the cervix

Uterus
The uterus is shaped like a flattened pear, and normally lies in the position shown here – bent forwards over an empty bladder

Round ligament of the uterus

Vesicouterine pouch
A pocket of peritoneal cavity between the bladder and the uterus. The peritoneal cavity is a potential space between the peritoneum lining the abdominal walls, and the abdominal and pelvic organs

Bladder

Pubic symphysis
Cartilage joint at front of pelvis; softens during pregnancy then widens slightly during childbirth

Endometrium
Lining of the uterus; innermost layer of endometrium is shed during menstruation. From the Greek for within the womb

Suspensory ligament of ovary
Carries ovarian arteries and veins to and from the ovary

Oviduct
Literally, egg-duct; each one is about 10cm (4in) long

Ovary
Means egg-place in Latin; each of the two ovaries lies on the pelvic side wall, in the angle between the internal and external iliac arteries

Sacrum

Perimetrium
The peritoneum (serous membrane lining the abdominal cavity) lies over the uterus

Rectouterine pouch
A pocket of peritoneal cavity between the rectum and the uterus

Body of uterus

Cavity of uterus

Posterior fornix of vagina

Rectum

Coccyx

Cervix of uterus
Literally, the neck of the uterus

Anterior fornix of vagina
Fornices are gutter-like areas that form as the cervix projects down into the vagina, known as anterior, lateral, and posterior fornices; fornix means arch or vault in Latin

Rectovaginal septum

Anal canal

External anal sphincter

Clitoris
Contains spongy, erectile tissue similar to that in the penis

Urethra

External urethral sphincter

Vagina
Tube, around 9cm (3½in) long, with walls of fibrous and muscular tissue

SAGITTAL SECTION / FEMALE

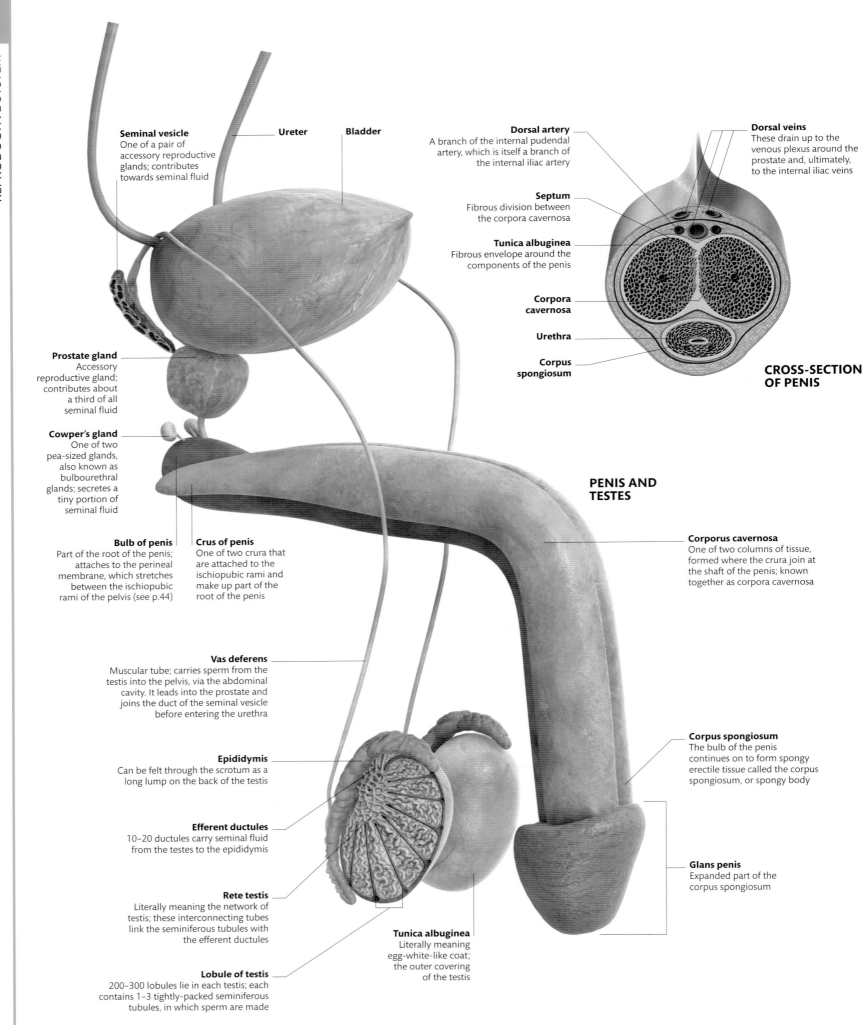

Seminal vesicle
One of a pair of accessory reproductive glands; contributes towards seminal fluid

Ureter

Bladder

Dorsal artery
A branch of the internal pudendal artery, which is itself a branch of the internal iliac artery

Dorsal veins
These drain up to the venous plexus around the prostate and, ultimately, to the internal iliac veins

Septum
Fibrous division between the corpora cavernosa

Tunica albuginea
Fibrous envelope around the components of the penis

Corpora cavernosa

Urethra

Corpus spongiosum

CROSS-SECTION OF PENIS

Prostate gland
Accessory reproductive gland; contributes about a third of all seminal fluid

Cowper's gland
One of two pea-sized glands, also known as bulbourethral glands; secretes a tiny portion of seminal fluid

PENIS AND TESTES

Bulb of penis
Part of the root of the penis; attaches to the perineal membrane, which stretches between the ischiopubic rami of the pelvis (see p.44)

Crus of penis
One of two crura that are attached to the ischiopubic rami and make up part of the root of the penis

Corporus cavernosa
One of two columns of tissue, formed where the crura join at the shaft of the penis; known together as corpora cavernosa

Vas deferens
Muscular tube; carries sperm from the testis into the pelvis, via the abdominal cavity. It leads into the prostate and joins the duct of the seminal vesicle before entering the urethra

Corpus spongiosum
The bulb of the penis continues on to form spongy erectile tissue called the corpus spongiosum, or spongy body

Epididymis
Can be felt through the scrotum as a long lump on the back of the testis

Efferent ductules
10–20 ductules carry seminal fluid from the testes to the epididymis

Glans penis
Expanded part of the corpus spongiosum

Rete testis
Literally meaning the network of testis; these interconnecting tubes link the seminiferous tubules with the efferent ductules

Tunica albuginea
Literally meaning egg-white-like coat; the outer covering of the testis

Lobule of testis
200–300 lobules lie in each testis; each contains 1–3 tightly-packed seminiferous tubules, in which sperm are made

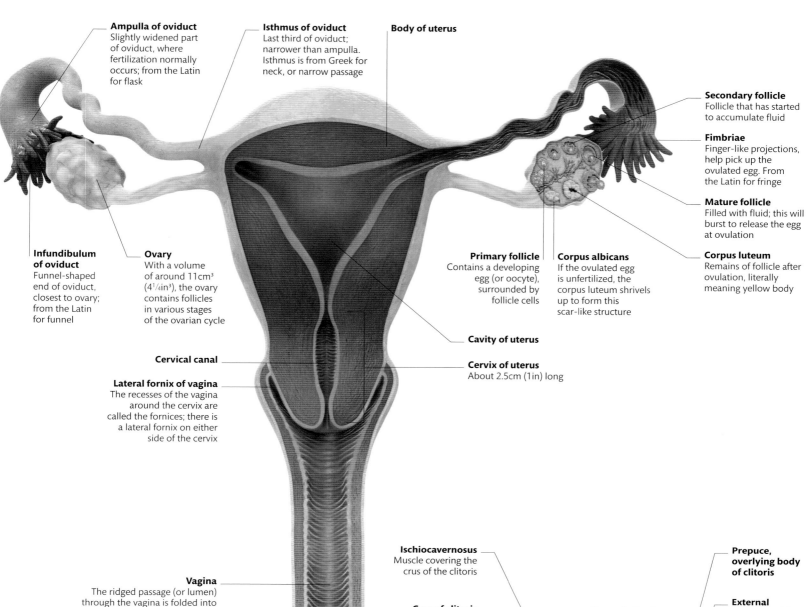

Ampulla of oviduct
Slightly widened part of oviduct, where fertilization normally occurs; from the Latin for flask

Isthmus of oviduct
Last third of oviduct; narrower than ampulla. Isthmus is from Greek for neck, or narrow passage

Body of uterus

Secondary follicle
Follicle that has started to accumulate fluid

Fimbriae
Finger-like projections, help pick up the ovulated egg. From the Latin for fringe

Mature follicle
Filled with fluid; this will burst to release the egg at ovulation

Infundibulum of oviduct
Funnel-shaped end of oviduct, closest to ovary; from the Latin for funnel

Ovary
With a volume of around 11cm³ (4¼in³), the ovary contains follicles in various stages of the ovarian cycle

Primary follicle
Contains a developing egg (or oocyte), surrounded by follicle cells

Corpus albicans
If the ovulated egg is unfertilized, the corpus luteum shrivels up to form this scar-like structure

Corpus luteum
Remains of follicle after ovulation, literally meaning yellow body

Cavity of uterus

Cervical canal

Cervix of uterus
About 2.5cm (1in) long

Lateral fornix of vagina
The recesses of the vagina around the cervix are called the fornices; there is a lateral fornix on either side of the cervix

Vagina
The ridged passage (or lumen) through the vagina is folded into an H-shape, enabling it to expand

UTERUS

ABDOMEN AND PELVIS

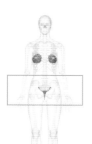

At a very fundamental level, the reproductive systems of man and woman must work together to allow eggs and sperm to meet. These views of the isolated organs and reproductive tracts show clearly how the anatomy is arranged to achieve this. The ovaries, where eggs (or ova) are produced, are deep inside the female pelvis. The eggs are collected from the ovaries by a pair of tubes, the oviducts, and it is usually here that fertilization takes place. The fertilized egg then moves along the oviduct, dividing into a ball of cells. The embryo eventually reaches the uterus, which is designed to accommodate and support the growing fetus. As well as forming a way for sperm to get in, the vagina provides the route for the baby to get out at birth.

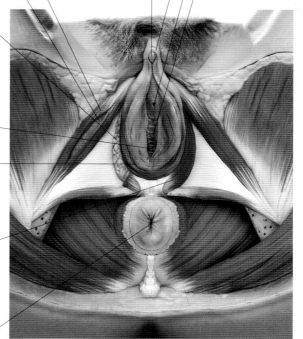

Ischiocavernosus
Muscle covering the crus of the clitoris

Crus of clitoris
Smaller in size than the crus of the penis; attached to the ischiopubic ramus of the bony pelvis

Glans of clitoris
Erectile organ, equivalent to the penis; the body of the clitoris comprises two corpora cavernosa

Prepuce, overlying body of clitoris

External urethral orifice

Vestibule
Area between the labia minora; Latin for entrance court

Bulb of vestibule
One of a pair of structures equivalent to the single bulb of the penis; made of spongy erectile tissue

Vaginal orifice

Labia minora
Folds of skin either side of the vestibule; singular is labium minus

Bulbospongiosus
Muscle covering the bulb of vestibule; helps to increase pressure in the underlying spongy tissue

Anus

EXTERNAL FEMALE GENITALIA

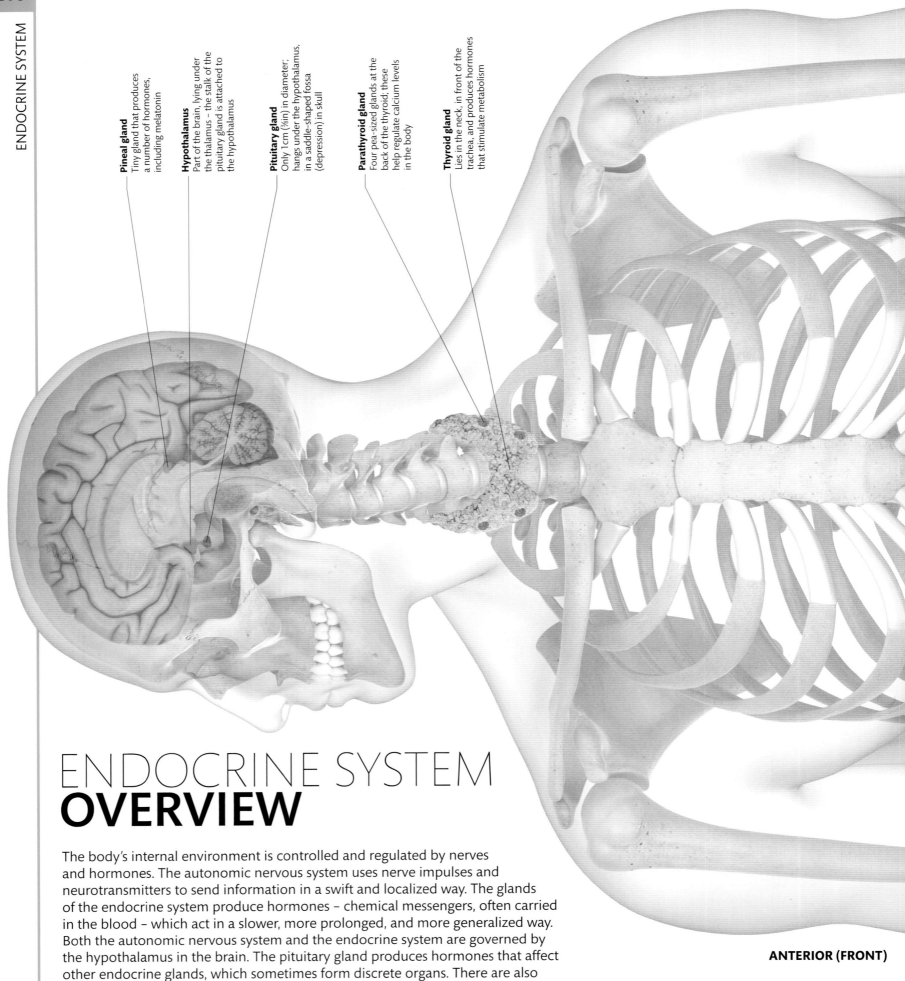

Pineal gland
Tiny gland that produces
a number of hormones,
including melatonin

Hypothalamus
Part of the brain, lying under
the thalamus – the stalk of the
pituitary gland is attached to
the hypothalamus

Pituitary gland
Only 1cm (⅜in) in diameter;
hangs under the hypothalamus,
in a saddle-shaped fossa
(depression) in skull

Parathyroid gland
Four pea-sized glands at the
back of the thyroid; these
help regulate calcium levels
in the body

Thyroid gland
Lies in the neck, in front of the
trachea, and produces hormones
that stimulate metabolism

ENDOCRINE SYSTEM
OVERVIEW

The body's internal environment is controlled and regulated by nerves
and hormones. The autonomic nervous system uses nerve impulses and
neurotransmitters to send information in a swift and localized way. The glands
of the endocrine system produce hormones – chemical messengers, often carried
in the blood – which act in a slower, more prolonged, and more generalized way.
Both the autonomic nervous system and the endocrine system are governed by
the hypothalamus in the brain. The pituitary gland produces hormones that affect
other endocrine glands, which sometimes form discrete organs. There are also
hormone-producing cells in the tissues of many other organs.

ANTERIOR (FRONT)

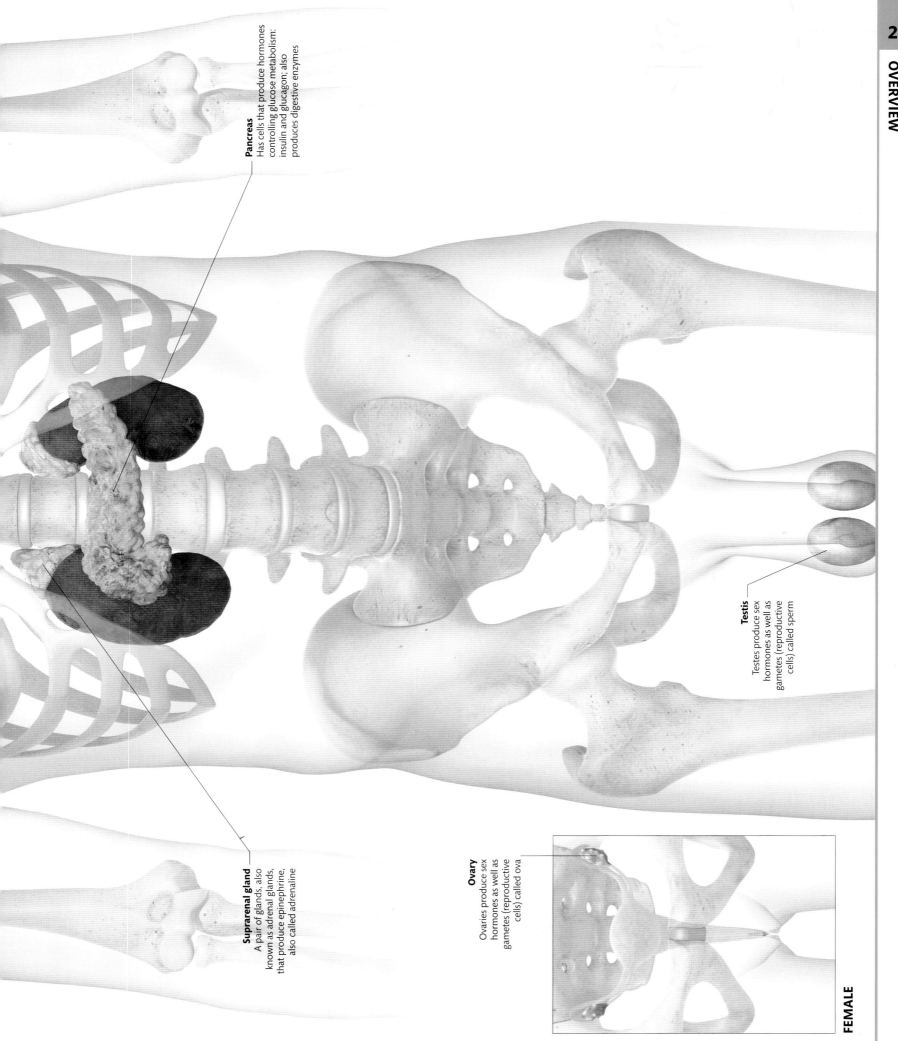

Pancreas
Has cells that produce hormones
controlling glucose metabolism:
insulin and glucagon; also
produces digestive enzymes

Suprarenal gland
A pair of glands, also
known as adrenal glands,
that produce epinephrine,
also called adrenaline

Testis
Testes produce sex
hormones as well as
gametes (reproductive
cells) called sperm

Ovary
Ovaries produce sex
hormones as well as
gametes (reproductive
cells) called ova

FEMALE

HEAD AND NECK

The insides of our bodies are regulated by the autonomic nervous and endocrine systems. There is overlap between these two systems, and their functions are integrated and controlled within the hypothalamus of the brain. The pituitary gland has two lobes; its posterior lobe develops as a direct extension of the hypothalamus. Both lobes of the pituitary gland secrete hormones into the blood stream, in response to nerve signals or blood-borne releasing factors from the hypothalamus. Many of the pituitary hormones act on other endocrine glands, including the thyroid gland in the neck, the suprarenal glands on top of the kidneys, and the ovaries or testes.

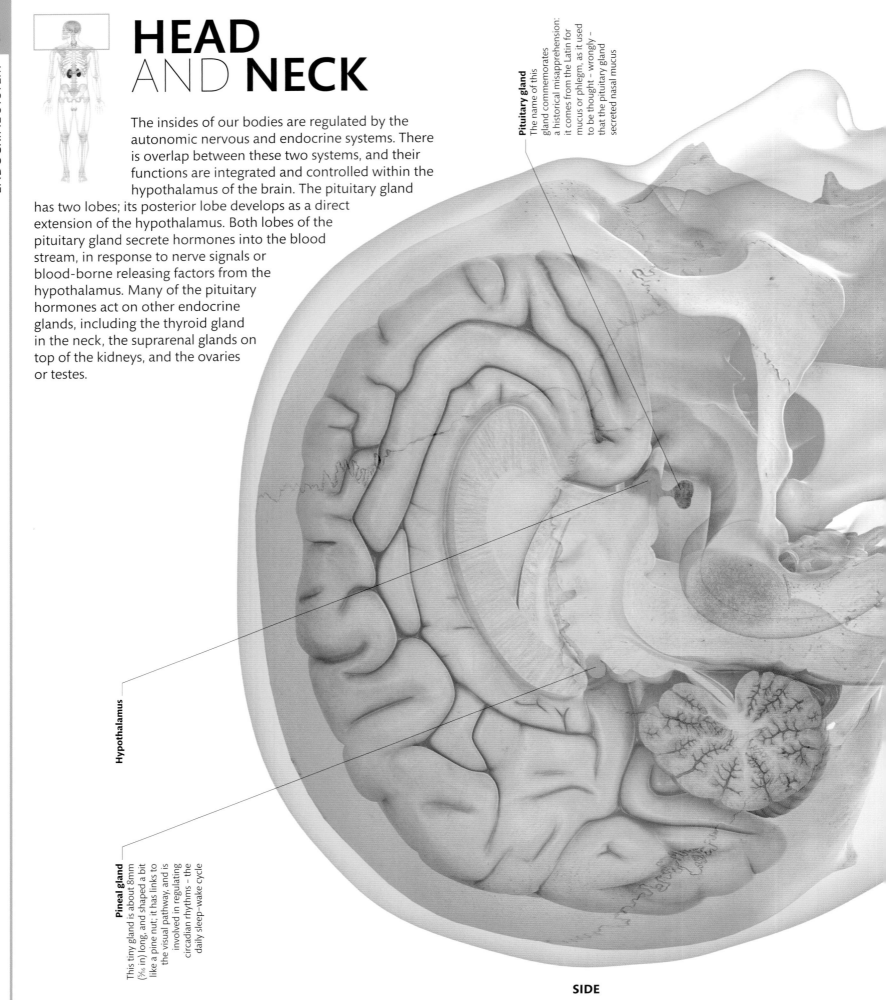

Pituitary gland
The name of this gland commemorates a historical misapprehension: it comes from the Latin for mucus or phlegm, as it used to be thought – wrongly – that the pituitary gland secreted nasal mucus

Hypothalamus

Pineal gland
This tiny gland is about 8mm (5/16 in) long, and shaped a bit like a pine nut; it has links to the visual pathway, and is involved in regulating circadian rhythms – the daily sleep–wake cycle

SIDE

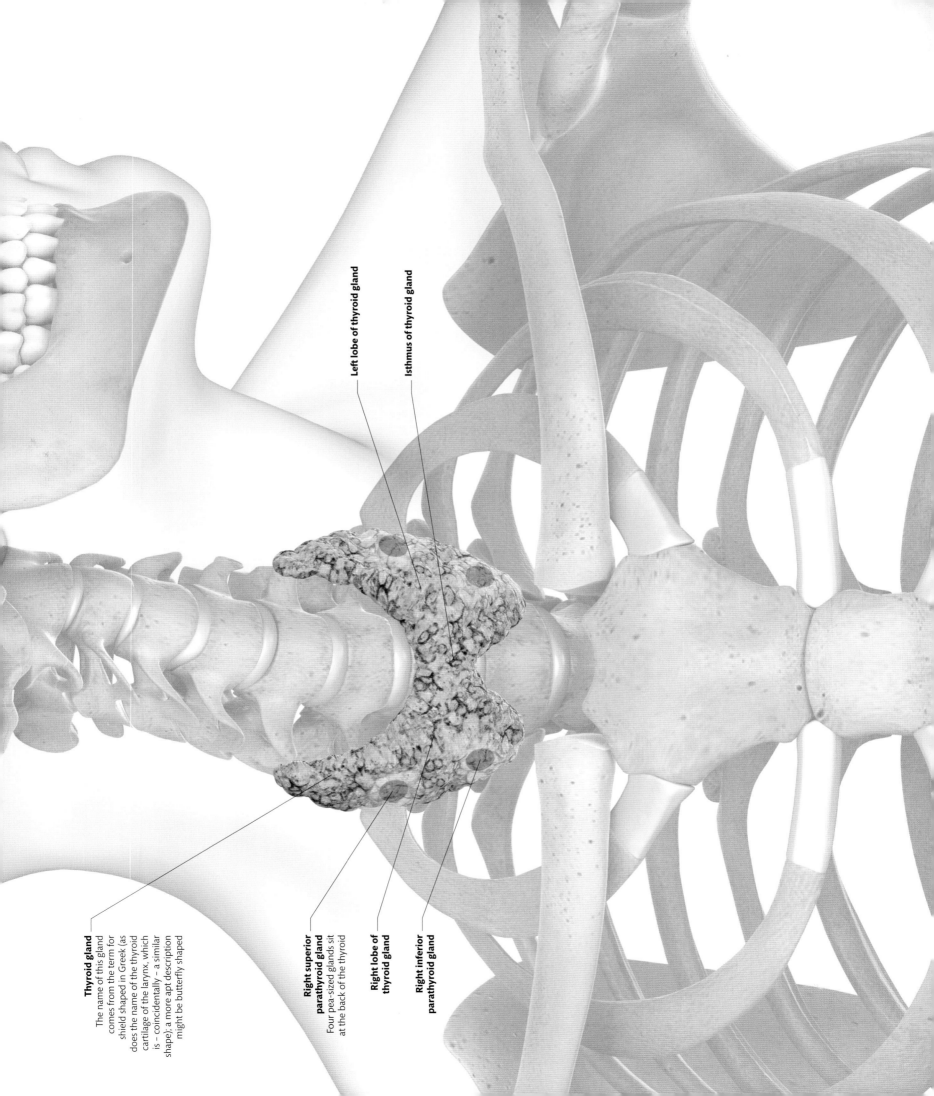

Thyroid gland
The name of this gland comes from the term for shield shaped in Greek (as does the name of the thyroid cartilage of the larynx, which is – coincidentally – a similar shape); a more apt description might be butterfly shaped

Right superior parathyroid gland
Four pea-sized glands sit at the back of the thyroid

Right lobe of thyroid gland

Right inferior parathyroid gland

Left lobe of thyroid gland

Isthmus of thyroid gland

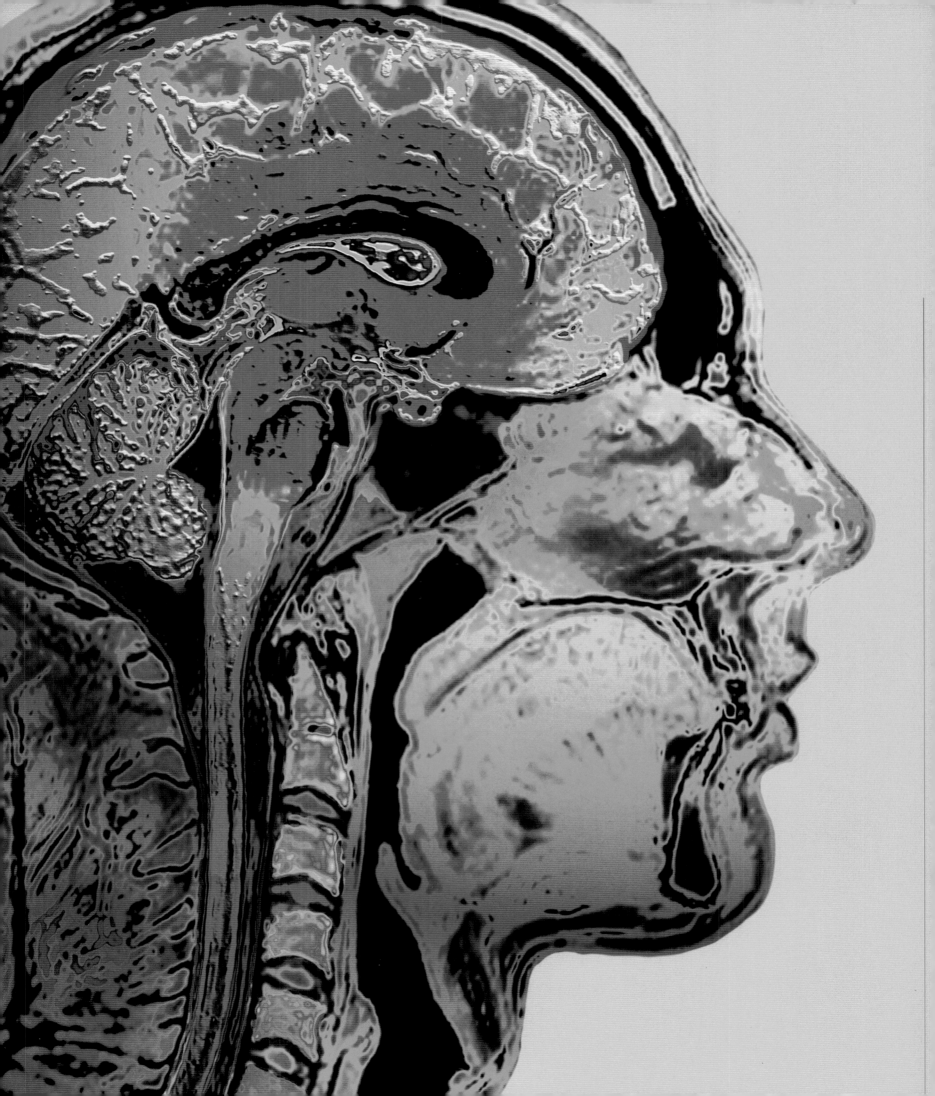

03 Imaging the Body

The human body is a "living machine" with many complex working parts. To understand how the body functions, and to cure the various ailments that afflict it, it is crucial for medical professionals to examine it in detail. Advances in technology have made it possible to view human anatomy without dissecting the body. Techniques such as magnetic resonance imaging (MRI) reveal the inside of the body with great accuracy and allow us to build up a complete picture of our anatomy from every possible angle.

222 Imaging techniques

224 Head and neck

226 Thorax

228 Abdomen and pelvis

230 Lower arm and hand

232 Lower limb and foot

IMAGING TECHNIQUES

Imaging is vital to diagnose illness, unravel disease processes, and evaluate treatments. Modern techniques provide detailed information with minimum discomfort to the patient and have largely replaced surgery in establishing the presence and extent of disease. Imaging has also helped advance biological research.

The invention of the X-ray in 1895 made the development of non-invasive medicine possible. Without the ability to see inside the body, many internal disorders could only be found after major surgery. Computerized imaging now helps doctors make early diagnoses, which at times greatly increase the likelihood of recovery. Computers process and enhance raw data to aid our visual ability. However, sometimes direct observation is essential. Viewing techniques have also become less invasive with the development of instruments such as the endoscope (see opposite).

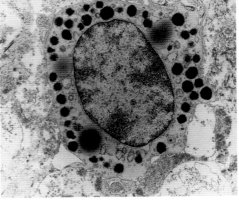

TEM of mast cell
Magnifications of several million times are possible using transmission electron microscopy (TEM). This image shows a mast cell with granules (dark purple) that it releases when it is damaged or is fighting microbes.

MICROSCOPY

Light microscopy (LM) uses magnifying lenses to focus light rays. The light passes through a thin section of material and enlarges it up to 2,000 times. Higher magnifications are achieved with beams of electrons (subatomic particles). In scanning electron microscopy (SEM), the beam runs across a specimen coated with gold film and bounces off the surface to create a three-dimensional image.

SEM of seminiferous tubule
This freeze-fracture image – in which the specimen is frozen and then cracked open before being scanned – shows sperm heads buried in Sertoli cells (orange) with the tails (blue) projecting into the tubule's lumen.

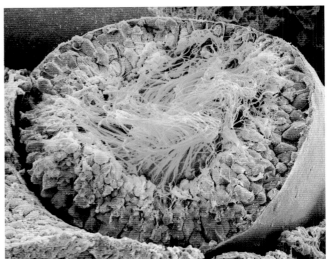

X-RAY

Like light rays, X-rays are electromagnetic energy, but of short wavelength. When passed through the body to strike photographic film, they create shadow images (radiographs). Dense structures, such as bone, absorb more X-rays and show up as white, while soft tissues, such as muscle, appear grey. Air-filled spaces, such as the lungs, appear black. The spaces inside the digestive tract, or within blood vessels, may be visualized by filling them with a contrast medium – such as iodine or barium – that absorbs X-rays. A contrast X-ray image of blood vessels is known as an angiogram.

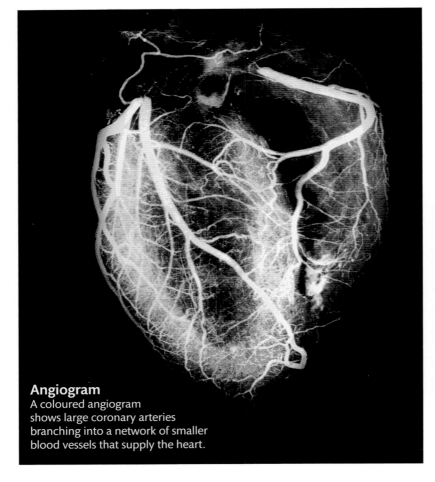

Angiogram
A coloured angiogram shows large coronary arteries branching into a network of smaller blood vessels that supply the heart.

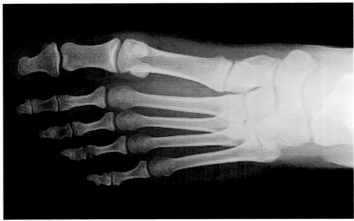

X-ray of foot
This X-ray image shows the foot bones of an adult from above. X-rays are especially useful for viewing dense tissue, such as bone.

RADIONUCLIDE AND PET SCANNING

In radionuclide imaging, a radioactive substance is injected into the body and is absorbed by the area to be imaged. As the substance decays, it emits gamma rays, which a computer forms into an image. Positron emission tomography (PET) is a type of radionuclide scanning where the injected chemical emits radioactive particles called positrons. PET gives data about how the brain functions rather than anatomy.

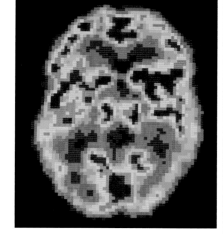

PET brain scan
This scan of the brain shows that the organ is active even in sleep mode. Areas in red, orange, and yellow represent high levels of activity.

ULTRASOUND

In ultrasound, a device called a transducer emits very high-frequency sound waves as it is passed over the body part being examined. The sound waves echo back to the device based on the density of the tissues they encounter. A computer analyses the reflected waves and creates an image.

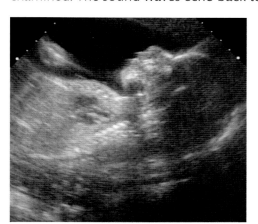

Fetal ultrasound
Low-intensity ultrasound is a safe way to monitor fetal development. In this scan the fetus's head can be seen clearly in profile on the right.

ENDOSCOPY

Telescope-like endoscopes are inserted through natural orifices or incisions to image the body's interior. They can be bent and controlled, and may carry instruments for other purposes as well, such as surgery and biopsy. Endoscopes have been designed to fit various body parts – bronchoscope for airways, gastroscope for the stomach, laparoscope for abdomen, and proctoscope for lower bowel.

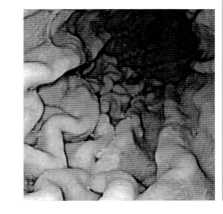

Endoscopic view of stomach
The gastric mucosa (inner lining) of a healthy stomach as seen through an endoscope. This procedure may be carried out to investigate upper digestive tract disorders.

MRI AND CT SCANNING

Computerized tomography (CT) and magnetic resonance imaging (MRI) detail various tissue types. In CT, a scanner using X-rays rotates around the patient as a computer records the levels of electromagnetic energy passing through tissues of different densities. A cross-section is built from layers of data. In MRI, a person lies in a magnetic chamber, which causes hydrogen atoms in the body to align. A pulse of radiowaves is released, throwing the atoms out of alignment. As they realign, they emit radio signals which are used to create an image.

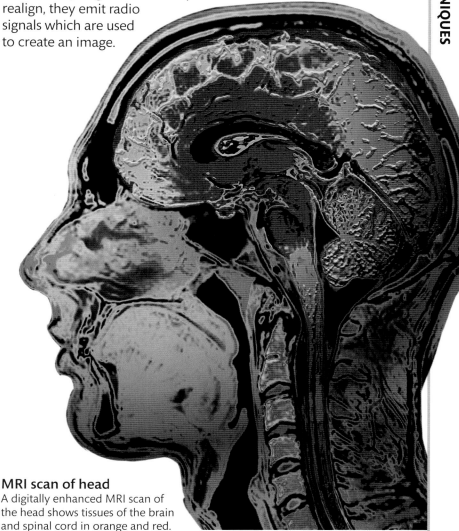

MRI scan of head
A digitally enhanced MRI scan of the head shows tissues of the brain and spinal cord in orange and red.

CT scan of lungs
In a horizontal slice through the chest, the spongy tissues and airways of the healthy lungs (oranges and yellows) show up clearly. The heart and major blood vessels between the lungs are mid-blue; the vertebrae, ribs, and sternum are dark blue.

ELECTRICAL ACTIVITY

Monitoring electrical activity in the body can reveal whether it is functioning normally. Signals coming from muscles and nerves are detected by applying sensor pads to the skin. The signals are sent to a computer, which coordinates, amplifies, and displays them as a real-time trace – usually a spiky or wavy line. Examples of this technique include electrocardiography (ECG) of the heart, electromyography (EMG) of skeletal muscles, and electroencephalography (EEG) of the brain's nerve activity.

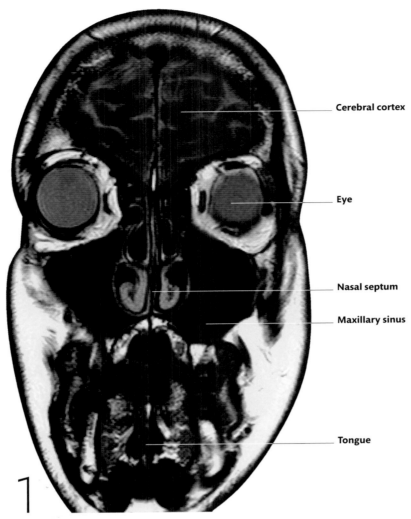

Cerebral cortex

Eye

Nasal septum

Maxillary sinus

Tongue

1

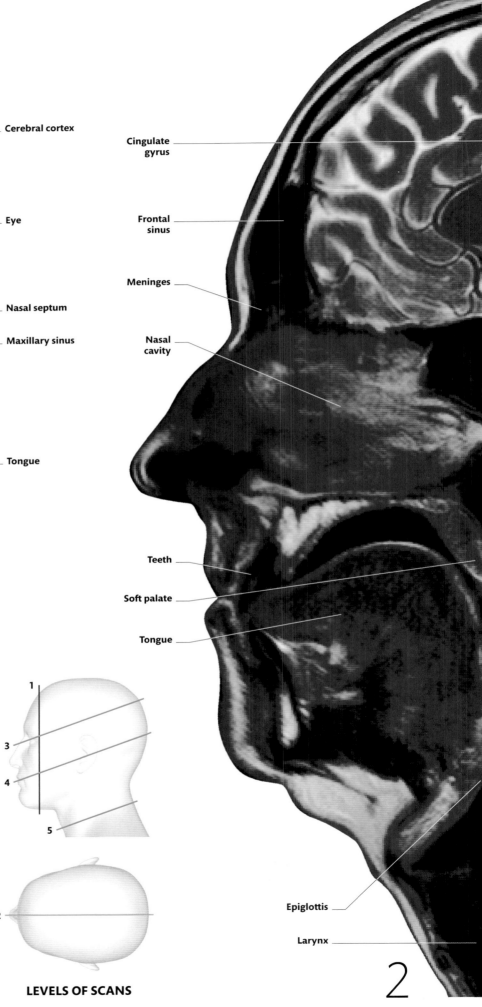

Cingulate gyrus

Frontal sinus

Meninges

Nasal cavity

Teeth

Soft palate

Tongue

Epiglottis

Larynx

2

HEAD AND NECK

The discovery of X-rays at the end of the 19th century suddenly created the possibility of looking inside the human body – without having to physically cut it open. Medical imaging is now an important diagnostic tool, as well as being used for the study of normal anatomy and physiology. In computed tomography (CT), X-rays are used to produce virtual sections or slices through the body. Another form of sectional imaging, using magnetic fields rather than X-rays to create images, is magnetic resonance imaging (MRI), as shown here. MRI is very useful for looking in detail at soft tissue, for instance, muscle, tendons, and the brain. Also seen clearly in these sections are the eyes (1 and 3), the tongue (1 and 2), the larynx, vertebrae, and spinal cord (2 and 5).

1

3

4

5

2

LEVELS OF SCANS

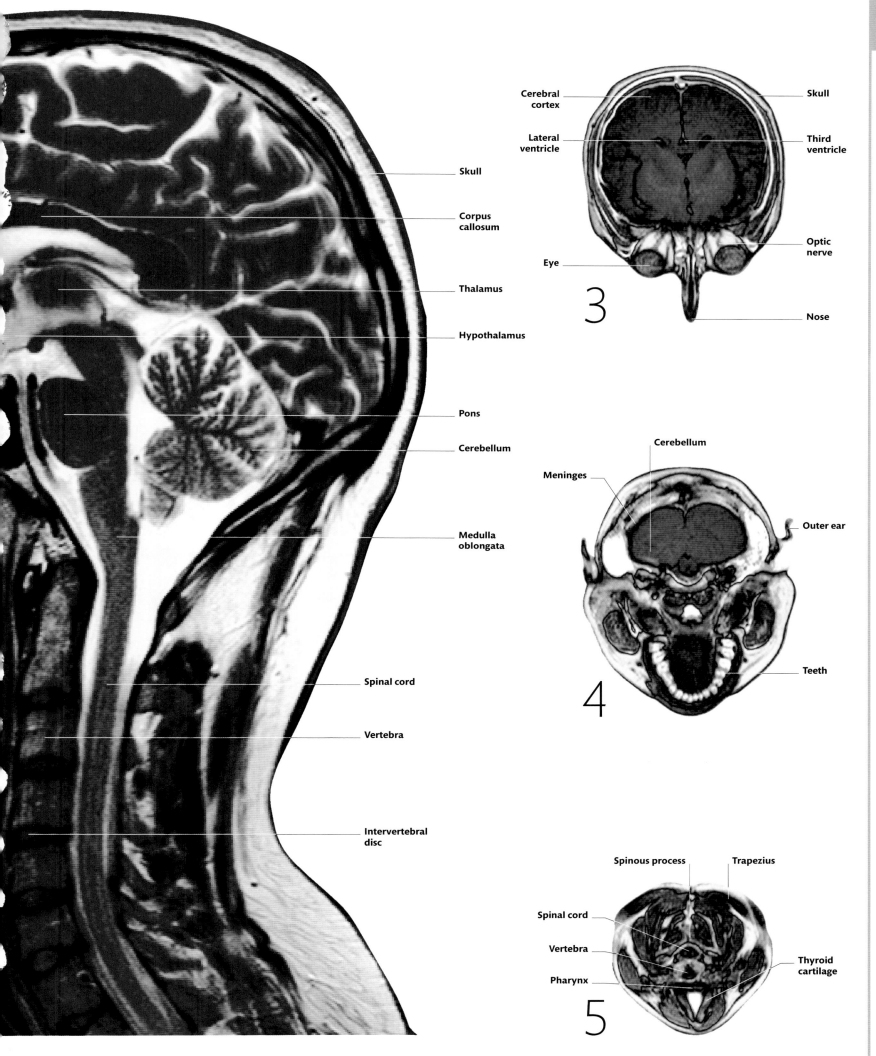

Skull

Corpus callosum

Thalamus

Hypothalamus

Pons

Cerebellum

Medulla oblongata

Spinal cord

Vertebra

Intervertebral disc

Cerebral cortex

Skull

Lateral ventricle

Third ventricle

Optic nerve

Eye

Nose

3

Cerebellum

Meninges

Outer ear

Teeth

4

Spinous process

Trapezius

Spinal cord

Vertebra

Thyroid cartilage

Pharynx

5

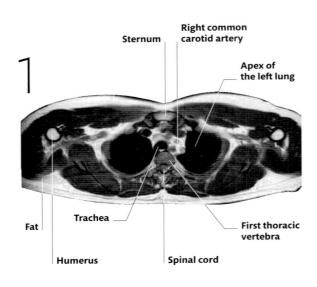

1

Sternum — Right common carotid artery

Apex of the left lung

Fat — Trachea — First thoracic vertebra — Spinal cord

Humerus

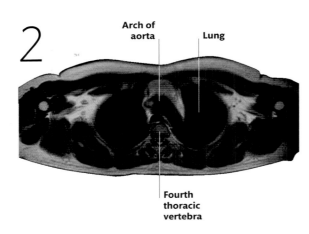

2

Arch of aorta — Lung

Fourth thoracic vertebra

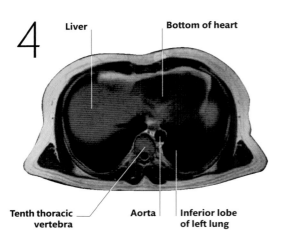

3

Lung — Superior vena cava — Left atrium — Sternum

Inferior lobe of right lung — Right inferior pulmonary artery — Seventh thoracic vertebra — Spinal cord

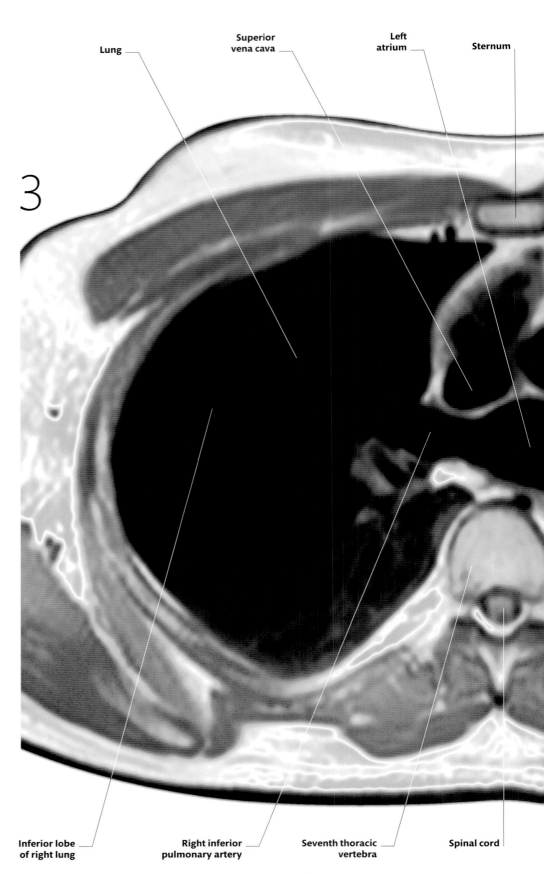

4

Liver — Bottom of heart

Tenth thoracic vertebra — Aorta — Inferior lobe of left lung

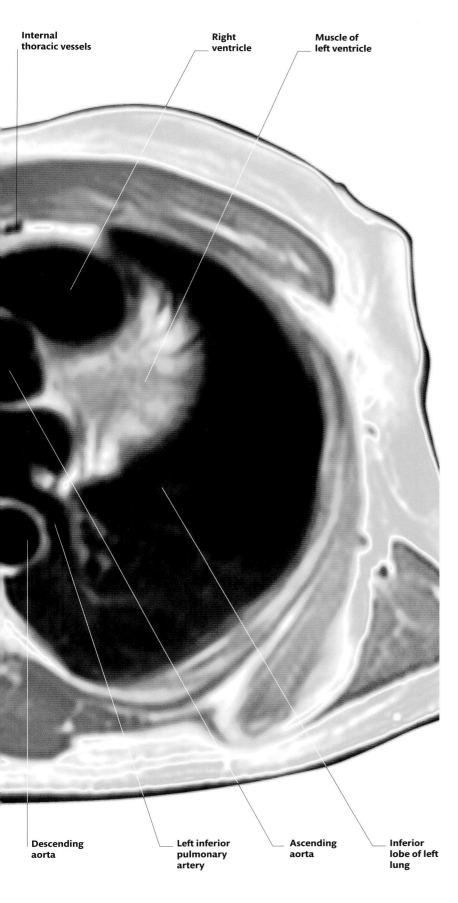

Internal thoracic vessels

Right ventricle

Muscle of left ventricle

Descending aorta

Left inferior pulmonary artery

Ascending aorta

Inferior lobe of left lung

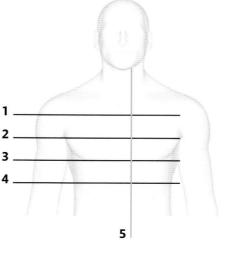

1

2

3

4

5

LEVELS OF SCANS

THORAX

The axial, or transverse, sections through the chest (sections 1–4) show the heart and large blood vessels lying centrally within the thorax, flanked by the lungs, and all set within the protective, bony casing of the ribcage. Section 1 shows the clavicles, or collarbones, joining the sternum at the front, the apex (top) of the lungs, and the great vessels passing between the neck and the thorax. Section 2 is lower down in the chest, just above the heart, while section 3 shows the heart with detail of its different chambers. The aorta appears to be to the right of the spine in this image, rather than to the left, but this is the usual way in which scans are viewed. You need to imagine yourself standing at the foot of the bed, looking down at the patient. This means that the left side of the body appears on the right side of the image as you view it. Section 4 shows the very bottom of the heart, and the inferior lobes of the lungs.

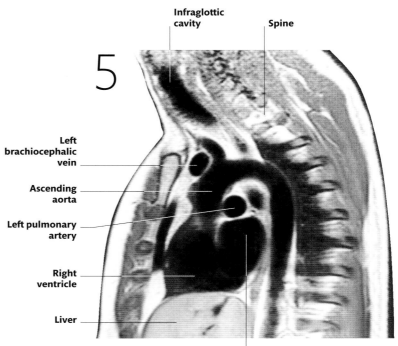

Infraglottic cavity

Spine

Left brachiocephalic vein

Ascending aorta

Left pulmonary artery

Right ventricle

Liver

5

Left atrium

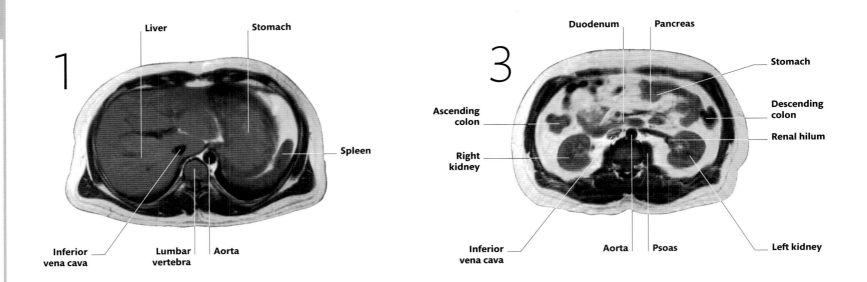

1
Liver
Stomach
Spleen
Inferior vena cava
Lumbar vertebra
Aorta

3
Duodenum
Pancreas
Stomach
Descending colon
Renal hilum
Ascending colon
Right kidney
Inferior vena cava
Aorta
Psoas
Left kidney

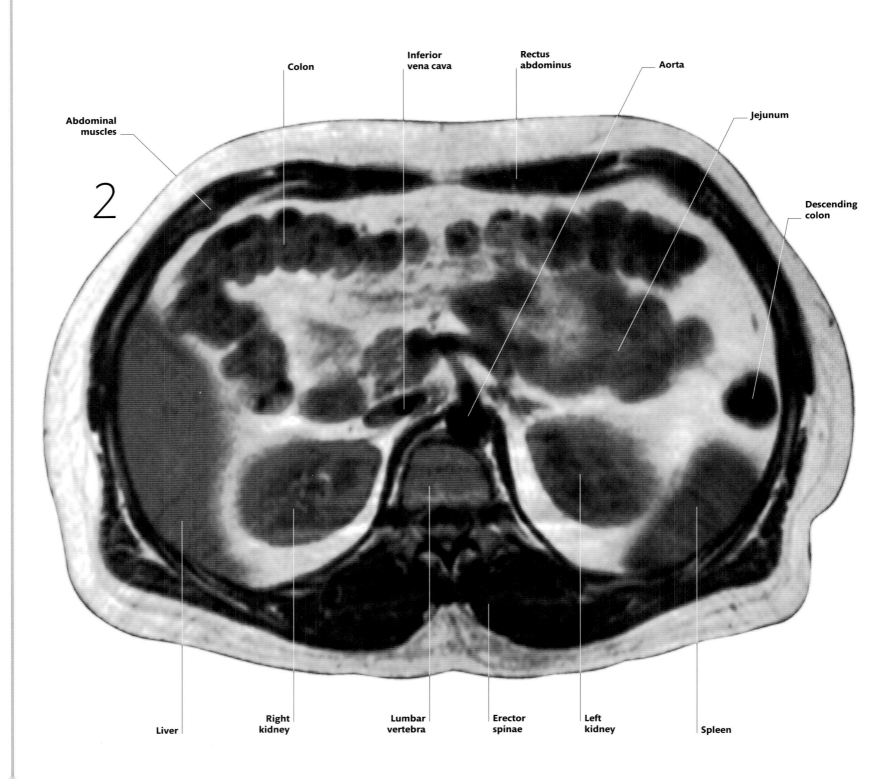

2
Colon
Inferior vena cava
Rectus abdominus
Aorta
Jejunum
Abdominal muscles
Descending colon
Liver
Right kidney
Lumbar vertebra
Erector spinae
Left kidney
Spleen

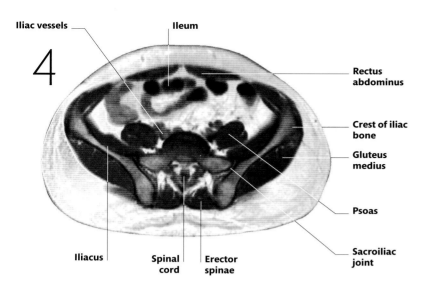

4

Iliac vessels
Ileum
Rectus abdominus
Crest of iliac bone
Gluteus medius
Psoas
Sacroiliac joint
Iliacus
Spinal cord
Erector spinae

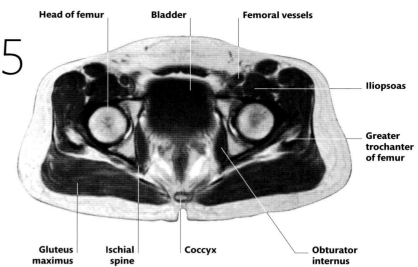

5

Head of femur
Bladder
Femoral vessels
Iliopsoas
Greater trochanter of femur
Obturator internus
Gluteus maximus
Ischial spine
Coccyx

ABDOMEN AND PELVIS

MRI is a useful way of looking at soft tissues – and for visualizing the organs of the abdomen and pelvis, which only appear as subtle shadows on a standard radiograph. In the series of axial or transverse sections through the abdomen and pelvis, we can clearly see the dense liver, and blood vessels branching within it (section 1); the right kidney lying close to the liver, and the left kidney close to the spleen (section 2); the kidneys at the level where the renal arteries enter them (section 3), with the stomach and pancreas lying in front; coils of small intestine, the ileum, resting in the lower part of the abdomen, cradled by the iliac bones (section 4); and the organs of the pelvis at the level of the hip joints (section 5). The sagittal view (section 6) shows how surprisingly shallow the abdominal cavity is, in front of the lumbar spine. In a slim person, it is possible to press down on the lower abdomen and feel the pulsations of the descending aorta – right at the back of the abdomen.

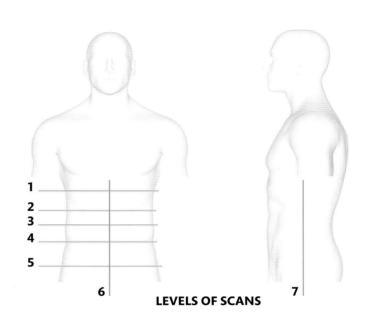

1
2
3
4
5
6
7

LEVELS OF SCANS

6

Intervertebral disc
Lumbar vertebra
Sacrum
Pubic symphysis

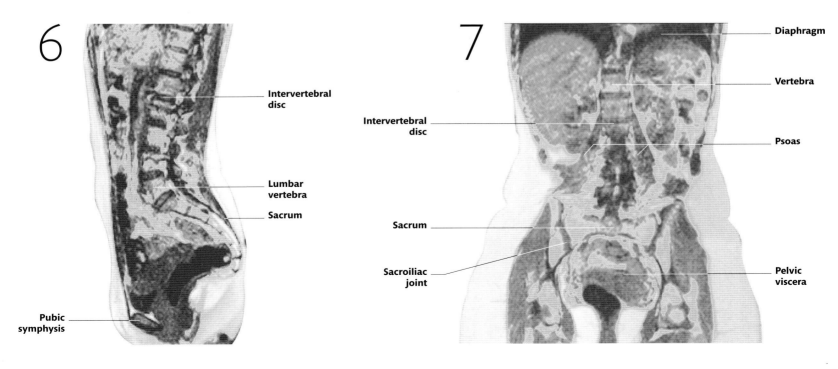

7

Diaphragm
Vertebra
Intervertebral disc
Psoas
Sacrum
Sacroiliac joint
Pelvic viscera

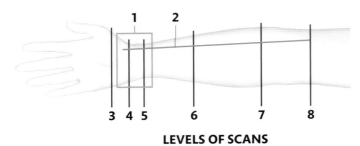

LEVELS OF SCANS

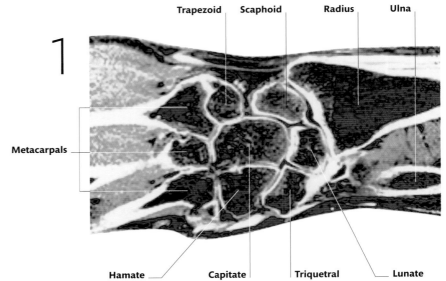

1

Trapezoid Scaphoid Radius Ulna

Metacarpals

Hamate Capitate Triquetral Lunate

LOWER ARM AND HAND

These scans of the arm, forearm, and hand show how tightly packed the structures are. Section 1 reveals the bones of the wrist – the carpals – interlocking like a jigsaw. The wrist joint itself is the articulation between the radius and the scaphoid and lunate bones. In section 2, part of the elbow joint is visible, with the bowl-shaped head of the radius cupping the rounded end of the humerus. Muscles in the forearm are grouped into two sets, flexors on the front and extensors behind the forearm bones and interosseous membrane. Compare sections 3–8 with sections through the leg (see pp.234–35) – both limbs have a single bone (humerus or femur) in the upper part, two bones in the lower part (radius and ulna in the forearm; tibia and fibula in the lower leg), a set of bones in the wrist and ankle (carpals and tarsals), fanning out to five digits at the end of the limb. Evolutionarily, these elements developed from the rays of a fish fin.

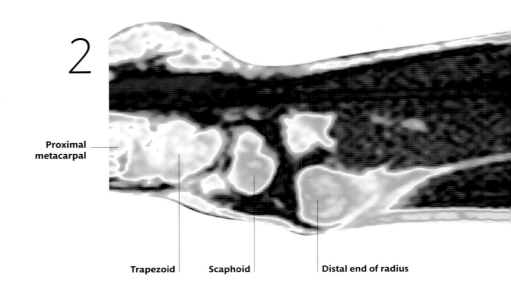

2

Proximal metacarpal

Trapezoid Scaphoid Distal end of radius

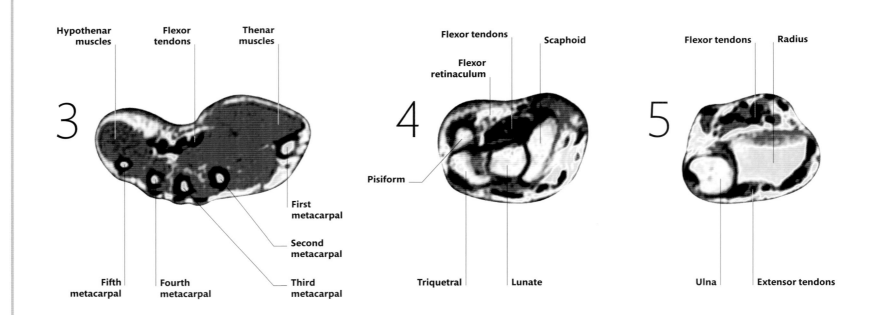

3

Hypothenar muscles Flexor tendons Thenar muscles

First metacarpal

Second metacarpal

Fifth metacarpal Fourth metacarpal Third metacarpal

4

Flexor tendons Scaphoid

Flexor retinaculum

Pisiform

Triquetral Lunate

5

Flexor tendons Radius

Ulna Extensor tendons

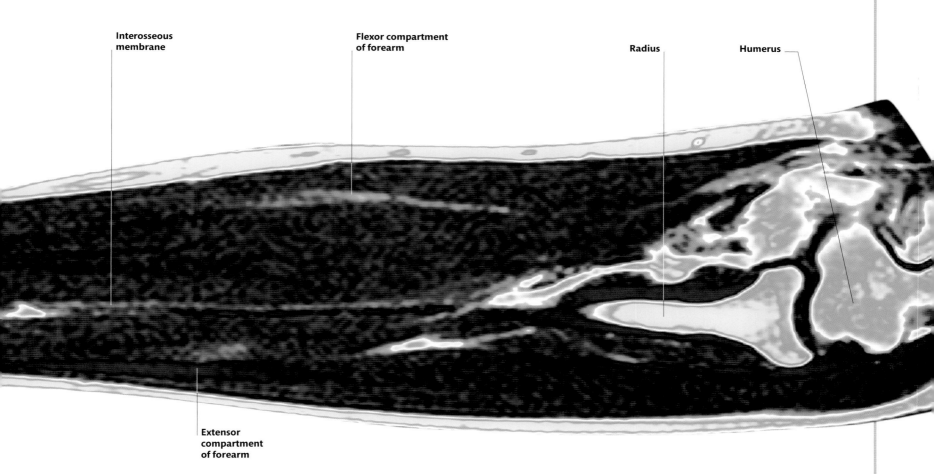

Interosseous membrane

Flexor compartment of forearm

Radius

Humerus

Extensor compartment of forearm

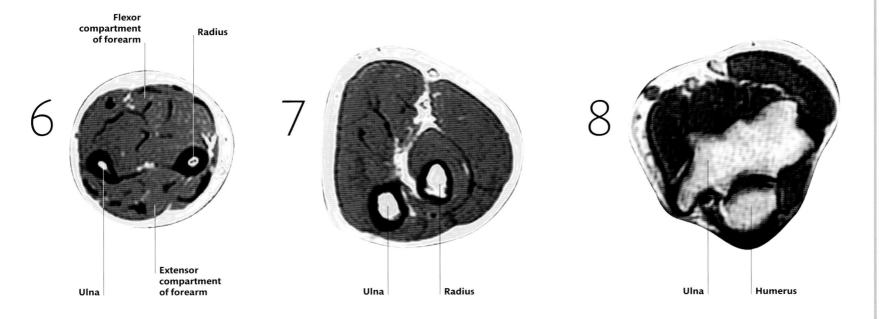

Flexor compartment of forearm

Radius

6

Ulna

Extensor compartment of forearm

7

Ulna

Radius

8

Ulna

Humerus

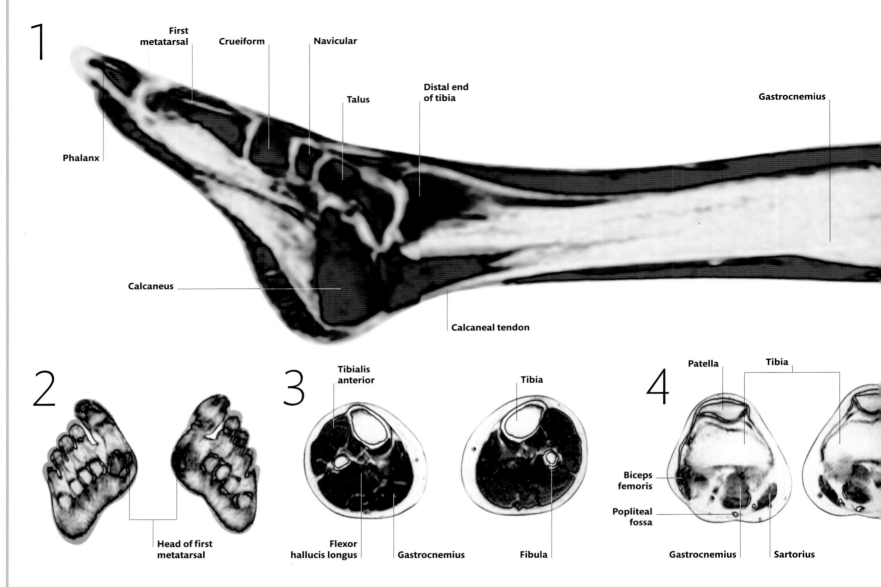

1

First metatarsal
Crueiform
Navicular

Talus

Distal end of tibia

Gastrocnemius

Phalanx

Calcaneus

Calcaneal tendon

2

Head of first metatarsal

3

Tibialis anterior

Tibia

Flexor hallucis longus
Gastrocnemius
Fibula

4

Patella
Tibia

Biceps femoris

Popliteal fossa

Gastrocnemius
Sartorius

LOWER LIMB AND FOOT

The sequence of axial and transverse sections through the thigh and lower leg show how the muscles are arranged around the bones. Groups of muscles are bound together with fascia – fibrous packing tissue – forming three compartments in the thigh (the flexor, extensor, and adductor muscles), and three in the lower leg (flexor, extensor, and peroneal or fibular muscles). Nerves and deep blood vessels are also packaged together in sheaths of fascia, forming "neurovascular bundles". Section 2 shows the bones of the forefoot, while the tightly packed muscles surrounding the tibia and fibula in the lower leg are visible in section 3. At the knee joint, shown in section 4, the patella can be seen to fit neatly against the reciprocal shape of the femoral condyles. The neurovascular bundle is clearly visible here, at the back of the knee, in a space known as the popliteal fossa – with the hamstring muscles on either side. Sections 5 and 6, through the middle and upper thigh, show the powerful quadriceps and hamstring muscles surrounding the thigh bone, or femur.

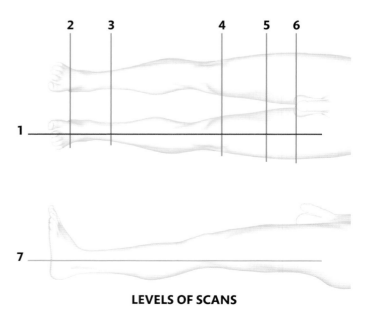

2   3   4   5   6

1

7

LEVELS OF SCANS

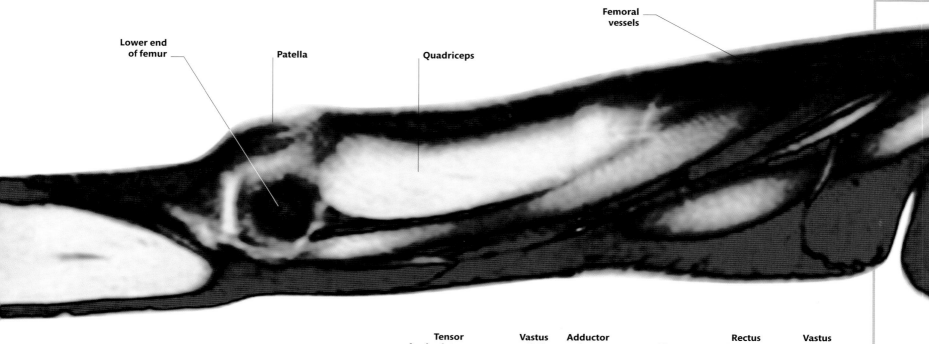

Femoral vessels

Lower end of femur

Patella

Quadriceps

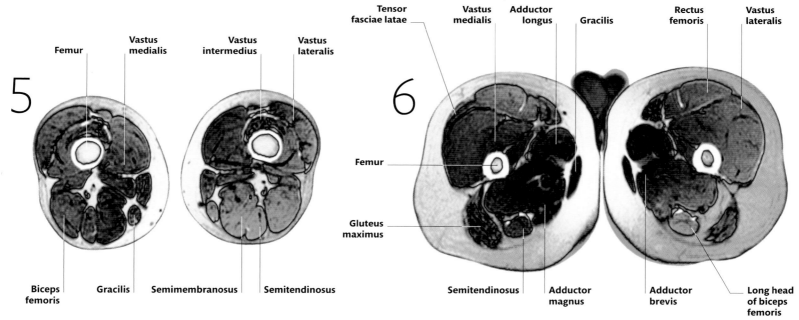

5

Femur

Vastus medialis

Vastus intermedius

Vastus lateralis

Biceps femoris

Gracilis

Semimembranosus

Semitendinosus

6

Tensor fasciae latae

Vastus medialis

Adductor longus

Gracilis

Rectus femoris

Vastus lateralis

Femur

Gluteus maximus

Semitendinosus

Adductor magnus

Adductor brevis

Long head of biceps femoris

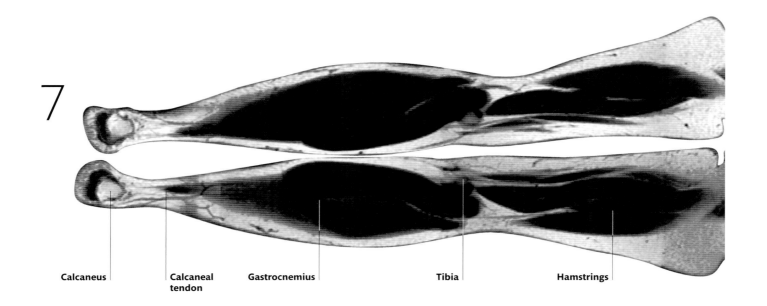

7

Calcaneus

Calcaneal tendon

Gastrocnemius

Tibia

Hamstrings

Glossary

Terms defined elsewhere in the glossary are in italics. All distinct terms are in **bold**.

abduction
The action of moving a limb further from the midline of the body. In muscle names, **abductor** indicates a muscle that has this action. See also *adduction*.

adduction
The action of moving a limb closer to the midline of the body. In muscle names, **adductor** indicates a muscle that has this action. See also *abduction*.

adipose tissue
Fat-storage *tissue*.

adrenal glands
See *suprarenal gland*.

adrenaline
See *epinephrine*.

afferent
In the case of blood vessels, carrying blood towards an organ, and in *nerves*, conducting impulses towards the *central nervous system*. See also *efferent*.

alveolus (pl. alveoli)
A small cavity; specifically, one of the millions of tiny air-sacs in the lungs where exchange of gases with the blood takes place; also the technical term for a tooth socket.

amino acid
Proteins are made from up to 20 different types of these small, nitrogen-containing *molecules*; amino acids also play various other roles in the body. See also *peptide*.

amnion
The *membrane* that encloses the developing *fetus* within the *uterus* (womb). The fluid inside it (amniotic fluid) helps to cushion and protect the fetus.

anastomosis
An interconnection between two otherwise separate blood vessels

(e.g. two *arteries*, or an artery and a *vein*).

angio-
A prefix relating to blood vessels.

angiography
In medical imaging: any technique for obtaining images of blood vessels in the living body.

anterior
Towards the front of the body, when considered in a standing position. **Anterior to** means in front of. See also *posterior*.

antibiotic
Any of various chemical compounds, natural or synthetic, that destroy or prevent the growth of micro-organisms (e.g. *bacteria*, yeasts, and fungi).

antibody
Defensive *proteins* produced by white blood cells that recognize and attach to particular "foreign" chemical components (*antigens*), such as the surface of an invading *bacterium* or *virus*. The body is able to produce thousands of different antibodies targeted at different invaders and toxins.

aorta
The body's largest *artery*, conveying blood pumped by the left *ventricle* of the heart. It extends to the lower abdomen, where it divides into the two common iliac arteries.

aponeurosis
A flattened, sheet-like *tendon*.

arteriole
A very small *artery*, leading into *capillaries*.

artery
A vessel carrying blood from the heart to the *tissues* and organs of the body. Arteries have thicker, more muscular walls than *veins*.

articular fat pad
In a joint bone, such as in the knee, the fatty tissue within the synovial membrane.

articular surface
The part of the long bone where the *epiphysis* forms the joint surface of the bone, which is covered in articular *cartilage*.

articulation
A *joint*, especially but not necessarily one allowing movement; also, a location within a joint where two bones meet in close proximity. A bone in a joint is said to **articulate with** the other bone(s) forming the joint.

ATP
Short for **adenosine triphosphate**, an energy-storing *molecule* used by all living *cells*.

atrium (pl. atria)
Either of the two smaller chambers of the heart that receive blood from the *veins* and pass it on to the corresponding *ventricle*.

autonomic nervous system
The part of the nervous system that controls non-conscious processes such as the activity of the body's *glands* and the muscles of the gut. It is divided into the **sympathetic** nervous system, the roles of which include preparing the body for "fight or flight", and the **parasympathetic nervous system**, which stimulates movement and secretions in the gut, produces erection of the penis during coitus, and empties the bladder.

axon
A wire-like extension of a *nerve cell* (*neuron*) along which electrical signals are transmitted away from the cell.

bacterium (pl. bacteria)
Any member of a large group of single-celled living organisms, some of which are dangerous *pathogens*. Bacterial *cells* are much smaller than animal and plant cells, and lack *nuclei*.

basal ganglia
Groups of *nerve cells* deep in the *cerebrum*; consists of the caudate nucleus, putamen, globus pallidus, and subthalamic nucleus. Functions include controlling movement.

belly (of muscle)
The widest part of a *skeletal muscle*, which bulges further when it contracts.

bile
A yellow-green fluid produced by the liver, stored in the *gallbladder*, and discharged into the intestine via the bile duct. It contains excretory products together with bile acids that help with fat digestion.

biopsy
A sample taken from a living body to test for infection, cancerous growth, etc; also the sampling process.

brachial
Relating to the arm.

brainstem
The lowest part of the brain, leading down from the rest of the brain to the *spinal cord*. In descending order, it consists of the *midbrain*, pons, and *medulla oblongata*.

bronchus (pl. bronchi)
The air tubes branching from the *trachea* and leading into the lungs; right and left main bronchi enter each lung respectively and divide into lobar bronchi, and eventually into much smaller tubes called **bronchioles**.

bursa
A pocket of **synovial fluid** that may lubricate the movement of tendons around joints, such as in the suprapatellar region of the knee joint.

caecum
The first part of the large intestine.
calcitonin. See *thyroid gland*.

cancer
An uncontrolled growth of *cells* with the potential to spread and form colonies elsewhere in the body. Cancer cells typically look different from their non-cancerous equivalents under the microscope. Cancers can arise in many different *tissues*.

capillaries
The smallest blood vessels, with a wall only one *cell* thick, supplied by

arterioles and draining into *veins*. Capillaries form networks, and are the sites where nutrients, gases, and waste products are exchanged between body *tissue* and blood.

carbohydrates
Naturally occurring chemical substances containing carbon, hydrogen, and oxygen atoms, e.g. *sugars*, *starch*, cellulose, and *glycogen*.

cardiac
Relating to the heart.

carpal
Relating to the wrist.

cartilage
A rubbery or tough supportive *tissue* (colloquially "gristle") found in various forms around the body.

cell
A tiny structure containing *genes*, a surrounding fluid (cytoplasm) that carries out chemical reactions, *organelles*, and an enclosing *membrane*. See also *nucleus*.

central nervous system
The brain and *spinal cord*, as distinct from the *nerves* that run through the rest of the body (the *peripheral* nervous system).

central osteonal canal
Also known as the Haversian canal, it is a channel in the centre of each *osteon* in a compact bone, containing blood and lymphatic vessels.

cerebellum
An anatomically distinct region of the brain below the back of the *cerebrum*, responsible for coordinating the details of complex bodily movements, and managing balance and posture.

cerebrospinal fluid
The clear fluid that fills the *ventricles* of the brain and surrounds the brain and *spinal cord*, helping to provide a constant environment and acting as a shock absorber.

cerebrum
The largest part of the brain and the locus of most "higher" *mental* activities; part of the forebrain in evolutionary

terms. It is divided into two halves called **cerebral hemispheres**.

cervical
1. Relating to the neck.
2. Relating to the *cervix* (neck) of the *uterus*.

cervix
The narrow "neck" of the *uterus*, opening into the upper end of the vagina; widens during childbirth.

cholesterol
A natural chemical that is an essential constituent of the body's *cell membranes* and is an intermediate *molecule* in the production of *steroid hormones*. It is a constituent of the plaques that cause the *arteries* to narrow in atherosclerosis.

chromosomes
The microscopic packages in the *nucleus* of a *cell* that contain genetic information in the form of *DNA*. Humans have 23 pairs of chromosomes, with a complete set present in nearly every cell of the body. Each chromosome consists of a single DNA *molecule* combined with various *proteins*.

cilium (pl. cilia)
A microscopic, beating, hair-like structure found in large numbers on the surfaces of some *cells* – for example in the air tubes of the lungs, where they help to remove foreign particles.

circadian rhythm
An *internal*, daily body rhythm. It is kept accurate by reference to external light and dark.

CNS
Short for *central nervous system*.

cochlea
The complex spiral structure in the *inner ear* that translates sound vibrations in the fluid it contains into electrical impulses to be sent to the brain.

collagen
A tough fibrous, structural *protein* that is widespread in the body (particularly in bone, *cartilage*, blood-vessel walls, and skin).

colon
The main part of the large intestine; comprises the ascending, transverse, and descending colon.

commissure
A link between two structures, especially any of several *nerve tracts* in the brain and *spinal cord* that crosses the midline of the body.

compartment (as in anatomical grouping or area)
In the case of muscles, used to define an anatomically and functionally discrete group of muscles, e.g. flexor compartment of the forearm.

condyle
A rounded, knuckle-like projection on a bone that forms part of a *joint*.

connective tissue
Any *tissue* comprising *cells* embedded in an acellular *matrix*; includes *cartilage*, bone, *tendon*, *ligament*, and blood.

cornea
The tough, transparent, protective layer at the front of the eye; helps focus light on the *retina*.

coronal section
A real or imagined section down the body that divides it from side to side; it is perpendicular to a *sagittal section*.

corpus callosum
A large *tract* of *nerve* fibres (*commissure*) that links the brain's two cerebral hemispheres.

cortex
The Latin word for bark, used for the outer parts of some organs, especially:
1. The **cerebral** or **cerebellar cortex** – the surface layers of *cells* (the "grey matter") of these parts of the brain.
2. The **suprarenal cortex** – the outer part of the *suprarenal glands*.

corticosteroid
Any of several *steroid hormones* produced by the suprarenal *cortex* (see *suprarenal glands*). Examples include **cortisone** and **cortisol** (*hydrocortisone*), which have many effects on the body's *metabolism*

and also suppress *inflammation*. The mineral-regulating hormone *aldosterone* is also a corticosteroid.

cranial
1. Relating to the *cranium*.
2. Towards the head.

cranial nerves
Pairs of *nerves* that lead directly from the brain rather than from the *spinal cord*. They mainly supply structures in the head and neck.

cranium
Together with the mandible (jaw), forms the skull.

CSF
Short for *cerebrospinal fluid*.

CT
Short for **computed tomography**, a sophisticated X-ray technique that produces images in the form of "slices" through the patient's body.

cutaneous
Relating to the skin.

cyst
A fluid-filled cavity in the body. Also, an old term for the bladder; hence **cystitis**.

dendrite
A branch-like outgrowth of a *nerve cell* (*neuron*) that carries incoming electrical signals to that cell. A neuron usually has many dendrites.

depressor
Term used in names of several muscles that act to pull down, e.g. depressor anguli oris (pulls down the angle of the mouth). See also *levator*.

diaphragm
A sheet of muscle that separates the *thorax* from the abdomen. When relaxed it is domed upwards; it flattens when contracted, to increase thoracic volume and draw air into the lungs. It is the most important muscle used in breathing.

diaphysis
A cylinder of compact bone around a central marrow cavity in a long bone.

diffusion
The net movement of *molecules* in a fluid (gas or liquid) from regions of high to lower concentration.

dilated
Opened or stretched wider.

distal
Relatively further away from the centre of the body or from the point of *origin*. See also *proximal*.

DNA
Short for **deoxyribonucleic acid**, a very long *molecule* made up of small individual units or nucleotides, containing one of four bases. DNA is found in the *chromosomes* of living *cells*; the order of the bases "spells out" the genetic instructions of the animal. See also *gene*.

dorsal
Relating to the back or back surface of the body, or to the top of the brain; also, relating to the back (**dorsum**) of the hand or the upper surface of the foot.

dorsal (sensory) root ganglion
Part of the *spinal cord* where cell bodies of sensory nerves cluster.

duodenum
The first part of the small intestine, leading out of the stomach.

efferent
In the case of blood vessels, carrying blood away from an organ; in the case of *nerves*, conducting impulses away from the *central nervous system*. See also *afferent*.

electrocardiography
Recording the electrical activity produced by the heart muscle, using electrodes applied to the patient's skin.

embryo
The earliest stage of a developing unborn individual in the *uterus*, from *fertilization* until 8 weeks of gestation (after which it is known as a *fetus*).

endocrine system
The system comprising *glands* that produce *hormones*.

endometrium
The inner lining of the *uterus*.

endosteal blood vessels
Blood vessels that travel inside a compact bone. See also *periosteal blood vessels*.

endothelium
The *cell* layer that forms the inner lining of blood vessels.

enzyme
Any of a large variety of different *molecules* (nearly always *proteins*) that catalyse a particular chemical reaction in the body.

epicondyle
A small bulge found on some bones near a *joint*, usually forming a site for muscle attachment.

epidermis
The outermost layer of skin, with a surface consisting of dead cells packed with the tough *protein keratin*.

epiglottis
A flexible flap of *cartilage* in the throat that helps to cover the *trachea* (windpipe) during swallowing.

epinephrine
A *hormone* released by the *suprarenal* glands in response to stressful situations. It prepares the body for a "fight or flight" response by increasing heart rate, diverting blood flow to muscles, etc.

epiphysis
The end of a bone that expands to form a joint surface. It is covered with a relatively thin shell of compact bone and is full of spongy or cancellous bone. See also *metaphysis*.

epithelium
Any *tissue* that forms the surface of an organ or structure. It may consist of a single layer of *cells*, or several layers.

erythrocyte
A red blood *cell*.

extension
The movement that increases the angle of, or straightens, a *joint*. The name **extensor** indicates a muscle

that has this action, e.g. extensor digitorum extends the fingers. See also *flexion*.

external
In anatomy: closer to the outer surface.

extracellular
Outside the *cell*; often used in reference to the fluid or *matrix* between cells of a *connective tissue*.

Fallopian tube
Another name for the oviduct or uterine tube; two oviducts attach to the *uterus*, extending to the *ovary* on each side; the *ovum* travels down this tube after *ovulation*.

fascia (pl. fasciae)
Layers of fibrous *tissue* between and around muscles, vessels, and organs.

fascicle
A bundle of muscle fibres, packed in connective tissue called **endomysium** and contained in a sheath of **perimysium**.

fertilization
The union of a *sperm* with an unfertilized egg (*ovum*), the first step in the creation of a new individual. See also *zygote*.

fetus
The unborn individual in the *uterus*, from 8 weeks after *fertilization*, when it begins to show a recognizably human appearance. See also *embryo*.

flexion
The bending movement at a *joint*. The name **flexor** indicates a muscle that has this action, e.g. flexor carpi ulnaris bends the wrist. See also *extension*.

follicle
A small cavity or sac-like structure: e.g. the hair follicle from which a hair grows.

foramen
An opening, hole, or connecting passage.

fossa
A shallow depression or cavity.

frontal
Relating to or in the region of the forehead; **frontal bone**, the skull bone of the forehead; **frontal lobe**, the foremost lobe of each cerebral hemisphere, lying behind the forehead.

gallbladder
The hollow organ into which *bile* (formerly known as gall) secreted by the liver is stored and concentrated before being transferred to the intestine.

gamete
A *sperm* or an *ovum* (egg). Gametes contain just one set of 23 *chromosomes*, whereas normal body *cells* have two sets (46 chromosomes). When sperm and egg combine during *fertilization*, the two-set condition is restored. See also *zygote*.

ganglion
1. A concentration of *nerve cell* bodies, especially one outside the *central nervous system*.
2. A swelling on a *tendon* sheath.

gastric
Relating to the stomach.

gene
A length of a *DNA molecule* that contains a particular genetic instruction. Many genes are blueprints for making particular *protein* molecules, while some have a role in controlling other genes. Between them, the thousands of different genes in the body provide the instructions for a fertilized egg to grow into an adult, and for all essential activities of the body to be carried out. Nearly every *cell* in the body contains an identical set of genes, although different genes are "switched on" in different cells.

genome
The complete set of *genes* found in a human or other living species. The human genome is thought to contain about 20,000–25,000 different genes.

gland
A structure in the body, the main purpose of which is to secrete particular chemical substances or

fluids. Glands are either **exocrine**, releasing their secretions through a duct onto an *external* or *internal* surface, such as the salivary glands, or **endocrine**, releasing *hormones* into the blood stream. See also *endocrine system*.

glial cells
Cells in the nervous system that are not *neurons* but play various supportive and protective roles within the nervous system.

gloss-, glosso-
Prefixes relating to the tongue.

glucagon
A *hormone* produced by the pancreatic islets (see *pancreas*) that increases *glucose* levels in the blood; its effect is opposite to that of *insulin*.

glucose
A simple *sugar* that is the main energy source used by the body's *cells*.

glycogen
A *carbohydrate* made up of long, branched chains of connected *glucose* molecules. The body stores glucose in the form of glycogen, especially in the muscles and liver; also called **animal starch**.

gonad
An organ that produces sex *cells* (*gametes*) – i.e. an *ovary* or a *testis*. A **gonadotropin** is a *hormone* that specifically affects the gonads.

grey matter
Part of the brain that contains cell bodies of *neurons*. See also *white matter*.

gyrus (pl. gyri)
One of the folds on the outer surface of the brain. See also *sulcus*.

haemoglobin
The red pigment within *erythrocytes* that gives blood its colour and carries oxygen to the *tissues*.

head (of a muscle)
Where a muscle has several *origins* or *proximal* attachments, these may be referred to as "heads", as in the long and short heads of biceps brachii.

hepatic
Relating to the liver.

homeostasis
The maintenance of stable conditions in the body, e.g. in terms of chemical balance or temperature.

hormone
A chemical messenger produced by one part of the body that affects other organs or parts. There also exist **local hormones** that affect only nearby *cells* and *tissues*. Chemically, most hormones are either *steroids*, *peptides*, or small *molecules* related to *amino acids*. See also *neurotransmitter*.

hypothalamus
A small but vital region at the base of the brain, which is the control centre for the *autonomic nervous system*, regulating processes such as body temperature and appetite. Also controls the secretion of *hormones* from the *pituitary gland*.

ileum
The last part of the small intestine, ending at the junction with the large intestine (*colon*) N.B: Not the same as ilium, one of the bones of the hip.

immune system
The *molecules*, *cells*, organs, and processes that are involved in defending the body against disease.

immunity
Resistance to attack by a *pathogen* (disease-causing organism); **specific immunity** develops as a result of the body's *immune system* being primed to resist a particular pathogen.

immunotherapy
Any of various treatments involving either the stimulation or suppression of the activity of the immune system.

inferior
Lower down the body, when considered in a standing position (i.e. nearer the feet). See also *superior*.

inflammation
An immediate reaction of body *tissue* to damage, in which the affected area becomes red, hot, swollen, and painful, as white blood cells (see *leukocyte*) accumulate at the site to attack potential invaders.

inguinal
Relating to, or in the region of, the groin.

inner ear
The fluid-filled innermost part of the ear, which contains the organs of balance (the semicircular canals) and the organs of hearing within the *cochlea*. See also *middle ear*.

insertion
The point of attachment of a muscle to the structure that typically moves when the muscle is contracted. See also *origin*.

insulin
A *hormone* produced by the pancreatic islets (see *pancreas*) that promotes the uptake of *glucose* from the blood, and the conversion of glucose to the storage *molecule*, *glycogen*.

integument
The *external* protective covering of the body.

internal
In anatomy: inside the body, distant from the surface. See also *external*.

internal elastic media
The layer between the *tunica media* and *tunica intima* that is prominent in large arteries, including the *aorta* and its main branches. This layer is absent from some *veins*, including those around the brain.

intra-
Prefix meaning within, as in **intracellular** or **intramuscular**.

islets of Langerhans
See *pancreas*.

-itis
Suffix meaning "inflammation", used in words such as **tonsillitis** and **laryngitis**.

joint
Any junction between two or more bones, whether or not movement is possible between them. See

also *articulation*, *suture*, *symphysis*, *synovial joint*.

keratin
A tough *protein* that forms the substance of hair and nails, gives strength to the skin, etc.

labia (sing. labium)
Either of the two paired folds that form part of the *vulva* in females: the outer **labia majora** and the more delicate inner **labia minora**.

labial
Relating to the lips, or to the *labia* of the female genitals.

lactation
Secretion of milk by the breasts.

larynx
The voicebox: a complex structure situated at the top of the *trachea* (windpipe). It includes the **vocal cords**, structures that function to seal off the trachea when necessary, as well as creating sound when their edges are made to vibrate during breathing.

lateral
Relating to or towards the sides of the body. See also *medial*.

leukocyte
A white blood *cell*. There are several types, acting in different ways to protect the body against disease as part of its *immune response*. Leukocytes are found in *lymph nodes* and other *tissues* generally, as well as in the blood.

levator
Term used in the names of several muscles whose action is to lift up, such as the levator scapula (lifts the shoulder blade). See also *depressor*.

ligament
A tough fibrous band that holds two bones together. Many ligaments are flexible, but they cannot be stretched. The term is also used for bands of *tissue* connecting or supporting some internal organs.

limbic system
Several connected regions at the base of the brain, involved in memory, behaviour, and emotion.

line of fusion
A line in the bone that shows the area of fusion of the cartilage growth plate with a long bone. The cartilage plate allows long bones to grow quickly in length during childhood and fuses by adulthood, but the line of fusion may still be evident for a few years.

lingual
Relating to the tongue.

lipid
Any of a large variety of fatty or fat-like substances that are found naturally in living things and are relatively insoluble in water.

lumbar
Relating to the lower back and sides of the body between the lowest ribs and the top of the hip bone. The **lumbar vertebrae** are the *vertebrae* that lie within this region.

lumen
The space inside a tubular structure, such as a blood vessel or glandular duct.

lymph node
A small lymphoid organ; lymph nodes serve to filter out and dispose of *bacteria* and debris, such as *cell* fragments.

lymphocyte
A specialized *leukocyte* that produces antibodies including *natural killer cells*, T-cells, and B-cells.

lymphoid tissue
The *tissue* of the lymphatic system, which has an immune function, including *lymph nodes*, the *thymus*, and the *spleen*.

M line
A fine line present in skeletal muscle, which connects the thick myosin filaments and holds them in place. See also *Z disc*.

macromolecule
A large *molecule*, especially one that consists of a chain of small similar "building blocks" joined together. *Proteins, DNA,* and *starch* are examples of macromolecules.

mammary
Of, or relating to, the breasts.

marrow
In anatomical contexts, usually short for **bone marrow**, the soft material located in the cavities of bones; in some areas this *tissue* is mainly fat; in others, it is blood-forming tissue.

matrix
The *extracellular* material in which the cells of *connective tissues* are embedded. It may be hard, as in bone; tough, as in *cartilage*; or fluid, as in blood.

meatus
A channel or passage. For example, the external **auditory meatus**, the ear canal.

medial
Towards the midline of the body. See also *lateral*.

medullary (marrow) cavity
Cavities of long bones that are filled with blood-forming red marrow at birth, but this is replaced with fat-rich yellow marrow by adulthood. Red marrow persists in other parts such as the skull, spine, ribs, and pelvis.

medulla
1. Short for **medulla oblongata**, the elongated lower part of the brain that connects with the *spinal cord*.
2. The central part or core of some organs such as the kidneys and *suprarenal glands*.

melanin
A dark brown naturally occurring pigment *molecule*, which occurs in greater amounts in tanned or darker skin, and protects deeper *tissues* from ultraviolet radiation.

melatonin
A *hormone* secreted by the pineal *gland* in the brain, which plays a role in the body's sleep–wake cycle (see *circadian rhythm*).

membrane
1. A thin sheet of *tissue* covering an organ, or separating one part of the body from another.
2. The outer covering of a *cell* (and similar structures within the cell).

A cell membrane is composed of a double layer of *phospholipid molecules* with other molecules such as *proteins* embedded in it.

meninges
Membranes that enclose the outside of the brain and *spinal cord*. **Meningitis** is *inflammation* of the meninges, usually resulting from infection.

meniscus (pl. menisci)
Crescent-shaped articular disc made of fibrocartilage present in the knee joint. A pair of menisci facilitates the complex movements of this joint.

menstrual cycle
The monthly cycle that takes place in the *uterus* of a non-pregnant woman of reproductive age. The *endometrium* (lining of the uterus) grows thicker in preparation for possible pregnancy; an egg is released from the ovary (*ovulation*); then, if the egg is not fertilized, the endometrium breaks down and is discharged through the vagina in a process known as **menstruation**.

mental
1. Relating to the mind (Latin *mens*).
2. Relating to the chin (Latin *mentum*).

mesentery
A folded sheet of *peritoneum* (the *membrane* lining the abdominal cavity and organs), forming a connection between the intestines and the back of the abdominal cavity.

metabolism
The chemical reactions taking place in the body. The **metabolic rate** is the overall rate at which these reactions are occurring.

metaphysis
Neck of a long bone where spongy bone starts to encroach on marrow cavity. See also *epiphysis*.

midbrain
The upper part of the *brainstem*.

middle ear
The air-filled middle chamber of the ear, between the inner surface of

the eardrum and the *inner ear*. See also *ossicles*.

molecule
The smallest unit of a chemical compound that can exist, consisting of two or more atoms joined together by chemical bonds. The water molecule is a simple example, consisting of two hydrogen atoms joined to one oxygen atom. See also *macromolecule*.

motor
Adjective relating to the control of muscle movements, as in **motor neuron, motor function**, etc. See also *sensory*.

MRI scan
Short for **magnetic resonance imaging scan**, a medical imaging technique based on the energy released when magnetic fields are applied then removed from the body; it can produce very detailed images of the soft *tissues* of the body.

mucosa (pl. mucosae)
A *membrane* that secretes *mucus*.

mucus
A thick fluid produced by some *membranes* of the body for protection, lubrication, etc. (Adjective mucous.)

muscle fibre
Cylindrical units in a skeletal muscle that range from a few millimeters to several centimetres in length. They are formed by the merging of many *cells*, and therefore contain many nuclei.

myelin
Fatty substance forming a layer around some *nerve axons*, called **myelinated** axons, insulating them and speeding their nerve impulses.

myelo-
1. Prefix relating to the *spinal cord*.
2. Prefix relating to bone *marrow*.

myo-
Prefix relating to muscle.

myofibril
Fibres in *skeletal muscle* that contain filaments made of contractile

proteins, mainly **actin** and **myosin**. The way these filaments are organized gives *skeletal muscle* a striped or striated appearance under a light microscope.

natural killer (NK) cell
A type of *lymphocyte* that can attack and kill *cancer cells* and *virus*-infected cells.

neocortex
All the *cortex* of the *cerebrum* except the region concerned with smell and the hippocampal formation.

nerve
A cable-like structure transmitting information and control instructions in the body. A typical nerve consists of *axons* of many separate nerve *cells* (*neurons*) running parallel to, but insulated from, each other; the nerve itself is surrounded by an overall protective sheath of fibrous *tissue*. Nerves may contain nerve fibres controlling muscles or *glands* (*efferent* fibres), while others contain fibres carrying *sensory* information back to the brain (*afferent* fibres); some nerves carry both types of nerve fibre.

neuron
A *nerve cell*. A typical neuron consists of a rounded cell body; branch-like outgrowths called *dendrites* that carry incoming electrical signals to the neuron; and a single, long, wire-like extension, called an *axon*, which transmits outgoing messages. There are many variations on this basic pattern, however.

neurotransmitter
Any of various chemical substances released at *synapses* by the ends of *nerve cells*, where they function to pass a signal on to another nerve cell or muscle. Some neurotransmitters act mainly to stimulate the action of other cells, others to inhibit them.

nucleus (pl. nuclei)
1. The structure within a *cell* that contains the *chromosomes*.
2. Any of various concentrations of *nerve* cells within the *central nervous system*.
3. The central part of an atom.

occipital
Relating to the back of the head. The **occipital bone** is the skull bone forming the back of the head. The **occipital lobe** is the hindmost lobe of each cerebral hemisphere, lying below the occipital bone.

oesophagus
The gullet: the tubular part of the alimentary canal between the *pharynx* and the stomach.

oestrogens
Steroid hormones produced predominantly by the *ovary*, and which regulate female sexual development and *physiology*. Artificial oestrogens are used in *oral* contraceptives and hormone replacement therapy.

olfactory
Relating to the sense of smell.

oligodendrocyte
A structure in nerve cell that manufactures the *myelin* sheath along the axons in the central nervous system.

optic nerve
The *nerve* that transmits visual information from the *retina* of the eye to the brain.

oral
Relating to the mouth.

orbit
The bony hollow in the skull within which the eye is contained.

organelle
Any of a variety of small structures inside a *cell*, usually enclosed within a *membrane*, which are specialized for functions such as energy production or secretion.

origin
The point of attachment of a muscle to the structure that typically remains stationary when the muscle is contracted. See also *insertion*.

ossi-, osteo-
Prefixes relating to bone.

ossicles
Three small bones of the *middle ear* that transmit vibrations caused by

sound waves from the eardrum to the *inner ear*.

osteocyte
Bone cells that lie in minute cavities between the concentric, cylindrical layers of bone mineral. The *cells* communicate with each other via thin processes, which run through microscopic canals in the mineral.

osteon
The basic unit in compact bone. It consists of concentric layers of *tissue*.

ovary
Either of the two organs in females that produce and release egg *cells* (*ova*). They also secrete sex *hormones*.

ovulation
The point in the *menstrual* cycle at which an egg *cell* (*ovum*) is released from the *ovary* and begins to travel towards the *uterus*.

ovum (pl. ova)
An unfertilized egg *cell*.

palate
The roof of the mouth, comprising the bony **hard palate** in the front and the muscular **soft palate** behind it.

pancreas
A large, elongated *gland* lying behind the stomach, with a dual role in the body. The bulk of its *tissue* secretes digestive *enzymes* into the *duodenum*, but it also contains scattered groups of cells called **pancreatic islets** or *islets of Langerhans* that produce important *hormones*, including *insulin* and *glucagon*.

parasympathetic nervous system
See *autonomic nervous system*.

parathyroid glands
Four small *glands* that are often embedded in but are separate from the *thyroid gland*. They produce parathyroid hormone, which regulates calcium *metabolism* in the body.

parietal
A term (derived from the Latin word for "wall") with various applications in anatomy. The **parietal bones** form the side

walls of the skull, and the **parietal lobes** of the brain lie beneath those bones. *Membranes* (such as the *pleura* and *peritoneum*) are described as parietal where they are attached to the body wall.

pathogen
Any disease-causing agent, including *bacteria* and *viruses*.

pathology
The study of disease; also, the physical manifestations of a disease.

pelvic girdle
The hip bones attach to the sacrum to form the pelvic girdle, linking the leg bones to the spine.

pelvis
1. The cavity enclosed by the *pelvic girdle*, or the area of the body containing the pelvic girdle.
2. The **renal pelvis** is the cavity in the kidney where the urine collects before passing down the *ureter*.

peptide
Any *molecule* consisting of two or more *amino acids* joined together, usually in a short chain. There are many types, some of which are important *hormones*. Proteins are polypeptides: long chains of amino acids.

peri-
Prefix meaning round or surrounding.

periosteal blood vessels
The blood vessels that run around the outside of the compact bone. See also *endosteal blood vessels*.

periosteum
The outer lining of bones; contains cells that can lay down or remove bone tissue.

peripheral
Towards the outside of the body or to the extremities of the body. The term **peripheral nervous system** refers to the whole of the nervous system except for the brain and *spinal cord*. See also *central nervous system*.

peritoneum
A thin, lubricated sheet of *tissue* that enfolds and protects most of the organs of the abdomen.

pharynx
The muscular tube behind the nose, mouth, and *larynx*, leading into the *oesophagus*.

phospholipid
A type of *lipid molecule* with a phosphate (phosphorus plus oxygen) group at one end. The phosphate group is attracted to water while the rest of the molecule is not. This property makes phospholipids ideal for forming *cell* membranes if two layers of molecules are situated back-to-back.

physiology
The study of the normal functioning of body processes; also, the body processes themselves.

pituitary gland
Also called the **hypophysis**, a complex pea-sized structure at the base of the brain, sometimes described as the body's "master *gland*". It produces various *hormones*, some affecting the body directly and others controlling the release of hormones by other glands.

placenta
The organ that develops on the inner wall of the *uterus* during pregnancy, allowing the transfer of substances, including nutrients and oxygen, between maternal and fetal blood. See also *umbilical cord*.

plasma
Blood minus its cellular components (red and white blood *cells*, and *platelets*).

platelets
Specialized fragments of *cells* that circulate in the blood and are involved in blood clotting.

pleura (pl. pleurae)
The lubricated *membrane* that lines the inside of the thoracic cavity and the outside of the lungs.

plexus
A network, usually in reference to *nerves* or blood vessels.

pneum-, pneumo-
1. Prefix relating to air.
2. Prefix relating to the lungs.

portal vein
The large *vein* carrying blood from the intestines to the liver; previously known as the **hepatic portal vein**.

posterior
Towards the back of the body, when considered in a standing position. **Posterior to**, behind. See also *anterior*.

process
In anatomy: a projection or extended part of a bone, *cell*, etc.

progesterone
A *steroid hormone* produced by the *ovaries* and *placenta*, which plays a role in the *menstrual cycle* and in the maintenance and regulation of pregnancy.

pronation
The rotation of the radius around the ulna in the forearm, turning the palms of the hand to face downwards or backwards. In muscle names, **pronator** indicates a muscle that has this action, e.g. pronator teres. See also *supination*.

prostate gland
A *gland* located below the male bladder; its secretions contribute to *semen*.

proteins
Large *molecules* consisting of long folded chains of small linked units (*amino acids*). There are thousands of different kinds in the body. Nearly all *enzymes* are proteins, as are the tough materials *keratin* and *collagen*. See also *peptide*.

proximal
Relatively closer to the centre of the body or from the point of *origin*. See also *distal*.

puberty
The period of sexual maturation between childhood and adulthood.
pulmonary
Relating to the lungs.

pyloric
Relating to the last part of the stomach, or pylorus. The muscle wall of the end of the pylorus is thickened to form the **pyloric sphincter**.

receptor
1. Any sense organ, or the part(s) of a sense organ that collects information.
2. A *molecule* in a *cell*, or on a cell's outer *membrane*, that responds to an outside stimulus, such as a *hormone* molecule attaching to it.

rectum
The short final portion of the large intestine, connecting it to the anal canal.

rectus
In muscle names, a straight muscle.

reflex
An involuntary response in the nervous system to certain stimuli, for example the "knee-jerk" response. Some reflexes, called **conditioned reflexes**, can be modified by learning.

renal
Relating to the kidneys.

respiration
1. Breathing.
2. Also called cellular respiration, the biochemical processes within cells that break down fuel *molecules* to provide energy, usually in the presence of oxygen.

retina
The light-sensitive layer that lines the inside of the eye. Light falling onto *cells* in the retina stimulates the production of electrical signals, which are transmitted to the brain via the *optic nerve*.

ribosomes
Particles within *cells* involved in *protein* synthesis.

RNA
Short for **ribonucleic acid**, a long *molecule* similar to *DNA*, but usually single- rather than double-stranded. RNA has many important roles including making copies of the DNA code for *protein* synthesis.

sacral
Relating to or in the region of the **sacrum**, the bony structure made up of fused *vertebrae* at the base of the spine that forms part of the *pelvic girdle*.

sagittal section
A real or imagined section down the body, or part of the body, that divides it into right and left sides.

sarcoplasm
The **cytoplasm** of muscle cell; contains many nuclei.

scrotum
The loose pouch of skin holding the *testes* in males.

sebum
An oily, lubricating substance secreted by sebaceous *glands* in the skin.

semen
The fluid released through the penis when the male ejaculates; it contains *sperm* and a mixture of nutrients and salts. Also called **seminal fluid**.

sensory
Concerned with transmitting information coming from the sense organs of the body.

serous membrane
A type of body *membrane* that secretes lubricating fluid and envelops various internal organs and body cavities. The pericardium, *pleura*, and *peritoneum* are all serous membranes.

sinus
A cavity; especially: 1. One of the air-filled cavities in the bones of the face that connect to the nasal cavity.
2. An expanded portion of a blood vessel, for example the carotid sinus and coronary sinus.

skeletal muscle
A type of muscle also known as *voluntary* or *striated muscle*, usually under voluntary control. Appears striped under the microscope. Many – but not all – skeletal muscles attach to the skeleton, and are important in movement of the body. See also *smooth muscle*.

smooth muscle
Muscle *tissue* that lacks stripes when viewed under a microscope, in contrast to *striated muscle*. Smooth muscle is found in the walls of internal organs and structures,

including blood vessels, the intestines, and the bladder. It is not under conscious control, but controlled by the *autonomic nervous system*.

somatic
1. Of or relating to the body, e.g. somatic cells.
2. Relating to the body wall.
3. Relating to the part of the nervous system involved in voluntary movement and sensing the outside world.

somatosensory
Related to sensations received from the skin and internal organs, including senses such as touch, temperature, pain, and awareness of *joint* position, or proprioception.

sperm
A male sex *cell* (*gamete*), equipped with a long moving "tail" (flagellum) to allow it to swim towards and fertilize an egg in the body of the female. Colloquially the word is also used to mean *semen*.

sphincter
A ring of muscle that allows a hollow or tubular structure in the body to be drawn closed (e.g. the *pyloric* sphincter and anal sphincter).

spinal cord
The part of the *central nervous system* that extends down from the bottom of the brain through the vertebral column, which protects it. Most *nerves* that supply the body originate in the spinal cord.

spinal nerve
A nerve in the *central nervous system* that is formed by the merging of the sensory and motor nerve rootlets.

spleen
A structure in the abdomen composed of *lymphoid tissue*. It has various roles, including blood storage.

starch
A plant *carbohydrate* made up of long, branched chains of *glucose molecules* linked together.

stem cell
A *cell* in the body that can divide to give rise to more cells. This could be

either more stem cells, or a range of more specialized types of cell. Stem cells contrast with highly specialized cells, which play specific roles in the body, and which may have lost the ability to divide completely – such as *nerve* cells.

steroids
Substances that share a basic molecular sturcture, consisting of four rings of carbon atoms fused together. Steroids, which may be naturally occurring or synthetic, are classified as *lipids*. Many of the body's *hormones* are steroids, including *oestrogen*, *progesterone*, *testosterone*, and cortisol.

striated muscle
A muscle with *tissue* that presents a striped appearance under a microscope. Striated muscle includes *skeletal muscles* and *cardiac* (heart) muscle. See also *smooth muscle*.

sucrose
See *sugar*.

sugar
1. Commonly used foodstuff, also called *sucrose*.
2. Any of a number of naturally occurring substances that are similar to sucrose. They are all *carbohydrates* with relatively small *molecules*, in contrast to other carbohydrates that are *macromolecules*, such as *starch*.

sulcus (pl. sulci)
One of the grooves on the folded outer surface of the brain. See also *gyrus*.

superficial
Near the surface; **superficial to**, nearer the surface than. (Opposite term **deep**.)

superior
Higher up the body, when considered in a standing position. See also *inferior*.

supination
The rotation of the radius around the ulna in the forearm, turning the palms of the hand to face upwards or forwards. The opposite to *pronation*. In muscle names, **supinator** indicates a muscle having this action, e.g. the supinator of the forearm.

suprarenal gland
Also called adrenal glands, a pair of glands found one on top of each kidney.Each gland consists of an outer **suprarenal cortex**, which secretes *corticosteroid* hormones, and an inner **suprarenal medulla**, which secretes *epinephrine*. See also *corticosteroid*.

suture
1. A stitched repair to a wound.
2. A rigid joint between two bones, as between the bones of the skull.

sympathetic nervous system
See *autonomic nervous system*.

symphysis
A cartilaginous *joint* between two bones, containing fibrocartilage.

synapse
A close contact between two nerve cells (neurons) allowing signals to be passed from the end of the first neuron on to the next. Synapses can either be electrical (where the information is transmitted electrically) or chemical (where neurotransmitters are released from one neuron to stimulate the next one). Synapses also exist between nerves and muscles.

synovial cavity
Cavity in a joint that is filled with a thin film of lubricating **synovial fluid**.

synovial joint
A lubricated, movable *joint*, such as the knee, elbow, or shoulder. In synovial joints the ends of the bones are covered with smooth *cartilage* and lubricated by a slippery liquid known as **synovial fluid**.

systemic
Relating to or affecting the body as a whole, not just one part of it. The **systemic circulation** is the blood circulation supplying all of the body apart from the lungs.

systole
The part of the heartbeat where the *ventricles* contract to pump blood.

tarsal
1. Relating to the ankle.
2. One of the bones of the tarsus, the

part of the foot between the tibia and fibula, and the metatarsals.

temporal
Relating to the temple – the area on either side of the head. The **temporal bones** are two bones, one on each side of the head, that form part of the *cranium*. The **temporal lobes** of the brain are located roughly below the temporal bones.

tendon
A tough fibrous cord that attaches one end of a muscle to a bone or other structure. See also aponeurosis.

testis (pl. testes)
Either of the pair of organs in men that produce male sex cells (sperm). They also secrete the sex hormone testosterone.

testosterone
A *steroid hormone* produced mainly in the testes, which promotes the development of and maintains male bodily and behavioural characteristics.

thalamus
Paired structures deep within the brain, forming a relay station for *sensory* and *motor* signals.

thick filament
Structure in the centre of the anisotropic or **A band** in skeletal muscle that is composed of the protein myosin. See also *M line*.

thin filament
Structure in the centre of the **A band** in skeletal muscle, which is mainly composed of protein actin. See also *tropomyosin*.

thorax
The chest region, which includes the ribs, lungs, heart, etc.

thymus
A *gland* in the chest composed of *lymphoid tissue*. Largest and most active in childhood, its roles include the maturation of T-lymphocytes.

thyroid gland
An endocrine *gland* located at the front of the throat, close to the

larynx (voicebox). Thyroid *hormones* such as **thyroxin** are involved in controlling *metabolism*, including regulating overall metabolic rate. The hormone *calcitonin*, which helps regulate the body's calcium, is also secreted by the thyroid.

tissue
Any type of living material in the body that contains distinctive types of cells, usually together with *extracellular* material, performing a specific function. Examples of tissues include bone, muscle, *nerve*, and *connective tissue*.

trachea
The windpipe: the tube leading between the *larynx* and the *bronchi*. It is reinforced by rings of *cartilage* to keep it from collapsing.

tract
An elongated structure or connection that runs through a certain part of the body. In the *central nervous system*, the term is used instead of *nerve* for bundles of nerve fibres that connect different body regions.

transmitter
See *neurotransmitter*.

tropomyosin
Actin-bonding protein that is present in the *thin filament* of skeletal muscle.

tunica adventitia
The outermost coat of a blood vessel, which is composed of connective tissue and elastic fibres. See also *tunica intima* and *tunica media*.

tunica intima
The innermost lining of an artery; made up of a single layer of flattened cells, also known as the *endothelium*. Also present in veins. See also *tunica media* and *tunica adventitia*.

tunica media
Middle layer of muscle cells that is thinner in veins than in arteries. See also *tunica intima* and *tunica adventitia*.

umbilical cord
The cord that attaches the developing *fetus* to the *placenta* of the mother, within the *uterus*. Blood from the fetus passes through blood vessels inside the cord, transporting nutrients, dissolved gases, and waste products between the placenta and the fetus.

urea
A small nitrogen-containing *molecule* formed in the body as a convenient way of getting rid of other nitrogen-containing waste products. It is excreted in the urine.

ureter
Either of two tubes that convey urine from the kidneys to the bladder.

urethra
The tube that conveys urine from the bladder to the outside of the body; in men it also conveys *semen* during ejaculation.

uterus
The womb, in which the *fetus* develops during pregnancy.

valve
In a *vein*, a pocket-like structure that allows deoxygenated blood to flow only towards the heart and prevents its backflow into cells. In the heart, it is present in each of the two *atria* and helps to direct the flow of blood in the chambers.

vascular system
The network of *arteries*, *veins*, and *capillaries* that conveys blood around the body.

vaso-
Prefix relating to blood vessels.

vein
A vessel carrying blood from the *tissues* and organs of the body back to the heart.

ventral
Relating to the front of the body, or the bottom of the brain.

ventricle
1. Either of the two larger muscular chambers of the heart. The right ventricle pumps blood to the lungs to be oxygenated, while the stronger-muscled left ventricle pumps oxygenated blood to the rest of the body. See also *atrium*.
2. One of the four cavities in the brain that contain *cerebrospinal fluid*.

venule
A very small vein, carrying blood away from *capillaries*.

vertebra (pl. vertebrae)
Any of the individual bones forming the **vertebral column** or spine.

villi (sing. villus)
Small, closely packed, fingerlike protrusions on the lining of the small intestine, giving the surface a velvety appearance and providing a large surface area, which is essential for the absorption of nutrients.

virus
A tiny parasite that lives inside *cells*, often consisting of only a length of *DNA* or *RNA* surrounded by *protein*. Viruses are much smaller than cells, and operate by "hijacking" cells to make copies of themselves. They are unable to replicate by themselves. Many viruses are dangerous *pathogens*.

viscera
Another term for organs. The adjective **visceral** applies to *nerves* or blood vessels, for example, that supply these organs.

vitamin
Any of a variety of naturally occurring substances that are essential to the body in small amounts, but which the body cannot make itself and so must obtain from the diet.

voluntary muscle
See *skeletal muscle*.

vulva
The outer genitalia of females, comprising the entrance to the vagina and surrounding structures.

white matter
Present in the brain and *spinal cord*, and made up of the axons of neurons. See also *grey matter*.

whole muscle
Part of the *skeletal muscle* that is made up of **fasciculi** and covered in a layer of *fascia* (fibrous tissue) called **epimysium**.

Z disc
Present in the centre of the isotropic or **I band** in skeletal muscle; it anchors the *thin filaments*. See also *M line*.

zygote
A cell formed by the union of two *gametes* at *fertilization*.

Index

Page numbers in **bold** indicate main treatments of a topic.

A

abdomen and pelvis
  arteries 155
  cardiovascular system
    166–7
  digestive system
    198–203
  immune and lymphatic
    system **186–7**
  muscles **82–5**
  nervous system **132–3**
  reproductive system
    212–5
  skeletal system **42–7**
  terminology 18
abducent nerve 122, 124
abduction 19
abductor digiti minimi
  muscle 94, 95, 109
abductor pollicis brevis
  muscle 95
abductor pollicis longus
  muscle 97
accessory nerve 122, 123,
  128, 129
acetabulum 58, 60
Achilles tendinitis 69
Achilles tendon 66,
  67, 107
acoustic meatus
  external 126, 127
  internal 30, 33
acromioclavicular
  ligament 52
acromion of scapula
  25, 48, 50, 52, 53,
  70, 86, 92
actin 68, 201
active diffusion, cell
  transport 13
"Adam's apple" (thyroid
  prominence) 149
adduction 19
adductor brevis muscle
  102
  MRI 233
adductor compartment of
  thigh 68, 69
adductor longus muscle 98,
  102
  MRI 233
adductor magnus muscle
  100, 102, 104
  MRI 233

adductor muscles 61, 232
adductor pollicis muscle
  97
adductor tendons 103
adductor tubercle 59,
  61
adenine, DNA 10–11
adenoids 183
adenosine diphosphate
  (ADP) 12
adenosine triphosphate
  (ATP) 12
adipocytes 15
adipose cells 14
adipose tissue 15
adrenal glands
  see suprarenal glands
ala of sacrum 43
alimentary canal
  see digestive system
alveolar bone 34
alveolar nerve, inferior
  123
alveolar process of mandible
  29
alveolar sac 151
alveoli 151
  mammary glands 208
amino acids
  cell transport 13
  protein synthesis 10
ampulla of oviduct 215
anal canal 193, 198,
  212, 213
anal sphincters 212, 213
anastomoses 170
anatomy
  abdomen and pelvis
    cardiovascular system
      166–7
    digestive system
      198–203
    immune and lymphatic
      system **186–7**
    muscular system **82–5**
    nervous system **132–3**
    reproductive system
      212–5
    skeleton **42–7**
  brain **112–23**
  cardiovascular system
    154–5
  cells **12–3**
  digestive system **192–3**
  eyes **124–5**
  head and neck
    cardiovascular system
      156–9
    digestive system
      194–5

endocrine system **218–9**
  lymphatic and immune
    system **182–3**
  muscular system **70–5**
  nervous system
    112–29
  respiratory system
    148–9
  skeletal system **26–33**
hip and thigh
  cardiovascular system
    174–7
  lymphatic and immune
    system **190–1**
  muscular system
    98–105
  nervous system **140–3**
  skeletal system **58–63**
lower arm and hand
  cardiovascular system
    172–3
  muscular system **94–7**
  nervous system **138–9**
  skeletal system **54–7**
lower leg and foot
  cardiovascular system
    178–9
  muscular system **106–9**
  nervous system **144–5**
  skeletal system **64–7**
lymphatic and immune
  system **180–1**
muscular system **68–109**
nervous system **110–1**
reproductive system
    208–9
shoulder and upper arm
  cardiovascular system
    168–71
  immune system **188–9**
  muscular system **86–93**
  nervous system **134–7**
  skeletal system **48–53**
skeletal system **24–5**
skin, hair and nail **22–3**
terminology **18–19**
thorax
  cardiovascular system
    160–3
  digestive system **196–7**
  immune and lymphatic
    systems **184–5**
  muscular system **76–81**
  nervous system **130–1**
  reproductive system
    208–11
  respiratory system
    152–3
  skeletal system **36–41**
  urinary system **204–5**

anconeus muscle 89, 93,
  95, 97
angiograms 222
  arteries of abdomen and
    legs 155
angular artery 156
angular gyrus 114
angular vein 157
anisotropic fibres, muscle 68
ankle
  bones 65
  joint 24, 66
annular ligament of the
  radius 53
annular tendon, common
  124
annulus fibrosus,
  intervertebral discs 43
anterior arch, vertebrae 40
anterior fissure, spinal cord
  133
antibodies 181
antihelix, ear 127
antitragus, ear 127
anus 193, 215
  sphincters 212, 213
aorta 155, 160, 166,
  196, 197
  abdominal 167, 204,
    167
  arch of the 154, 155, 161,
    162, 163
  ascending 161, 162, 164
  bifurcation of the 167
  descending 154, 155,
    161
  MRI 226, 227, 228
aortic nodes 180, 186, 191
aortic valve, heart 165
apex
  heart 160, 162, 163
  lung 150, 151, 152, 153
aponeuroses 68
appendix 193, 198, 201
aqueous humour 125
arachnoid granulation
  121
arachnoid mater
  brain 120, 121
  spinal cord 133
arcuate ligaments
  lateral 196
  medial 196
  median 196
arcuate line 83
areola 210
arm
  anterior surface of 18
  cardiovascular system
    154, **168–71**, **172–3**

extensor compartment
  69
flexor compartment
  68, 69
lymphatic and immune
  system **188–9**
muscles **86–93**, **94–7**
nervous system 111,
  134–7, **138–9**
posterior surface of 18
skeletal system 25,
  48–53, **54–5**
armpit 18
arrector pili muscle 23
arteries **155**
  around brain 158
  external arteries of head
    156
  structure 155
  see also artery and specific
    arteries
arterioles 23, 181
artery,
  angular 156
  anterior cerebral 158
  anterior circumflex
    humeral 168, 170
  anterior interventricular
    162, 163, 165
  anterior tibial 154, 155,
    178, 179
  anterior tibial recurrent
    178
  auricular, posterior 156
  axillary 154, 155, 168,
    170
  basilar 158
  brachial 154, 155, 169,
    170, 172, 173
    deep 169, 171
  buccal 156
  carotid 155
    common 129, 154, 156,
      158, 160, 161, 163, 226
    external 154, 155, 156,
      158
    internal 154, 155, 156,
      158
  cerebellar
    posterior inferior 158
    superior 158
  cerebral
    anterior 158
    middle 158
    posterior 158
  circumflex 163
  common carotid 129, 154,
    156, 158, 160, 161, 163
  common iliac 154, 155,
    166, 167, 205, 207

communicating
  anterior 158
  posterior 158
coronary
  CT scan 223
  left 165
  right 162, 163, 165
deep brachial 169, 171
dorsal 214
dorsalis pedis 179
digital 173
external carotid 154, 155,
  156, 158
external iliac 154, 155,
  166, 167, 174, 206
facial 156
femoral 140, 154
genicular
  lateral inferior 175, 177,
  178
  lateral superior 175, 177
  medial inferior 175,
  178
  medial superior 175,
  177
gluteal
  right superior 166
gonadal 155, 166, 167
hepatic 166
humeral
  anterior circumflex
  168, 170
  posterior circumflex
  168, 170
ileocolic 166
iliac
  common 154, 155, 166,
  167, 205
  external 154, 155, 166,
  167, 174, 206
  internal 154, 155, 166,
  167, 176, 177, 206
inferior labial 156
inferior mesenteric 154,
  155, 167
inferior ulnar collateral
  169, 171
infraorbital 156
internal carotid 154, 155,
  156, 158
internal iliac 154, 155,
  166, 167, 176, 177, 206
internal thoracic 160
interosseous 172
intercostal 160, 161
interventricular
  anterior 162, 163, 165
  posterior 163, 165
labial
  inferior 156
  superior 156

lateral inferior genicular
  175, 177, 178
lateral plantar 179
lateral superior genicular
  175, 177
left coronary 165
marginal 162
maxillary 156
medial inferior genicular
  175, 178
medial plantar 179
medial superior genicular
  175, 177
mental 156
mesenteric 166
  inferior 154, 155, 167
  superior 154, 155,
  167
middle cerebral 158
occipital 156
ophthalmic 158
palmar digital 173
perforating 176, 177
peroneal (fibular) 154,
  178
plantar
  lateral 179
  medial 179
pontine 158
popliteal 154, 155, 175,
  177, 178, 179
posterior auricular 156
posterior cerebral 158
posterior circumflex
  humeral 168, 170
posterior inferior
  cerebellar 158
posterior interventricular
  163, 165
posterior tibial 154, 178,
  179
pulmonary 160, 161
  left 152, 161, 163, 227
  left inferior 227
  right 153, 160, 162, 163,
  164
  right inferior 226
radial 154, 155, 169, 171,
  172, 173
radial collateral 169, 171
radial recurrent 169,
  171
rectal, superior 167
renal 154, 166, 167, 204,
  206, 207
right coronary 162, 163,
  165
right superior gluteal
  166
spinal, anterior 158
splenic 167

subclavian 154, 155, 160,
  161, 163, 168
submental 156
subscapular 168, 170
superior cerebellar 158
superior labial 156
superior mesenteric 154,
  155, 167
superior ulnar collateral
  169, 171
sural 175, 177
temporal, superficial 156
thoracic, internal 160
thoracoacromial 168
thyroid, superior 156
tibial
  anterior 154, 155, 178,
  179
  posterior 154, 178, 179
tibial recurrent
  anterior 178
ulnar 154, 155, 169, 171,
  172, 173
ulnar collateral
  inferior 169, 171
  superior 169, 171
ulnar recurrent 169, 171
vertebral 156, 158
articular cartilage, knee
  joint 63
articular eminence 31
articular processes,
  vertebrae 41
arytenoid cartilage 149
association areas, brain
  114
asterion 28
atlas 25, 40
  intervertebral discs 43
atria, heart 160, 163, 164
  MRI 226, 227
atrium, nose 148
auricle, ear 126, 127
auricles, heart 160, 161,
  162, 163
auricular artery, posterior
  156
auricular nerve
  greater 123
  posterior 123
auricular vein, posterior
  157
auriculotemporal
  nerve 123
autonomic nervous system
  (ANS) 110, 216
  and smooth muscle 201
axilla 18
  see also armpit
axillary artery 154, 155,
  168, 170

axillary nerve 110, 111, 134,
  136
axillary nodes 180, 181,
  184, 185
  anterior 189
  apical 188
  central 188
  lateral 188
axillary tail 210
axillary vein 168, 170
axis 25, 40
  intervertebral discs 43
axons 14
  length 111
  spinal cord 133
azygos vein 155, 160,
  161, 197

B

babies, skull of 34
backbone see spine
bacteria
  in stomach 201
basal cells, epidermis 23
basal ganglia 120
bases, DNA 10, 11
basilar artery 158
basilic vein 154, 168, 169,
  171, 172, 173
basiocciput 30
biceps aponeurosis 87, 94
biceps brachii muscle 68,
  87
  long head of 86
  short head of 86
biceps femoris muscle 101,
  106
  long head of 100
  MRI 232, 233
  short head of 105
biceps tendon 53, 87, 94
bicuspid valves, heart 164,
  165
bile 12, 193, 201, 202
bile duct 202, 203
bladder 204, 205, 206,
  207, 214,
  female anatomy 213
  male anatomy 212
  MRI 229
blind spot, retina 125
blood
  cardiovascular system 17,
  164-5
  kidneys 206
  respiratory system 146
blood cells see red blood
  cells

blood pressure
  pulmonary circulation
  155
  systemic circulation
  155
blood vessels
  in bone 24
  hair 22
  lymph nodes 181
  structure **155**
  see also cardiovascular
  system
body composition
  14-5
body systems **16-7**
bone marrow
  blood cell production
  181
  stem cells 14
bones
  blood vessels 24
  composition 24
  radionuclide scan 223
  spongy bone 15
  structure **24**
  see also skeletal system
  and specific bones
bowels see colon; large
  intestine
brachial artery 154, 155,
  169, 170, 172, 173
  deep 169, 171
brachial plexus 110, 111,
  129, 134, 136
brachial veins 154, 155,
  169, 171
brachialis muscle 69, 87,
  90, 91, 93, 94, 96
brachiocephalic trunk 154,
  155, 160, 163
brachiocephalic veins 154,
  155, 160, 161
  MRI 227
brachioradialis muscle 68,
  87, 89, 91, 92,
  93, 94, 95, 96
brain
  anatomy 112-23
  appearance 114-5
  basal ganglia 120
  blood vessels 154
  cardiovascular system
  158-9
  caudate nucleus 120
  fornix 120
  hippocampus 120
  internal capsule 120
  lentiform nucleus
  120
  lobes 112
  motor cortex 120

nerve tissue 15
optic radiation 120
PET scan 223
size of 112
brainstem
anatomy 111, 112
medulla oblongata 113
breastbone see sternum
breasts 208, **210–1**
lymphatic system 184, 185
breathing
control of 81
mechanics of 16
muscles 16
see also respiratory system
bregma 26
bronchi 147, 150, 151
bronchioles 150, 151
bronchus
left main 151, 152
right main 150, 153
superior lobar 153
brow ridge 26
buccal artery 156
buccal node 182
bulb, hair 22
bulbospongiosus 215
bursas, knee joint 63, 105, 108
buttocks
muscles 85
terminology 18

C

caecum 193, 198, 201
calcaneal tendon 66, 67
MRI 232, 233
calcaneal tuberosity 67
calcaneocuboid ligament 66
calcaneofibular ligament 66, 67
calcaneonavicular
ligaments 66
plantar 67
calcaneus bone 24, 25, 65, 66, 67, 107, 109
MRI 232, 233
calcium, cell storage 13
calf 18
calf muscle 68
calyx, kidney 206, 207
cambial layer, sutures 34
cancellous bone 24
canine teeth 195

capillaries
alveolar sac (lung) 150
network 151
structure **155**
capitate bone 54, 55, 56, 57
MRI 230
capsular layer, sutures 34
carbon dioxide
cell metabolism 12
respiratory system 16, 147
cardiac impression, lung 152
cardiac muscle 15, 165
cardiac notch
lung 151, 152
stomach 200
cardiac veins
great 162
middle 163
small 162, 163
cardiovascular system 15, 17
abdomen and pelvis **166–7**
anatomy **154–5**
blood vessels **155**
head and neck **156–9**
hip and thigh **174–7**
lower arm and hand **172–3**
lower leg and foot **178–9**
muscles and 17
shoulder and upper arm **168–71**
thorax **160–3**
tissues 15
see also blood vessels; heart
carotid arteries 155, 226
common 154, 129, 156, 158, 160, 161, 163, 226
external 154, 155, 156, 158
internal 154, 155, 156, 158
carotid canal 31
carpal ligaments
palmar 57
radiate 57
carpal tunnel syndrome 138
carpals 25
MRI 230
carpometacarpal joint 56, 57
carpometacarpal ligaments 56
cartilage 15, 17, 24
ear 126
growth plate 24

intervertebral discs 40
joints 43
nose 70
photomicrograph 222
structure **24**
cauda equina 110, 111
caudate lobe of liver 202
caudate nucleus 120, 121
cavernous sinus 159
cells **12–3**
anatomy **12–3**
body composition **14**
capillary walls 155
cardiac muscle 165
cell division 13
cell transport 13
DNA 10–1
lymphocytes 180, 181
metabolism 12
muscle **68**
neurons 113
retina 14
tissues **15**
types of **14**
see also red blood cells
cellular respiration 12
cement, tooth 34, 195
central sulcus 112, 113, 114
centrioles, cells 12
centromeres 13
cephalic vein 154, 155, 168, 170, 172, 173, 188
accessory 172, 173
cerebellar arteries
posterior inferior 158
superior 158
cerebellar hemisphere 116, 117
cerebellar vermis 117
cerebellum 111, 112, 113, 115, 119
fissures 116, 117
MRI 225
cerebral aqueduct 119, 121
cerebral arteries
anterior 158
middle 158
posterior 158
cerebral cortex, MRI 224, 225
cerebral fossa, lateral 115
cerebral hemispheres 112, 114
cerebral peduncle 115
cerebral veins, great 159
cerebrospinal fluid (CSF)
brain 119, 120, 121
spinal cord 133
cerebrum 111, 116, 119
longitudinal fissure 116, 117

cervical canal 215
cervical nerves 110, 111, 128, 129
cervical nodes 181
deep 180
superficial 180, 182
cervical spine 28, 40
curvature 40
cervical vertebrae 25, 27, 40, 75, 129
cervix 209, 213, 215
cheek bone see zygomatic bone
chemicals
body composition 15
chest see thorax
chewing, muscles 72
child, thymus gland 185
chin 27
choana 31
cholesterol 12
chondrocytes 15, 25
chordae tendineae 164
choroid layer of eye 125
choroid plexus 119, 120
chromatin 10
chromosomes
cell division 13
DNA 10
egg 14
human genome **11**
karyotype 11
cilia, bronchi 151
ciliary body 125
ciliary ganglion 124
cingulate gyrus 118
MRI 225
cingulate sulcus 114
circadian rhythms 218
circle of Willis 158
circular folds, small intestine 201
circumflex artery 163
cisterna chyli 180, 186, 187
clavicle 24, 25, 27, 36, 48, 50, 52, 76, 86, 134, 136, 150, 151
MRI 226
clitoris 213, 215
coccygeal cornua 41
coccyx 25, 40, 41, 42, 44, 212, 213
facet joint 41
MRI 229
cochlea 126, 127
cochlear nerve 127
coeliac nodes 187
coeliac trunk 155, 166, 167
collagen 15

collar bone see clavicle
collateral ligaments
elbow 53
fibular 108
hand 56
lateral 108
medial 108
colliculus
inferior 119
superior 119
colon 193
ascending 198, 201, 228
descending 199, 228
hepatic flexure of 198
MRI 228
sigmoid 199
splenic flexure of 199
transverse 198
commissure, anterior 118
communicating arteries
anterior 158
posterior 158
communication, facial muscles 70
compact bone 24
compound joints 63
concha
ear 127
nasal cavity 149
inferior 148
middle 148
superior 148
condyles 28
cone cells, retina 14
confluence of sinuses, brain 159
conjunctiva 124, 125
connective tissue 15
fascia 68
constrictor muscle, inferior 75
coracoacromial arch 53
coracoacromial ligament 52, 53
coracoclavicular ligament 52
coracoid process 48, 52, 53
cornea 125
coronal plane 19
coronal suture 26, 28
coronary arteries
right 162, 163
coronary ligament 202
coronary sinus 162, 163, 164
coronoid fossa 49
coronoid process 49, 54
corpus albicans 215

corpus callosum 116, 117, 120, 121
  body of 118
  genu of 118, 120
  MRI 225
  splenium of 119
corpus cavernosum 212, 214
corpus luteum 215
corpus spongiosum 212, 214
cortex
  brain 114
  hair 22
costal cartilage 24, 36, 77
costal surface, lung 152, 153
costodiaphragmatic recess 150
Cowper's gland 214
cranial bones 34–5
cranial nerves 110, 111, 114, 122, 123, 129
cranium 24, 25
  sutures 26, 28
  see also skull
cribriform plate of ethmoid bone 30, 148
cricoid cartilage 74, 148, 149
cricopharyngeus muscle 75
cricothyroid membrane 149
crista galli 30
crown, tooth 195
cruciate ligaments
  anterior 63
  posterior 63
cruciform bone 232
crus of clitoris 215
crus of diaphragm 80, 81, 196
crus of penis 214
cubital fossa 18
cubital vein, median 169, 171, 172
cuboid bone 65, 66
cuneiform bone
  intermediate 65
  lateral 65
  medial 65, 107
cutaneous nerves
  arm 134, 136
  forearm 134, 136
  lateral sural 144
  thigh 132, 140
cuticle
  hair 22
  nails 22

cystic duct, gallbladder 203
cytoplasm 12, 13
cytosine, DNA 10, 11
cytoskeleton, cells 12, 13

D

daughter cells, cell division 13
deep palmar venous arch 173
defecation 83
deltoid ligament 67
deltoid muscle 68, 69, 86, 88
  anterior fibres 90
  middle fibres 90
  posterior fibres 92
dendrites, neurons 111
dens (odontoid peg), vertebrae 40
dense connective tissue 15
dentine, tooth 195
deoxyribose 10
depressor anguli oris muscle 71, 73
depressor labii inferioris muscle 71, 73
dermal root sheath, hair 22
dermis 23
detrusor muscles 207
diaphragm 81, 147, 150, 162, 197
  central tendon of 80, 196
  left crus of 81, 196
  MRI 229
  muscular part of 80, 196
  right crus of 80, 196
  sternal part 196
diffusion
  cell transport 13
digastric muscle
  anterior belly of 73
  posterior belly of 73
digastric notch 31
digestive system 17
  abdomen and pelvis **198–203**
  anatomy **192–3**
  head and neck **194–5**
  thorax **196–7**
digital arteries, palmar 173
digital nerve, dorsal 145

digital vein
  dorsal 172
  palmar 173
discs see intervertebral discs
diseases and disorders
  gene therapy 11
  stem cells 14
dislocation, shoulder 53
DNA **10–1**, 12, 15
  cell division 13
  human genome 11
  "junk DNA" 10
dorsalis pedis artery 179
dorsum of foot 18
  artery of 154, 155
duodenum 200, 202, 203
  functions 193, 201
  MRI 228
dura mater
  brain 121
  spinal cord 133

E

ear **126–7**
  bones 34, 35
eardrum (tympanic membrane) 126, 127
ectoderm 14
efferent ductules 214
egg (ovum) 14, 208, 209, 215
elbow **53**
  bones 25
  joint 230, 231
electrical activity
  nervous system 14
embryo 14
embryonic stem cells (ESCs) 14
enamel, tooth 195
endocrine system 16, **216–7**
  head and neck **218–9**
endoderm 14
endometrium 213
endoplasmic reticulum 12, 13
endothelium, blood vessels 155
energy
  adipose tissue 15
  cell metabolism 12
  digestive system 17
enzymes
  cell metabolism 12, 13
  pancreas 202
  saliva 195
  stomach 201

epicondyles 25, 139
epicranial aponeurosis 70
epidermis 23
epididymis 209, 212, 214
epidural space 75
epigastric region 18
epiglottis 74, 146, 148, 149, 183, 192, 194
  MRI 224
epimysium 68
epithelial cells 14
epithelial root sheath, hair 22
epithelial tissue 15
erector spinae muscles 69, 78, 84
  MRI 228, 229
erythrocytes see red blood cells
ethmoid bone
  cribriform plate of 30, 148
  crista galli 30
  orbital plate of 35
ethmoid sinus 148, 149
Eustachian (pharyngotympanic) tubes 74, 126, 127
  opening of 183
exons 11
expressions, facial 70
extension 19
extensor carpi radialis brevis muscle 95, 97
extensor carpi radialis longus muscle 92, 93, 95, 97
extensor carpi ulnaris muscle 95, 97
extensor compartment
  of arm 69, 231
  of leg 68, 69
  of thigh 68
extensor digiti minimi muscle 95
extensor digitorum brevis muscle 109
extensor digitorum longus muscle 107, 108, 109
extensor digitorum longus tendon 107, 109
extensor digitorum muscle 95
extensor digitorum tendons 94
extensor hallucis brevis muscle 107
extensor hallucis longus muscle 108
extensor hallucis longus tendon 107, 109

extensor indicis muscle 96
extensor muscles 230, 232
extensor pollicis brevis muscle 96
extensor pollicis longus muscle 97
extensor retinaculum muscle 94, 96
  inferior 107, 109
  superior 107, 109
extensor tendons, MRI 230
external urethral orifice 207, 212, 215
external urethral sphincter 212, 213
exytosis, cells 12
eyelashes 124
eyelids 124
eyes **124–5**
  eye-socket 26
  MRI 224, 225
  muscles 124
  nerves of the orbit 124
  photoreceptor cells 14

F

face
  blood vessels 154
  bones 26–9, 32
  expressions 70
  muscles 68, **70–5**
facet joints, vertebrae 44
facial artery 156
facial nerve 122, 123, 128
  buccal branch 123
  cervical branch 123
  marginal mandibular branch 123
  temporal branch 123
  zygomatic branch 123
facial vein 157
facilitated diffusion, cell transport 13
faeces
  defecation 83
  storage in rectum 193, 201
falciform ligament 202
Fallopian tubes (oviducts)
  anatomy 208, 209, 213, 215
false vocal cord 148, 149
falx cerebri 121
fascia, muscles 68
fascicles, muscles 68

fat cells 14
adipose tissue 15
subcutaneous fat 75
feedback loops 16
feet *see* foot
female reproductive
system **208–9**, 212, **213**
breasts 208, **210–1**
femoral artery 140, 154,
155, 174, 175, 176
deep 154, 155, 174, 176,
177
lateral circumflex 174,
176
left 167
medial circumflex 174,
176
right 166
femoral condyles 59, 61,
63, 232
femoral cutaneous nerves
intermediate 141, 142
lateral 141, 143
medial 141, 142
femoral nerves 110, 111,
132, 140, 142
femoral veins 154, 155,
175, 176
left 167
right 166
femur 24, 25, 44, 58, 174,
176
condyles 59, 61, 63
epicondyles 59, 61
greater trochanter of 62,
104, 140, 142
head 58, 60, 229
intertrochanteric line of
62
knee joint 63
lesser trochanter 62
linea aspera 105
MRI 232, 233
neck of 58, 60, 140, 142
patellar surface 59
popliteal surface of 105,
143
shaft of 59, 61, 140, 142
fetus
ultrasound scan 223
fibres, muscle 68
fibroblasts 15
fibrous capsule
elbow joint 53
knee joint 63
fibrous joints 34
fibula
anatomy 24, 25, 66, 67,
178
fibrous joints 34
head of 64, 144

interosseous border 64
knee joint 63
MRI 232
neck of 64
shaft 64, 65
fibular collateral ligament
63, 108
fibular muscles 69
fibular (peroneal) nerve
141
common 110, 111, 143,
144, 145
deep 110, 144, 145
superficial 110, 145
fibularis (peroneus) brevis
muscle 107
fibularis (peroneus) brevis
tendon 66
fibularis (peroneus) longus
muscle 106, 107, 108
fibularis (peroneus) longus
tendon 109
fibularis (peroneus) tertius
muscle 109
filaments, skeletal muscle
68
filiform papillae 195
fimbriae, Fallopian tubes
209, 215
fingers
bones 24, 25, 54, 55, 97
joints **56**
muscles 96
see also hand
flexion 19
flexor carpi radialis muscle
94
flexor carpi ulnaris muscle
68, 93, 96
flexor compartment
of arm 68, 69, 231
of leg 69
of thigh 69
flexor digiti minimi brevis
muscle 95
flexor digitorum
profundus tendon 95
flexor digitorum
superficialis muscle 94,
95
flexor digitorum
superficialis tendon
95
flexor hallucis longus
muscle 109
MRI 232
flexor muscles 230, 232
flexor pollicis brevis
muscle 95
flexor pollicis longus
muscle 96

flexor retinaculum muscle
94, 96
MRI 230
flexor tendons 230
floating ribs 36
folia 117
foliate papillae 195
follicles
ovarian 215
food **192–3**, 195
swallowing 195, 197
foot
blood vessels 154, 155
cardiovascular system
178–9
dorsum of 18
eversion 109
MRI 232
muscles **106–9**
nervous system **145**
skeletal system 24, 25,
64–7
tendons 107
toe nails 22
venous arch 179
X-ray 222
foramina, skull 122
foramen caecum 30,
195
foramen lacerum 30, 31
foramen magnum 30, 31
foramen ovale 30, 31
foramen rotundum 30
foramen spinosum 30, 31
forearm
anterior surface 18
blood vessels 154
bones 24, 25, **54–5**
cardiovascular system
172–3
extensor compartment
69
flexor compartment
68, 69
MRI 230, 231
nervous system **138–9**
posterior surface 18
see also arm
fornix 121
frontal bone 25, 26, 28,
32, 34
orbital part of 30
zygomatic process 26
frontal gyrus
inferior 112, 114
middle 112, 114
superior 112, 114, 118
frontal lobe 112, 116
frontal nerve 124
frontal pole 112, 114, 115,
116

frontal sinus 32, 148, 149
MRI 224
frontal sulcus
inferior 114
superior 114
fundus of uterus 209
fungiform papillae 195

G

gallbladder 193, 202, 203
fundus of 198, 203
gametes *see* egg; sperm
ganglion impar 110
gastric veins 203
gastrocnemius muscles
69
lateral head 101, 106
medial head 101, 106
MRI 232, 233
gemellus muscles
inferior 104
superior 104
genetics and genes **10–1**
cell types 14
DNA **10–1**
gene therapy 11
genetic engineering 11
human genome **11**
genicular arteries
lateral inferior 175, 177,
178
lateral superior 175,
177
medial inferior 175, 178
medial superior 175,
177
genioglossus muscle 74
geniohyoid muscle 74,
194
genitofemoral nerve 132
gingivae 195
glabella 26
gladiolus 24
glands **216–7**
see also specific glands
glans penis 209, 212, 214
glenohumeral ligaments
52
glenoid cavity (fossa)
48, 50
glossopharyngeal nerve
122, 123, 128, 129
glottis 75
glucose, cell metabolism
12
gluteal artery, right
superior 166
gluteal muscles 58

gluteal nerves 143
superior 133, 142
gluteal region 18
gluteal tuberosity 60
gluteus maximus muscle
69, 84, 85, 100, 143
MRI 229, 233
gluteus medius muscle
68, 84, 102, 104
MRI 229
Golgi complex 12, 13
gomphosis, joints 34
gonadal arteries 155, 166,
167
gonadal veins 155, 166,
167
gracilis muscle 98, 100,
102
MRI 233
gracilis tendon 63
grey matter
brain 114
spinal cord 133
groin 18
"groin pulls" 103
growth plate, bones 24
guanine, DNA 10–1
gums 195
gyrus
angular 114
cingulate 118
inferior frontal 112,
114
inferior temporal 113,
115
middle frontal 112, 114
middle temporal 113
occipitotemporal
fusiform 115
parahippocampal 115
postcentral 113, 114
precentral 113, 114
straight 115
superior frontal 112, 114,
118
superior temporal 113
supramarginal 114

H

haemoglobin
red blood cells 14
hair 23
follicles 23
structure **22**
hair cells, in ear 126
hamate bone 54, 55, 56, 57
hook of 57
MRI 230

hamstrings 69, 100, 101, 143
  injuries 69
  MRI 232, 233
hand
  bones 25, **54-5**
  cardiovascular system
    172-3
  joints **56-7**
  muscles **94-7**
  nails 22
  nervous system **138-9**
  palmar surface of 18
  radiographs 57
  tendons 94
hard palate 74, 148, 194, 195
haustra 201
Haversian canal 24
head and neck
  arteries 155
  brain **116**
  cardiovascular system
    154, **156-9**
  digestive system **194-5**
  endocrine system **218-9**
  lymphatic and immune
    system **182-3**
  muscular system **70-5**
  MRI 223, 224, 225
  nervous system 111,
    112-29
  respiratory system **148-9**
  skeletal system 24, 25,
    26-33
heart 15, 147, 154, 155, **162-5**
  angiogram 222
  cardiac muscle 165
  ECG 223
  MRI 226, 227
  nervous system 16
  position in chest 160
  tissues 15
heel-bone 24, 25, 65
helix
  DNA 10-1
  ear 127
hemispheres, cerebral
  112, 114
hepatic arteries 166
hepatic ducts 203
hepatic portal vein 203
hepatic veins 154, 155, 203
hilum
  lung 152, 153
  renal 206, 228
hinge joints 63
hip
  cardiovascular system
    174-7
  joint 59
  lymphatic and immune
    system **190-1**

muscles **98-105**
  nervous system **140-3**
  skeletal system 25, **58-62**
hippocampus 120
histone 10, 11
horizontal fissure, lung 150,
  153
hormones
  endocrine system 16,
    216-7
  pancreas 202, 203
  pituitary gland 218
  see also sex hormones
human genome **11**
humeral arteries
  anterior circumflex 168,
    170
  posterior circumflex
    168, 170
humerus 24, 25, 49, 52, 53
  capitulum 49, 54
  head 136
  lateral epicondyle 49, 53,
    54, 55, 93, 95, 135, 137
  medial epicondyle 49,
    53, 54, 55, 87, 89, 91,
    94, 96, 135, 137
  MRI 226, 230, 231
  neck 48, 134, 136
  olecranon fossa 55
  shaft 49, 51, 92, 135, 137
  spiral groove 50
  trochlea 49, 54
hyaline cartilage
  joints 63
  photomicrograph 222
  spine 43
hydrochloric acid 193, 201
hyoid bone 25, 29, 74,
  149, 194
hypochondrial region 18
hypodermis 23
hypoglossal canal 30, 31,
  33
hypoglossal nerve 122,
  123, 128, 129
hypothalamus 118, 121,
  216, 218
  endocrine system 216
  MRI 225
hypothenar muscles 230

I band, muscle 68
ileal vein 203
ileocolic artery 166
ileum 193, 198, 201
  MRI 229

iliac arteries
  common 154, 155, 166,
    167, 205, 207
  external 154, 166, 167,
    174, 206
  internal 154, 155, 166,
    167, 176, 177, 206
iliac crest 42, 44, 46, 47,
  82, 85, 133
  MRI 229
iliac fossa 42
iliac nodes
  common 181, 187
  external 180, 181, 186
  internal 180, 181, 187
iliac region 18
iliac spine
  anterior superior 43
  posterior superior 44
iliac veins
  common 154, 155, 166,
    167, 205, 206
  external 154, 166, 167,
    206
  internal 154, 155, 166,
    167, 203, 206
iliacus muscle 102
  MRI 229
iliocostalis muscle 84
iliofemoral ligament 62
iliohypogastric nerve 132
ilioinguinal nerve 132
iliopsoas muscle 68, 98
  MRI 229
iliotibial tract 63, 98, 100
ilium 25, 42, 62
  gluteal surface 44
  imaging 222-3
immune system
  abdomen and pelvis
    186-7
  anatomy **180-1**
  head and neck **182-3**
  hip and thigh **190-1**
  lymphatic system 16
  lymphocytes 181
  shoulder and upper arm
    188-9
  thorax **184-5**
impingement syndrome
  90
incisive fossa 31
incisors 194, 195
incus 35, 126
inferior, definition 18, 19
inferior margin of liver
  202
infraclavicular nodes 188
infraglottic cavity 227
infraorbital artery 156
infraorbital foramen 26

infraorbital nerve 123
infraorbital vein 157
infrapatellar bursas
  deep 63
infrapatellar fat pad, knee
  joint 63
infraspinatus muscle 78, 79,
  88, 92
infraspinous fossa 50
infundibulum of oviduct
  215
inguinal ligament 82, 98
inguinal nodes 180, 181
  deep 190
    distal superficial 186, 190
    proximal superficial 186
inguinal region 18
injuries
  "groin pulls" 103
interatrial septum 164
intercalated discs, cardiac
  muscle 165
intercarpal ligaments 56
intercondylar fossa 61
intercostal arteries 160,
  161
intercostal membrane,
  internal 80
intercostal muscles 68, 77,
  78, 93, 131, 146, 210
  external 77, 81
  internal 77, 81
intercostal nerves 110, 111,
  130, 131, 133, 160
intercostal nodes 181, 184
intercostal veins 160, 161
intermaxillary suture 31
internal capsule 120
internal urethral orifice
  207
interosseous artery 172
interosseous membrane
  55, 230, 231
interosseus muscles (hand)
  dorsal 94, 96
  palmar 97
interosseus muscles (foot)
  dorsal 107, 109
interosseous nerve,
  posterior 138, 139
interpalatine suture 31
interpeduncular fossa 115
interphalangeal joints
  56, 57
intertendinous connections
  94
interthalamic adhesion 119
intertragic notch, ear 127
intertrochanteric crest 60
intertrochanteric line 58, 59
intertubercular region 18

interventricular arteries
  anterior 162, 163
  posterior 163
interventricular foramen
  121
interventricular septum
  164, 165
interventricular veins 165
intervertebral discs 40, 43
  MRI 225, 229
intervertebral foramina 40,
  130
intestines
  blood vessels 154
intraparietal sulcus 114
introns 11
involuntary muscle see
  smooth muscle
iris 124, 125
ischial spine 44
  MRI 229
ischial tuberosity 42, 45,
  58, 62, 104, 142
ischiocavernosus 215
ischiopubic ramus 42, 45,
  46, 47, 58, 62
ischium 42, 62
  body of 42
isotropic band, muscle 68
isthmus of oviduct 215
isthmus of thyroid gland
  219

J

jaws
  biting and chewing food
    72
  see also mandible; maxilla
jejunal vein 203
jejunum 193, 199, 201
  MRI 228
joints
  cartilage 15, 43
  fibrous joints 34
  sutures 34
  synovial joints 63
  see also specific joints
jugular foramen 30, 31
jugular veins 180
  external 154, 155, 157
  internal 129, 154, 155,
    157, 158, 159, 160, 161,
    182
jugulo-omohyoid node
  182
jugulodigastric nodes 181,
  182
"junk DNA" 10

K

karyotype, chromosomes
11
kidneys 204, 206, 207
blood vessels 154, 166
functions 204
MRI 228
renal hilum 206, 228
knee
anterior surface of 18
blood vessels 154
bones **63**
joint 63, 232
knee jerk reflex 99
patella 24, 25
popliteal fossa 18, 232
kyphosis 40

L

labia
majora 215
minora 215
labial arteries
inferior 156
superior 156
labial veins
inferior 157
superior 157
lacrimal bone 28, 35
lacrimal caruncle 124
lacrimal gland 124
lacrimal nerve 124
lacrimal papilla 124
lactiferous ducts 208, 210,
211
lactiferous sinus 210
lambda 26
lambdoid suture 26, 28, 31,
33
lamina
vertebrae 40, 41
large intestine 193, 199, 201,
203
laryngopharynx 148, 183,
194
larynx 129, 146, 147, 183,
192, 194
anatomy 149
MRI 224
lateral, definition 18, 19
lateral cord 134, 136
lateral mass, vertebrae 40
lateral sulcus 112, 113,
116
latissimus dorsi muscle 69,
79, 84, 85, 88, 89, 90

leg
adductor compartment
of thigh 68
anterior surface 18
calf 18, 68
cardiovascular system
154, 155, **174–7**, **178–9**
extensor compartment
68, 69
flexor compartment 69
lymphatic system **190–1**
MRI 232–3
muscles **98–105**, **106–9**
nervous system 110, 111,
140–3, **144–5**
posterior surface 18
skeletal system 24, 25,
58–61, **64–7**
lens 125
lentiform nucleus 120, 121
leukocytes *see* white blood
cells
levator labii superioris
alaeque nasi muscle
70, 72
levator labii superioris
muscle 70, 73
levator scapulae muscle
70, 71, 73, 75
ligaments 17, 34
connective tissue 15
ligamentum teres 202
ligamentum venosum 202
light
photoreceptor cells 14
limbs *see* arm; leg
linea alba 82
linea aspera 60, 61, 105
linea semilunaris 82
lingual nerve 123
lingual node, proximal
superficial 190
lingual tonsil 183
lingula 151, 152
lips 194
liver 193, 196, 198, 199
anatomy **202–3**
blood supply 166
cells 12
functions 202
MRI 226, 227, 228
lobes
brain 112
lung 150, 151, 152, 153
lobules
ear 127
mammary glands 208
long bones 24
longissimus muscle 84
longitudinal (cerebral)
fissure 114, 115, 116, 117

longitudinal ligament,
anterior 80
longus colli muscle 75, 81
loose connective tissue
15
lordosis 40
lower pole of kidney 206
lumbar plexus 110, 111,
133, 140, 141
lumbar region 18
lumbar spinal nerves 110,
111, 132
lumbar spine 41
curvature 41
lumbar triangle 85
lumbar vertebrae 25, 41,
42, 45
MRI 228, 229
lumbosacral joint 45
lumbosacral trunk 133
lumbrical muscles 95
lunate bone 54, 55, 57
MRI 230
lungs 16
anatomy 146–7, **150–3**
CT scan 223
epithelial cells 14
MRI 226, 227
lunula, nail 22
lymph nodes 180, 181
blood vessels 181
lymphatic ducts 184, 185,
197
lymphatic system 16
abdomen and pelvis
186–7
anatomy **180–1**
head and neck **182–3**
hip and thigh **190–1**
shoulder and upper arm
188–9
thorax **184–5**
lymphatics 180, 181
lymphocytes 180, 181
lymphoid tissue 181
lysosomes 12

M

M line, muscle 68
magnetic resonance
imaging *see* MRI scans
male reproductive system
209
anatomy **212**
malleolus
lateral 65, 66, 67
medial 65, 67, 109,
145

malleus 35, 126
handle of 127
lateral process 127
mammary glands 208,
210–1
mammillary bodies 115,
118, 121
mandible 24, 27
alveolar process of 29, 35
angle of 29, 35
body of 29, 35
coronoid process 28
ramus 27, 29, 35
tooth sockets 34
mandibular fossa 31
mandibular node 182
manubriosternal joint 37
manubrium 24, 37
marginal artery 162
marginal sinus 159
marginal veins
lateral 179
medial 179
marrow *see* bone marrow
marrow cavity, bone 24
masseter muscle 71, 73
mast cell, TEM image 222
mastication, muscles 72
mastoid foramen 30
mastoid nodes 182
mastoid process 29, 31, 35
matrix
nail 22
maxilla 27, 29, 34, 35
alveolar process 27, 35
frontal process 26
orbital surface 35
tooth sockets 34
zygomatic process 26, 31
maxillary artery 156
maxillary sinus 149
MRI 224
maxillary vein 157
meatus, nasal cavity
inferior 148
middle 148
superior 148
medial, definition 19
medial cord 134, 136
median nerve 110, 111,
135, 137, 138, 139
digital branches 138, 139
median vein of the forearm
172
mediastinal nodes,
posterior 185
medulla
hair 22
medulla oblongata 113,
116, 117, 119
MRI 225

medullary cavity, bone 24
melanocytes 22
membranes
cells 12, 13
meninges
brain 121, 224, 225
spinal cord 133
meniscus 63
lateral 63
medial 63
mental artery 156
mental foramen 27, 29
mental nerve 123
mental protuberance 27
mental vein 157
mentalis muscle 71, 73
mesenteric arteries 167
inferior 154, 155, 167
superior 154, 155, 167
mesenteric nodes 187
mesenteric veins
inferior 167, 203
superior 154, 155, 166,
203
mesentery, attachment of
201
mesoappendix 201
mesoderm 14
messenger RNA (mRNA) 10
metabolism, cell 12
metacarpal bones 97
metacarpal ligaments, deep
transverse 57
metacarpals 24, 25, 54, 55,
56, 57
MRI 230
metacarpophalangeal
joints 56, 57, 95, 97
metaphysis, bones 24
metatarsal bones 24, 25,
65, 66, 67, 107, 109
MRI 232
metatarsal ligaments
deep transverse 66
dorsal 66
microfilaments, cells 13
microscopy 222–3
microtubules 12
microvilli 12
micturition (urination) 83
midbrain 119
tectum 119
tegmentum 119
midclavicular line 18
milk
mammary glands
208, 211
mitochondria 12, 13
in cardiac muscle 165
in muscle cells 201
mitosis 13

molars 195
molecules, body
    composition 15
motor cortex 120
motor nerves
    rootlets 133
mouth
    digestive system 192,
    194–5
    mastication 72
    tongue 195, 224
    see also teeth
MRI (magnetic resonance
    imaging) scans **220–33**
    abdomen and pelvis
    228–9
    body 6–7
    head and neck 224–5
    liver 226–7, 228
    lower limb 232–3
    stomach 228
    thorax 226–7
    upper limb 230–1
mucosa
    intestines 201
    nasal cavity 149
mucus, nasal 149
multifidus muscle 75
muscle
    abductor digiti minimi
    94, 95, 109
    abductor pollicis
    brevis 95
    abductor pollicis longus
    97
    adductor brevis 102,
    233
    adductor longus 98,
    102, 233
    adductor magnus 100,
    102, 104, 233
    adductor pollicis 97
    anconeus 89, 93, 95, 97
    anterior scalene 71, 73,
    75, 77, 81
    arrector pili 23
    biceps brachii 68, 87
    biceps femoris 100, 101,
    105, 106, 232, 233
    brachialis 69, 87, 89, 90,
    91, 93, 94, 96
    brachioradialis 68, 87, 91,
    92, 93, 94, 95, 96
    calf 68
    constrictor, inferior 75
    deltoid 68, 69, 86, 88,
    90
    depressor anguli oris 71,
    73
    depressor labii inferioris
    71, 73

detrusor 207
digastric 73
erector spinae 69, 78, 84,
    228, 229
extensor carpi radialis
    brevis 95, 97
extensor carpi radialis
    longus 92, 93, 95, 97
extensor carpi ulnaris 95,
    97
extensor digiti minimi 95
extensor digitorum 95
extensor digitorum brevis
    109
extensor digitorum longus
    107, 108, 109
extensor hallucis brevis
    107
extensor hallucis longus
    108, 109
extensor indicis 96
extensor pollicis brevis 96
extensor pollicis longus
    97
extensor retinaculum 94,
    96
    inferior 107, 109
    superior 107, 109
external obliques 68, 76,
    79, 82, 85
fibular 69
fibularis (peroneus) brevis
    107, 109
fibularis (peroneus) longus
    106, 107, 108
fibularis (peroneus) tertius
    109
flexor carpi radialis 94
flexor carpi ulnaris 93, 96
flexor digiti minimi brevis
    95
flexor digitorum
    superficialis 94, 95
flexor hallucis longus 109,
    232
flexor pollicis brevis 95
flexor pollicis longus 96
flexor retinaculum 94, 96,
    230
gastrocnemius 69, 101,
    106, 233
gemellus
    inferior 104
    superior 104
genioglossus 74
geniohyoid 74, 194
gluteus maximus 69, 84,
    85, 100, 143, 229, 233
gluteus medius 68, 84,
    102, 104, 229
gracilis 98, 100, 102, 233

hypothenar 230
iliacus 102, 229
iliocostalis 84
iliopsoas 68, 98, 229
inferior extensor
    retinaculum 107, 109
inferior gemellus 104
infraspinatus 78, 79, 88,
    92
intercostal 68, 77, 78, 81,
    93, 131, 146, 210
internal obliques 77, 83,
    84
latissimus dorsi 69, 79, 84,
    85, 88, 89, 90
levator labii superioris 70,
    73
levator labii superioris
    alaeque nasi 70, 72
levator scapulae 70, 71,
    73, 75
longissimus 84
longus colli 75, 81
lumbrical 95
masseter 71, 73
medial rectus 124
mentalis 71, 73
middle scalene 73, 75,
    81
multifidus 75
mylohyoid 74, 194
nasalis 70, 72
obliques
    external 68, 76, 79, 82,
    85
    internal 77, 83, 84
obturator internus 104,
    229
occipitofrontalis 69, 70,
    72
omohyoid 71, 73, 75, 77
opponens digiti minimi
    97
opponens pollicis 97
orbicularis oculi 70, 72
orbicularis oris 71, 73
pectineus 98, 102
pectoralis major 68, 76,
    82, 86, 210, 211
pectoralis minor 68, 77,
    90, 210
peroneus (fibularis) brevis
    107
peroneus (fibularis) longus
    106, 107, 108
peroneus (fibularis) tertius
    109
piriformis 69, 84
platysma 68, 75
popliteus 105
posterior scalene 73

pronator teres 94, 95
psoas 58, 102, 196, 228
quadratus femoris 104
quadriceps femoris 61,
    68, 69, 99, 103, 141
rectus
    lateral 124, 125
    medial 124, 125
    superior 124
rectus abdominis 68, 76,
    82, 83, 228, 229
rectus femoris 63, 99, 233
risorius 71, 73
sartorius 98, 106, 141,
    232
scalene
    anterior 71, 73, 75, 77,
    81
    middle 73, 75, 81
    posterior 73
semimembranosus 101,
    105, 106, 233
semispinalis capitis 70, 75
semispinalis cervicis 75
semitendinosus 100, 106,
    233
serratus anterior 68, 69,
    76, 82
serratus posterior inferior
    69, 78, 84
soleus 107
spinalis 78, 84
splenius capitis 70, 73, 75
sternocleidomastoid 70,
    71, 73, 75, 76, 129
sternohyoid 71, 73, 75
sternothyroid 73, 75
stylopharyngeus 75
subclavius 77, 90
subscapularis 90
superior extensor
    retinaculum 107, 109
superior gemellus 104
superior oblique (eye)
    124
superior rectus 124
supinator 91, 97
supraspinatus 92
temporalis 69, 70, 72
teres major 78, 79, 88,
    90, 92
teres minor 78, 92
thenar 230
thyrohyoid 73, 75
tibialis anterior 106, 232
tibialis posterior 109
transversus abdominis
    68, 69
trapezius 68, 69, 70, 71,
    73, 75, 79, 85, 86, 88,
    129

triceps brachii 69, 87, 88,
    89, 95, 97
vastus intermedius 103,
    233
vastus lateralis 63, 99,
    100, 103, 104, 108, 233
vastus medialis 63, 99,
    103, 108, 233
zygomaticus major 70,
    73
zygomaticus minor 70
muscular system **68–9**
    abdomen and pelvis
    82–5
    breathing 16
    cells 14
    eyes 124
    head and neck **70–5**
    hip and thigh **98–105**
    lower arm and hand
    94–7
    lower leg and foot **106–9**
    muscle tissue 14, 15
    muscular system 17
    names of muscles 69
    shoulder and upper arm
    86–93
    skeletal muscle tissue 15
    structure **68–9**
    thorax **76–81**
    see also muscle and
    specific muscles
musculocutaneous nerve
    110, 111, 134, 136, 139
myelin sheath 111
mylohyoid muscle 74, 194
myocardium 164, 165
myofibrils 68, 165
myometrium 213
myosin 68, 201

nails
    structure **22**
naris (nostril) 146
nasal bone 25, 26, 28, 32,
    35
nasal cavity 146, 147, 149
    tonsils 183
    MRI 224
nasal concha
    inferior 26, 32
    middle 32
    superior 32
nasal crest 26
    anterior 32
nasal septum 149
    MRI 224

nasalis muscle 70, 72
nasociliary nerve 124
nasolacrimal duct 148
nasopharynx 148, 183, 194
navicular bone 65, 66
  MRI 232
neck
  nerves **128–9**
  see also head and neck
neck, tooth 195
nerve
  alveolar, inferior 123
  abducent 122, 124
  accessory 122, 123, 128, 129
  auricular
    greater 123
    posterior 123
  auriculotemporal 123
  axillary 110, 111, 134, 136
  cochlear 127
  cutaneous, lateral sural 144
  dorsal digital 145
  facial 122, 123, 128
    buccal branch 123
    cervical branch 123
    marginal mandibular branch 123
    temporal branch 123
    zygomatic branch 123
  femoral 110, 111, 132, 140, 142
  femoral cutaneous
    intermediate 141, 142
    lateral 141, 143
    medial 141, 142
    posterior 142
  fibular (peroneal) 141, 143
    common 110, 111, 144, 145
    deep 110, 144, 145
    superficial 110, 145
  frontal 124
  genitofemoral 132
  glossopharyngeal 122, 123, 128, 129
  gluteal 143
    superior 133, 142
  greater auricular 123
  hypoglossal 122, 123, 128, 129
  iliohypogastric 132
  ilioinguinal 132
  infraorbital 123
  intercostal 110, 111, 130, 131, 133, 160
  interosseous, posterior 138, 139

lacrimal 124
lateral plantar 145
lateral sural cutaneous 144
lingual 123
medial pectoral 134, 136
medial plantar 145
median 110, 111, 135, 137, 138, 139
mental 123
musculocutaneous 110, 111, 134, 136, 139
nasociliary 124
obturator 110, 111, 132, 140, 141, 142
oculomotor 122, 124
olfactory 122, 148
ophthalmic 123, 124
optic 112, 116, 122, 123, 124, 125
peroneal *see* fibular
phrenic 129, 130, 162
plantar
  lateral 145
  medial 145
posterior auricular 123
posterior interosseous 138, 139
pudendal 140, 142
radial 110, 111, 135, 137, 138, 139
recurrent laryngeal 162
sacral 111
saphenous 110, 140, 141, 142, 144, 145
sciatic 110, 111, 133, 140, 142, 143
subcostal 130, 133
supraorbital 124
supratrochlear 124
sural 144, 145
tibial 110, 111, 141, 143, 144, 145
  calcaneal branch 145
trigeminal nerve 122, 123, 128
trochlear 122, 124
ulnar 110, 111, 135, 137, 138, 139
  palmar branch 139
  palmar digital branches 138, 139
vagus 122, 123, 128, 129, 130, 162
vestibular 127
vestibulocochlear 122, 126, 127
nerve cells 14
nerves
  nerve tissue 14, 15
  plexuses 111

structure **111**
see also nerve and specific nerves
nervous system 16
  abdomen and pelvis **132–3**
  anatomy **110–1**
  brain **112–29**
  ear **126–7**
  eye **124–5**
  hip and thigh **140–3**
  lower arm and hand **138–9**
  lower leg and foot **144–5**
  neck **128–9**
  shoulder and upper arm **134–7**
  thorax **130–1**
  see also specific nerves
neurons 114
  structure **111**
neurovascular bundles 232
nipples 208, 210, 211
node of Ranvier 111
nodes, lymphatic system 180, 181
nose
  bones 26–7
  cartilage 70
  MRI 225
  respiratory system **148–9**
nostrils 146, 148
nuchal line
  inferior 31
  superior 31
nuclear membrane 12
nucleolus 12
nucleoplasm 12
nucleosomes 11
nucleotides 10, 11
nucleus 12
  muscle cells 201
nucleus pulposus, intervertebral discs 43
nutrients, digestion 12

O

oblique cord 53
oblique fissure, lung 150, 151, 152, 153
oblique muscle (eye), superior 124
oblique muscles
  external 68, 76, 79, 82, 85
  internal 77, 83, 84
  aponeurosis of 83
obturator foramen 43, 45, 58, 140, 142

obturator internus muscle 104
  MRI 229
obturator membrane 62
obturator nerve 110, 111, 132, 140, 141, 142
occipital artery 156
occipital bone 25, 26, 28, 31, 33, 34
occipital condyle 31
occipital lobe 112, 117, 120
occipital nodes 181, 182
occipital pole 112, 114, 115, 117
occipital protuberance
  external 31, 33
  internal 30
occipital region 18
occipital vein 157
occipitofrontalis muscle 69, 70
  frontal belly 72
  occipital belly 72
occipitomastoid suture 28
occipitotemporal fusiform gyrus 115
oculomotor nerve 122, 124
odontoid peg, vertebrae 40
oesophageal veins 203
oesophagus 196
  anatomy 74, 146, 192, 194, 195, 200
  endoscopic view 223
  muscles 75, 197
  thoracic part of 196, 197
olecranon 51, 53, 55, 89, 93, 95
olecranon fossa 51, 55
olfactory bulb 112, 115, 116, 122
olfactory nerve 122, 148
olfactory tract 115, 116, 122
olfactory trigone 115
oligodendrocytes 111
omohyoid muscle 75, 77
  inferior belly 71, 73
  superior belly 71, 73
ophthalmic artery 158
ophthalmic nerve 123, 124
ophthalmic veins
  inferior 159
  superior 159
opponens digiti minimi muscle 97
opponens pollicis muscle 97
optic canal 30
optic chiasma 115, 116, 118
optic disc 125

optic nerve
  anatomy 112, 116, 122, 123, 124, 125
  MRI 225
optic radiation 120
opposable thumb 56
oral cavity 194
orbicularis oculi muscle 70, 72
orbicularis oris muscle 71, 73
orbit 26
  medial wall of 124
orbital fissures
  inferior 26
  superior 26, 124
orbital sulcus 115
organelles 12, 13
organs 15
  connective tissue 15
  see also specific organs
orbital gyri 115
oropharynx 148, 183, 194
ossicles 34, 126
osteons 24
oval fossa 164
oval window, ear 126
ovaries 212, 213, 215
  anatomy 208, 209, 215
  blood vessels 166
  hormones 217, 218
  suspensory ligaments 213
oviducts *see* Fallopian tubes
ovum *see* egg
oxygen
  body composition 14
  cardiovascular system 17
  cell metabolism 12
  cell transport 13
  red blood cells 14
  respiratory system 16, 147

P

palate
  hard 74, 148, 194, 195
  soft 74, 75, 183, 195
palatine bone 32
palatine foramina
  greater 31
  lesser 31
palatine tonsil 74, 183
palatoglossal fold 74
palatomaxillary suture 31
palatopharyngeal fold 74
palmar aponeurosis 95
palmar arch
  deep 173
  superficial 173

palmar carpal ligament 57
palmar digital artery 173
palmar digital vein 173
palmar interosseous
    muscles 97
palmar ligament 57
palmar metacarpal ligament
    57
palmar surface of hand 18
palmar venous arch
    deep 173
    superficial 173
palmaris longus tendon 94
pancreas
    anatomy 193, 199, 202,
        203
    blood vessels 166
    functions 202, 217
    MRI 228
pancreatic ducts 203
pancreatic juice 202
papilla, hair 22
papillae, tongue 195
papillary muscles, heart
    164, 165,
para-aortic nodes 180
parahippocampal gyrus
    115
paramammary node 184
paranasal sinuses 148
parasternal nodes 180, 181,
    184, 185
parathyroid glands 216,
    219
paratracheal nodes 182
parietal bone 25, 26, 28,
    33, 34
parietal lobe 112, 117
parietal lobules
    inferior 114
    superior 114
parietal pleura 147, 150,
    153
parieto-occipital sulcus 112,
    114
parietomastoid suture 28
parotid (Stensen's) duct 192
parotid gland 192, 194, 195
parotid nodes 182
patella
    anatomy 24, 25, 59, 63, 64,
        68, 103, 106, 108, 141
    apex of 59
    base of 59
    knee joint 63
    MRI 232, 233
    quadriceps tendon 99
patellar ligament 63, 99, 106,
    108
patellar retinaculum, lateral
    63

pectineus muscle 98, 102
pectoral nerve, medial 134,
    136
pectoral region 18
pectoralis major muscle
    68, 76, 82, 86, 210, 211
pectoralis minor muscle
    68, 77, 90, 210
pelvis 24, 25
    blood vessels 154
    cardiovascular system
        166–7
    digestive system **198–203**
    immune and lymphatic
        system **186–7**
    muscles **82–5**
    pelvic brim 46, 47
    reproductive system
        212–5
    skeletal system **42–7**, 60
penis 209, 212, 214
    bulb of 212, 214
    corpus cavernosum 212,
        214
    corpus spongiosum 212,
        214
    crus 214
    glans penis 212, 214
perforated substance,
    anterior 115
perforating artery 176, 177
pericarditis 162
pericardium 162
perimetrium 213
perimysium 68
periodontal ligament 34,
    195
periosteal blood vessels
    24
periosteum 24
peroneal (fibular) artery
    154, 178
peroneal muscles 69
peroneal (fibular) nerves
    141, 143
    common 110, 111, 144,
        145
    deep 110, 144, 145
    superficial 110, 145
peroneus (fibularis) brevis
    muscle 107
peroneus (fibularis) longus
    muscle 106, 107, 108
peroneus (fibularis) longus
    tendon 109
peroneus (fibularis) tertius
    muscle 109
peroxisomes 13
petrosal sinus
    inferior 159
    superior 159

phalanges (foot) 24, 232
    distal 66, 109
    middle 66
    proximal 66, 67, 109
phalanges (hand) 25
    distal 54–7, 65
    joints 57
    middle 54–7, 65
    proximal 54–7, 65, 95
phalanx see phalanges
pharyngeal muscles 75
pharyngeal raphe 75
pharyngeal tonsil 183
pharyngeal tubercle 31
pharyngobasilar fascia 75
pharyngotympanic tubes
    see Eustachian tubes
pharynx
    anatomy 74, 146, 147,
        183, 192
    inferior constrictor 75
    middle constrictor 75
    MRI 225
    superior constrictor
        73, 75
    swallowing food 195
phosphates
    cell metabolism 12
    DNA 10, 11
photoreceptor cells 14
phrenic nerves 129, 130,
    162
pia mater
    brain 121
    spinal cord 133
pineal gland
    anatomy 119, 216, 218
piriform aperture 26
piriformis muscle 69, 84,
    104
pisiform bone 54, 57, 138,
    139
    MRI 230
pituitary fossa 30, 32
pituitary gland
    anatomy 115, 116, 118,
        216, 218
plantar arteries
    lateral 179
    medial 179
plantar ligaments
    long 66, 67
    short 66
plantar nerves
    lateral 145
    medial 145
platysma muscle 68, 75
pleura 150, 152, 153
pleural cavity 147, 160
plexuses, nerves 111
plica semilunaris 124

pons
    anatomy 113, 115, 116,
        119, 122
    MRI 225
pontine arteries 158
popliteal artery 154, 155,
    175, 177, 178, 179
popliteal fossa 18, 143, 145
    MRI 232
popliteal nodes 180, 181,
    191
popliteal surface, femur 61
popliteal vein 154, 155, 175,
    177, 178, 191
popliteus muscle 105
portal vein 154, 155, 166,
    202, 203
postcentral gyrus 113, 114
postcentral sulcus 113, 114
posterior cord 134, 136
posterior nasal spine 31
posterior rectus sheath 68
prepatellar bursa 99, 108
pretracheal nodes 180
preaortic nodes 180, 181
preauricular nodes 180
precentral gyrus 113, 114
precentral sulcus 113, 114
prelaryngeal nodes 180,
    182
premolar teeth 195
preoccipital notch 113
prepatellar bursa 63
prepuce, clitoris 215
presymphyseal node 190
pretracheal nodes 180, 181,
    182
pronator teres muscle 94,
    95
prostate gland 205, 208,
    209, 212, 214
proteins
    cell membrane 13
    cell metabolism 12
    genes and 10–1
    synthesis 11
proximal, definition 19
proximal superficial nodes
    190
psoas muscle 58, 102, 196
    MRI 228, 229
pterion 28
pterygoid hamulus 31
pterygoid plates
    lateral 31
    medial 31
pterygoid processes 32
pterygoid venous plexus
    157, 159
pubic ramus
    superior 42, 45, 46, 47, 62

pubic symphysis 43, 46, 47,
    82, 98
    female pelvis 213
    joint 43
    male pelvis 212
    MRI 229
pubic tubercle 43, 83
pubis, body of 42, 44, 62
pubofemoral ligament 62
pudendal nerve 140, 142
pulmonary arteriole 151
pulmonary arteries
    anatomy 160, 161
    left 152, 162, 163
    right 153, 162, 163
pulmonary circulation 155
pulmonary ligament 152,
    153
pulmonary trunk 161, 162,
    164
pulmonary valve 164, 165
pulmonary veins 163, 164
    inferior 152, 153
    superior 152, 153
pulmonary venule 151
pulp cavity, tooth 195
pulse 173
pupil 125
pyloric sphincter 200
pylorus 200
pyramid 122

quadrate lobe of liver 202
quadratus femoris muscle
    104
quadriceps femoris muscle
    61, 68, 69, 99, 103, 141
    MRI 232, 233
quadriceps femoris tendon
    63, 99, 103

radial artery 154, 155, 169,
    171, 172, 173
radial collateral artery 169,
    171
radial fossa 49
radial nerve 110, 111, 135,
    137, 138, 139
radial recurrent artery 169,
    171
radial vein 172, 173
radiocarpal ligaments 56,
    57

radiographs 222
  hand 57
  head 149
radioulnar joints 55
radius
  anatomy 24, 25, 49, 53,
    138, 172
  annular ligament 53
  head 51, 54, 55
  interosseous border 54,
    55
  MRI 230, 231
  radial tuberosity 51, 54,
    55
  shaft 51, 54, 55
  styloid process 54, 56
receptors
  photoreceptor cells 14
rectal artery, superior
  167
rectal veins
  inferior 203
  middle 203
  superior 203
rectouterine pouch 213
rectovaginal septum 213
rectovesical pouch 212
rectum
  anatomy 193, 198, 199,
    201, 212, 213
  blood supply 154
rectus abdominis muscle
  68, 76, 82, 83
  MRI 228, 229
rectus femoris muscle 63,
  99
  MRI 233
rectus femoris tendon 62
rectus muscle
  lateral 124, 125
  medial 124, 125
  superior 124
rectus sheath 77
  posterior layer of 83
recurrent laryngeal nerve
  162
red blood cells 14
  haemoglobin 14
  spleen 187
reflexes
  knee jerk 99
renal arteries 154, 166, 167,
  204, 206, 207
renal cortex 207
renal hilum 206
  MRI 228
renal medullary pyramid
  207
renal pelvis 207
renal veins 154, 166, 167,
  204, 206, 207

reproductive system 17
  abdomen and pelvis
    212–5
  anatomy **208–9**
  female 208, 209, 215
  male 208, 209, 212, 214
  thorax **210–1**
respiratory system 16, **146–7**
  head and neck **148–9**
  thorax **152–3**
rete testis 214
retina 125
retromandibular vein 157
rhomboid muscles 69
  major 70, 78
  minor 70, 78
ribosomes 10, 12, 13
ribs
  anatomy 24, 25, 27, 36–9,
    77, 84, 133, 146
  costal cartilages 24, 36
  intercostal muscles 131, 146
  joints 38, 40
  nerves 130–1
risorius muscle 71, 73
RNA (ribonucleic acid) 10
rod cells, retina 14
root
  nail 22
  teeth 195
root sheath, hair 22
rotator cuff muscles 50, 90,
  92
rough endoplasmic
  reticulum 13
round window, ear 127
rugae 200

S

saccule 126
sacral curvature 41
sacral foramina
  anterior 41, 43, 133
  posterior 45
sacral nerves 111
sacral plexus 110, 111, 132,
  141, 143
sacral promontory 46, 47
sacral spinal nerves 110
sacroiliac joint 42, 46, 47, 59
  MRI 229
sacrospinous ligament 44
sacrotuberous ligament 44
sacrum 24, 41, 42, 44, 212,
  213
  ala of 43
  facet joint 41
  MRI 229

sagittal plane 19
sagittal sinus
  inferior 159
  superior 121, 159
sagittal suture 26
saliva
  chewing food 192
  functions 195
salivary glands 192, 195
saphenous nerve 110, 140,
  141, 142, 144, 145
saphenous veins
  accessory 175, 176
  great (long) 154, 175, 177,
    178, 190
  small (short) 154, 179, 191
sarcoplasm 68
sartorius muscle 98, 141,
  106
  MRI 232
sartorius tendon 63
scalene muscles
  anterior 71, 73, 75, 77, 81
  middle 73, 75, 81
  posterior 73
scaphoid bone 54, 56
  MRI 230
scapula 24, 25, 36, 48,
  52, 53
  acromion of 25, 48, 50,
    52, 53, 70, 86, 92
  coracoid process 48, 52,
    53
  inferior angle 50, 78
  spine of 50, 70, 78, 88, 92
  vertebral (medial) border
    78, 92
scapular ligament,
    transverse 52
scapular muscles 69
sciatic foramen, greater
  143
sciatic nerve 110, 111, 133,
  140, 142, 143
sciatic notch, greater 46, 47
sclera 124, 125
screening see tests
scrotum 208, 209, 212
sebaceous glands 22, 23
sebum 23
secretory lobules, mammary
  glands 208, 210, 211
secretory vesicles, cells 12
semicircular canals 126
  anterior 127
  lateral 127
  posterior 126
semilunar cusps 165
semimembranosus muscle
  101, 105, 106
  MRI 233

seminal vesicles 208, 209,
  212, 214
seminiferous tubule 222
semispinalis capitis muscle
  70, 75
semispinalis cervicis muscle
  75
semitendinosus muscle 100,
  106
  MRI 233
semitendinosus tendon 63
senses
  vision **124–5**
sensory nerves
  ganglion 133
  rootlets 133
septum
  penis 214
  interatrial 164
septum pellucidum 118, 120,
  121
serous lining, small intestine
  201
serous pericardium 162, 164
serratus anterior muscle 68,
  69, 76, 82
serratus posterior inferior
  muscle 69, 78, 84
sex cells see egg; sperm
sex hormones 208, 217
shin bone see tibia
"shin splints" 69
shoulder
  cardiovascular system
    168–71
  joint 48, **52**
  lymphatic and immune
    system **188–9**
  muscles **86–93**
  nervous system **134–7**
  skeletal system **48–53**
shoulder blade see scapula
sight see vision
sigmoid colon 199
sigmoid sinus 159
sinuses, nasal 32, 148, 149
"six pack" 83
skeletal muscle 15, 68
skeletal system 17
  abdomen and pelvis
    42–7
  anatomy **24–5**
  bone and cartilage **24**
  head and neck **26–33**
  hip and thigh **58–62**
  knee **63**
  lower arm and hand
    54–7
  lower leg and foot **64–7**
  shoulders and upper arm
    48–53

thorax **36–41**
  see also bones
skin
  epithelial cells 14
  structure **22–3**
skull 24, **26–33**
  bones 34–5
  foramina 122
  interior **32–3**
  internal surface of base
    of **30**
  MRI 225
  sutures 34
  underside of **31**
sleep-wake cycle 218
small intestine
  anatomy 193, 199, 201
smooth endoplasmic
  reticulum 13
smooth muscle 201
  cells 14
  tissue 15
soft palate 74, 75, 183,
  195
  MRI 224
soleal line 64
soleus muscle 107
somatic nervous system
  110
speech
  larynx 149
sperm 14, 208, 209, 212,
  214, 215
sphenoid bone 34, 35
  greater wing of 28
  lesser wing of 30
sphenoidal sinus 32, 148
sphenoparietal sinus 159
sphincters
  anal 212, 213
  urethral 212, 213
spinal artery, anterior 158
spinal cord
  abdomen and pelvis
    133
  anatomy 75, 110, 111,
    113, 115, 116, 117, 119
  MRI scan 225, 226
  neck 129
  nerve tissue 15
  structure 133
spinal ganglia 133
spinal nerves
  anatomy 130, 131, 133
  cervical 110, 111
  lumbar 110, 111
  neck 129
  sacral 110, 111
  thoracic 110, 111
spinal process 225
spinalis muscle 78, 84

spindle, cell division 13
spine
  intervertebral discs 40, 43
  length of 40
  MRI scans 227, 229
  skeletal system **40–1**
  vertebral column 24, 25
  see also vertebrae
spleen
  anatomy 180, 181, 186, 187, 199
  blood vessels 166
  MRI 228
splenic artery 167
splenic vein 167, 203
splenium of corpus callosum 119
splenius capitis muscle 70, 73, 75
spongy bone 15, 24
"spring ligament" 67
squamosal suture 28, 33
stapes 35, 126
stem cells 13, 14
Stensen's (parotid) duct 192
sternocleidomastoid muscle 70, 73, 75, 76, 129
  clavicular head 71
  sternal head 71
sternohyoid muscle 71, 73, 75
sternothyroid muscle 73, 75
sternum 24, 37, 77, 160
  MRI 226
stomach 200
  anatomy 193, 197, 199, 201
  fundus of 197, 200
  greater curvature 200
  lesser curvature 200
  MRI 228
straight gyrus 115
straight sinus 159
stroma, breasts 210
stylohyoid ligament 75
styloid processes
  radius 54, 56
  skull 29, 31, 33
  ulna 54, 56, 57
stylomastoid foramen 31
stylopharyngeus muscle 75
subarachnoid space 75, 121, 133
subclavian arteries 154, 155, 160, 161, 163, 168
  grooves for 152, 153

subclavian veins 154, 155, 160, 161, 180
subclavius muscle 77, 90
subcostal nerve 130, 133
subcutaneous fat 75
sublingual gland 192, 194, 195
submandibular duct 192, 194
submandibular gland 192, 194, 195
submandibular nodes 182
submental artery 156
submental nodes 181, 182
submental vein 157
subparotid nodes 180
subpubic angle 46, 47
subscapular artery 168, 170
subscapularis muscle 90
sulcus terminalis 195
superciliary arch 26
superficial fascia 210
superior, definition 19
superior lobar bronchus 153
supinator crest 55
supinator muscle 91, 97
supraclavicular nodes 184, 185
supracondylar line
  lateral 61
  medial 61
supramarginal gyrus 114
supraorbital foramen 26
supraorbital nerve 124
supraorbital ridge 26
suprapatellar bursa 63
suprapubic region 18
suprarenal glands 204, 206, 217, 218
  blood vessels 166
supraspinatus muscle 92
supraspinous fossa 50
supratrochlear nerve 124
supratrochlear nodes 181, 189
sural artery 175, 177
sural nerve 144, 145
suspensory ligaments
  eye 125
sutures, skull 26, 28, 34
swallowing 75, 197
sweat 22
sweat glands 23
sympathetic ganglia 110, 132
sympathetic nervous system 110
  thorax 130

sympathetic trunks 110, 129, 132
synaptic knob 111
synovial cavity, knee joint 63
synovial joints 63
systemic circulation 155
systems of the body 15

T

T lymphocytes
  thymus gland 185
taenia coli 201
talofibular ligaments
  anterior 66
  posterior 67
talus 65, 66, 67
  MRI 232
tarsals 24, 25, 65, 230
tarsometatarsal ligaments
  dorsal 66
  plantar 66
taste buds 195
tears 124
  nasolacrimal duct 148
teeth 192, 194, 195
  eruption 195
  MRI 224, 225
  periodontal ligament 34
temporal artery, superficial 156
temporal bone 25, 28, 34, 35, 126
  petrous part 30, 35
  squamous part 33
  tympanic part 28, 31
  zygomatic process 35
temporal gyrus
  inferior 113, 115
  middle 113
  superior 113
temporal lobe 112, 116
temporal pole 112, 115, 116
temporal sulcus
  inferior 113, 115
  superior 113, 114
temporal vein, superficial 157
temporalis muscle 69, 70, 72
tendinous cords 165
tendinous intersection 82
tendons 68
  bursas 63
  foot 107
  hand 94, 95
  intertendinous connections 94

skeletal system 17
tissues 15
toes 109
wrist 95
tensor fasciae latae 98
  MRI 233
teres major muscle 78, 79, 88, 90, 92
teres minor muscle 78, 92
terminology, anatomical **18–9**
testes 208, 209
  anatomy 209, 212, 214
  blood vessels 166
  hormones 217, 218
  lobules 214
testicles see testes
thalamus 119, 120, 121
  MRI 225
thenar muscles 230
thigh
  adductor compartment 68
  anterior surface 18
  blood vessels 154
  cardiovascular system **174–7**
  extensor compartment 68
  flexor compartment 69
  lymphatic and immune system **190–1**
  muscles **98–105**
  nervous system 110, 111, **140–3**
  posterior surface 18
  skeletal system 24, 25, **58–61**
thoracic curvature 40
thoracic duct 180, 181, 184, 185, 186, 187
thoracic nerves 110, 111, 128, 130
thoracic spine 40
thoracic vertebrae 25, 37, 41, 131, 133
  MRI 226
thoracic vessels, MRI 227
thoracoacromial artery 168
thoracolumbar fascia 85
thoracromial artery 170
thorax
  cardiovascular system **160–3**
  digestive system **196–7**
  immune and lymphatic systems **184–5**
  MRI 226, 227
  muscular system **76–81**

nervous system **130–1**
reproductive system **210–1**
respiratory system **152–3**
skeletal system **36–41**
thumb
  bones 54, 55, 95, 97
  joints 56
  muscles 95, 96, 97
  opposable 56
thymine 10, 11
thymus gland
  lymphocytes 181, 185
thyrohyoid muscle 73, 75
thyroid artery, superior 156
thyroid cartilage 74, 75, 148, 149
  MRI 225
thyroid gland
  anatomy 74, 216, 218, 219
thyroid prominence ("Adam's apple") 149
thyroid vein, superior 157
tibia
  anatomy 24, 25, 59, 66, 67, 108, 141, 143, 178
  anterior border 65
  condyles 64
  fibrous joints 34
  interosseous border 64
  knee joint 63
  lateral condyle 64
  medial condyle 64
  medial surface 65, 107
  MRI 233
  nutrient foramina 64
  shaft 65
  tibial tuberosity 64
tibial arteries
  anterior 154, 155, 178
  posterior 154, 178, 179
tibial collateral ligament 63
tibial nerve 110, 111, 141, 143, 144, 145
  calcaneal branch 145
tibial recurrent artery, anterior 178
tibial veins
  anterior 154, 155, 178
  posterior 154, 155, 178
tibialis anterior muscle 106
  MRI 232
tibialis posterior muscle 109
tibialis tendons
  anterior 67
  posterior 67
tibiofibular ligaments

anterior 66
posterior 67
tissue types **14-5**
toes
bones 24, 65
tendons 107, 109
tongue
anatomy 183, 192, 194, 195
MRI 224
muscles 75
oral part of 195
pharyngeal part of 195
tonsils
anatomy 74, 183
touch sensors 23
trabeculae carnae 164, 165
trachea
anatomy 74, 146, 147, 148, 150, 161
bifurcation of 161
cartilage 149
MRI 226
tracheobronchial nodes 185
tragus 127
transcription, protein synthesis 10
translation, protein synthesis 10
transpyloric plane 18
transverse abdominis muscle 69
transverse colon 198
transverse foramen, vertebrae 40
transverse plane 19
transverse processes, vertebrae 38, 40, 41
transverse sinus 159
transversus abdominis muscle 68, 69
trapezium 54
trapezius muscle
anatomy 68, 69, 70, 71, 73, 75, 79, 85, 86, 88, 129
MRI 225
trapezoid 54
MRI 230
triangular ligaments 202
triceps brachii muscle 69, 87, 88, 89, 95, 97
medial head 91, 92, 93
triceps tendon 89, 93
tricuspid valve 164, 165
trigeminal nerve 123, 128
motor root 122
sensory root 122
trigone

bladder 207
olfactory 115
triquetral bone 54, 56
MRI 230
trochanter
greater 44, 58, 60, 229
lesser 44, 58, 60
trochlear nerve 122
tropomyosin 68
tuber cinereum 115
tubercle
greater 48
lesser 48
tunica adventitia 155
tunica albuginea 214
tunica intima 155
tunica media 155
tympanic membrane (eardrum) 126, 127

U

ulna
anatomy 24, 25, 49, 51, 53, 138, 172
head of 54
interosseous border 54, 55
MRI 230, 231
olecranon 51, 53, 55, 89, 93, 95
radial notch 54
shaft 51, 54, 55
styloid process 54, 56, 57
tuberosity 54
ulnar artery 154, 155, 169, 171, 172, 173
ulnar collateral arteries
inferior 169, 171
superior 169, 171
ulnar collateral ligament 53
ulnar nerve 110, 111, 135, 137, 138, 139
palmar branch 139
palmar digital branches 138
ulnar recurrent artery 169, 171
ulnar vein 172
umbilical cord
ligamentum venosum 202
umbilical region 18
umbilicus 82
uncinate process of pancreas 203
uncus 115

upper pole of kidney 206
ureter 204, 205, 206, 207, 214
ureteric orifice 207
urethra 204, 205, 206, 207, 208, 209
female anatomy 205, 213
male anatomy 205, 212, 214
urethral orifice
external 207, 212, 215
internal 207
urethral sphincter, external 212, 213
urinary system 17, **204-5**
urination (micturition) 83
urine 17, 204
uterus
anatomy 208, 209, 212, 213, 215
body of 213
cavity of 213
cervix 209, 213, 215
fundus 213
round ligament 213
utricle 126

V

vacuoles, cells 13
vagina
anatomy 208, 209, 212, 213
anterior fornix 213
cervix 215
lateral fornix 215
posterior fornix 213
vestibule 215
vagus nerve 122, 123, 128, 129, 130, 162
vallate papillae 195
valves
veins 155
vas deferens 208, 209, 212, 214
vastus intermedius muscle 103
MRI 233
vastus lateralis muscle 63, 99, 100, 103, 104, 108
MRI 233
vastus medialis muscle 63, 99, 103, 108
MRI 233

vein
accessory saphenous 175, 176
angular 157
anterior tibial 154, 155, 178
auricular
posterior 157
axillary 168, 170
azygos 155, 160, 161, 197
basilic 154, 168, 169, 171, 172, 173
brachial 154, 155, 169, 171
brachiocephalic 154, 155, 160, 161
MRI 227
cardiac
great 162
middle 163
small 162
cephalic 154, 155, 168, 170, 172, 173, 188
accessory 172, 173
common iliac 154, 155, 166, 167, 205
cubital, median 169, 171, 172
digital, dorsal 172
dorsal 214
external iliac 154, 166, 167, 206
external jugular 129, 154, 155, 157
facial 157
femoral 154, 155, 175, 176, 177
deep 155
left 167
right 166
gastric 203
gonadal 155, 166, 167
great cardiac 162
great (long) saphenous 154, 175, 177, 178, 190
hepatic 154, 155, 203
hepatic portal 203
ileal 203
iliac
common 154, 155, 166, 167, 205, 206
external 154, 166, 167
internal 154, 155, 166, 167
infraorbital 157
inferior labial 157
inferior mesenteric 167, 203
inferior ophthalmic 159
inferior rectal 203

infraorbital 157
intercostal 160, 161
internal iliac 154, 155, 166, 167, 206
internal jugular 129, 154, 155, 157, 158, 159, 160, 161, 182
interventricular 165
jejunal 203
jugular 180
external 154, 155, 157
internal 129, 154, 155, 157, 158, 159, 160, 161, 182
labial
inferior 157
superior 157
lateral marginal 179
marginal
lateral 179
medial 179
maxillary 157
medial marginal 179
median vein of the forearm 172
mental 157
mesenteric
inferior 167, 203
superior 154, 155, 166, 203
middle cardiac 163
middle rectal 203
occipital 157
oesophageal 203
ophthalmic
inferior 159
superior 159
palmar digital 173
popliteal 154, 155, 175, 177, 178, 191
portal 154, 155, 166, 203
posterior auricular 157
posterior tibial 154, 178
pulmonary 163, 164
inferior 152, 153
superior 152, 153
radial 172, 173
rectal
inferior 203
middle 203
superior 203
renal 154, 166, 167, 204, 206, 207
retromandibular 157
right gastric 203
saphenous
accessory 175, 176, 177
great (long) 154, 175, 177, 178, 190
small (short) 154, 155, 178, 179, 191

small cardiac 162
small (short) saphenous
154, 178, 179, 191
splenic 167, 203
subclavian 154, 155, 160,
161, 180
submental 157
superior labial 157
superior mesenteric 154,
155, 166, 203
superior ophthalmic
159
temporal, superior
157
thyroid, superior 157
tibial
anterior 154, 155, 178
posterior 154, 178
ulnar 172
veins
anatomy **155**
around brain 159
external veins of head
157
muscles and 17
structure **155**
valves 155
see also vein *and specific*
veins

vena cava
inferior 154, 155, 160,
163, 164, 166, 196,
202, 203, 204, 206,
228
superior 154, 155, 160,
162, 163, 164, 180,
226
venous arch
deep palmar 173
superficial 173
venous network, dorsal
172
venous sinuses, brain 158,
159
ventricles
brain 119, 120, 121, 225
heart 160, 161, 162, 163,
164, 165, 227
venules 155
in skin 23
vertebrae
anatomy **40-1**
anterior arch 40
articular processes 40,
41
body 40, 41
cervical 25, 27, 28, 39
dens (odontoid peg) 40

intervertebral discs 40,
43
joints 39, 40, 44
lamina 40, 41
lateral mass 40
lumbar vertebrae 25,
42, 45
MRI 225
number of 40
posterior arch 40
spinal nerves 130
spinous processes 40,
41
thoracic 25, 37, 131,
133
transverse foramen 40
transverse processes 37,
39, 40, 41
vertebral foramen 40,
41
vertebral arteries 156,
158
vertebral column 24
vesicles
cells 12
vesicouterine pouch 213
vestibular apparatus 127
vestibular cord 148
vestibular nerve 127

vestibule
ear 127
nose 148
vagina 215
vestibulocochlear nerve
122, 126, 127
villi
microvilli 12
visceral pleura 147, 150,
153
vision **124-5**
vitreous humour 125
vocal cords (ligament) 74,
75, 148, 149
voice box *see* larynx
voluntary muscle *see*
skeletal muscle
vomer 31, 35

water
cell metabolism 12
cell transport 13
white matter
spinal cord 133
windpipe *see* trachea

"wisdom teeth" 195
womb *see* uterus
wrist
bones 24, 25, 54
joints **56-7**, 230
tendons 95

X-rays 222
see also radiographs
xiphisternal joint 37
xiphoid process 24, 37, 196

Z disc, skeletal muscle 68
zygapophyseal joints 41, 43
zygomatic arch 28, 31
zygomatic bone 28, 35
zygomaticus major muscle
70, 73
zygomaticus minor
muscle 70

Acknowledgments

Dorling Kindersley would like
to thank the following people for
their help in the preparation of
this book: Alison Sturgeon for
proofreading and Jane Parker
for the index.

Medi-Mation would like to thank:
Senior 3D artists: Rajeev Doshi,
Arran Lewis, 3D artists: Owen
Simons, Gavin Whelan, Gunilla Elam.
Antbits Ltd would like to thank:
Paul Richardson, Martin Woodward,
Paul Banville, and Rachael Tremlett.

The publisher would like to
thank the following for their
kind permission to reproduce
their photographs:

(Key: a-above; b-below/bottom;
c-centre; f-far; l-left; r-right; t-top)

**6-7 Robert Steiner MRI Unit,
Imperial College London**. **11
Getty Images:** CMSP (bl). **Science
Photo Library:** Makoto Iwafuji (br).
14 Corbis: MedicalRF.com (bl). **15
Corbis:** Photo Quest LTD (Dense
Connective); Photo Quest Ltd /
Science Photo Library (Spongy
Bone). **Getty Images:** Dr Gladden
Willis (smooth tissue). **Peter Hurst,
University of Otago, NZ:** (Skeletal
Muscle, Nerve Tissue). **Science
Photo Library:** Biophoto
Associates (Loose Connective); M.I.
Walker (Cartilage); Steve
Gschmeissner (Epithelial Tissue,
Adipose Tissue). **15 Getty Images:**
SPL / Pasieka (bl). **19 Robert
Steiner MRI Unit, Imperial
College London:** (r). **57 Corbis:**
Lester Lefkowitz (bl). **66 Science
Photo Library:** Gustoimages (bl).

67 Science Photo Library: Living
Art Enterprises (tr). **181 Science
Photo Library:** PRJ Bernard / CNRI
(t); Susumu Nishinaga (tr). **124 Getty
Images:** Image Source (t). **127
Getty Images:** Image Source (tr).
Wellcome Images: (br). **149 Getty
Images:** Nick Veasey (l). **222
Science Photo Library:** (tr); (bl);
Susumu Nishinaga (cr); Living Art
Enterprises (tr). **223 Science Photo
Library:** Hank Morgan (ca);
Sovereign, ISM (cr); Zephyr (cb).
Getty Images: Peter Dazeley (cl).
Science Photo Library: David M.
Martin, M.D. (bc). **224-225 Robert
Steiner MRI Unit, Imperial
College London:** (all). **226-227
Robert Steiner MRI Unit,
Imperial College London:** (all).
**228-229 Robert Steiner MRI
Unit, Imperial College London:**

(all). **230-231 Robert Steiner MRI
Unit, Imperial College London:**
(all). **232-233 Robert Steiner MRI
Unit, Imperial College London.**

Endpapers: Photo Quest Ltd /
Science Photo Library (Spongy
Bone)

All other images © Dorling
Kindersley

For further information see:
www.dkimages.com

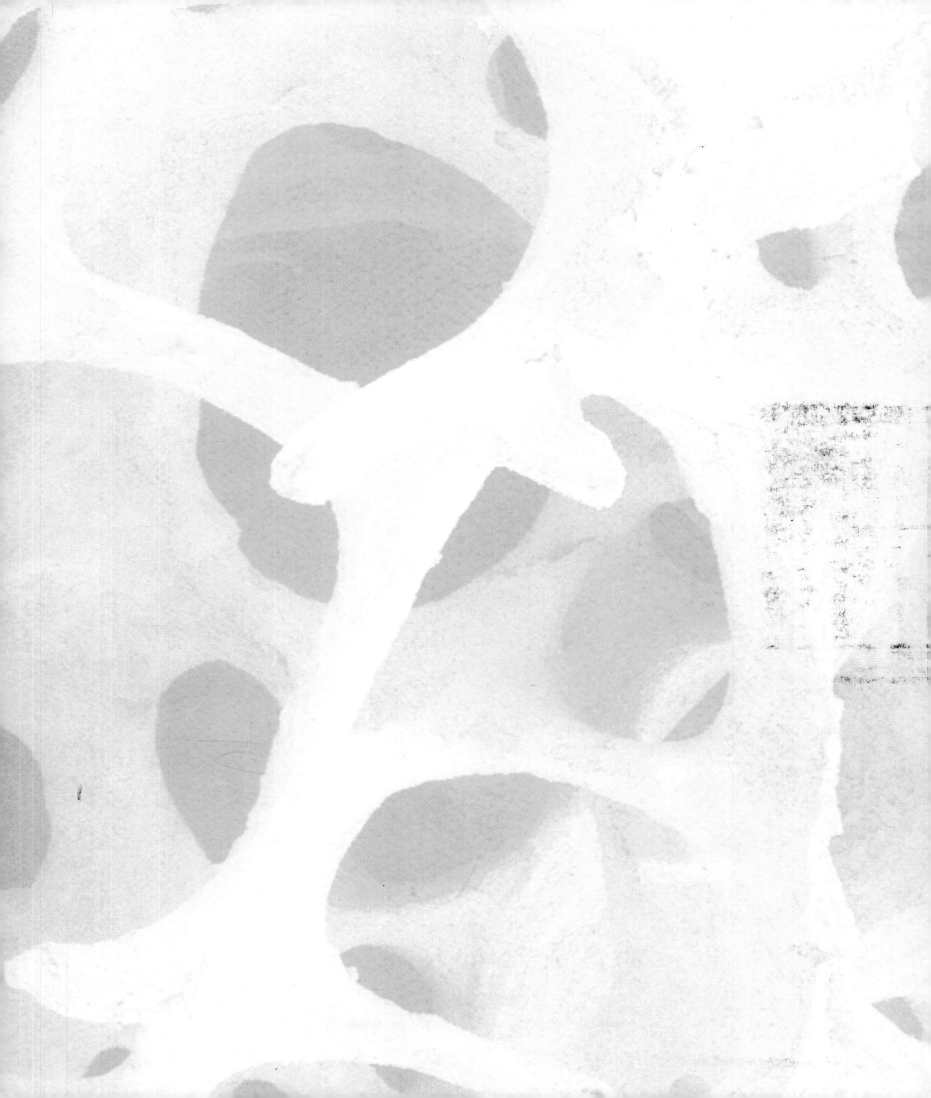